FARMING THE NORTH SEA COAST,
900–2000

Boydell Studies in Rural History

Series Editor
Professor Richard W. Hoyle

This series aims to provide a forum for the best and most influential work in agricultural and rural history, and on the cultural history of the countryside. Whilst it is anchored in the rural history of Britain and Ireland, it also includes within its remit Europe and the colonial empires of European nations (both during and after colonisation). All approaches and methodologies are welcome, including the use of oral history.

Proposals or enquiries are welcomed. They may be sent directly to the editor or the publisher at the email addresses given below.

Richard.Hoyle@reading.ac.uk

Editorial@boydell.co.uk

Previous titles in the series are listed at the back of the volume.

Farming the North Sea Coast, 900–2000

Managing Water, Reclaiming the Land

Piet van Cruyningen

THE BOYDELL PRESS

© Piet van Cruyningen 2025

All Rights Reserved. Except as permitted under current legislation no part of this work may be photocopied, stored in a retrieval system, published, performed in public, adapted, broadcast, transmitted, recorded or reproduced in any form or by any means, without the prior permission of the copyright owner

The right of Piet van Cruyningen to be identified as the author of this work has been asserted in accordance with sections 77 and 78 of the Copyright, Designs and Patents Act 1988

First published 2025
The Boydell Press, Woodbridge

ISBN 978 1 83765 262 4

The Boydell Press is an imprint of Boydell & Brewer Ltd
PO Box 9, Woodbridge, Suffolk IP12 3DF, UK
and of Boydell & Brewer Inc.
668 Mt Hope Avenue, Rochester, NY 14620-2731, USA
website: www.boydellandbrewer.com

A CIP catalogue record for this book is
available from the British Library

The publisher has no responsibility for the continued existence or accuracy of URLs for external or third-party internet websites referred to in this book, and does not guarantee that any content on such websites is, or will remain, accurate or appropriate

In memory of Jan Bieleman (1949–2021), agrarian historian

CONTENTS

	Illustrations	viii
	Preface	xi
	Glossary	xiii
	Introduction	1
1	The Great Reclamation: Transformation of the Wetlands, c. 900–1300	19
2	Pioneering in the Wetlands: Water Management and Agriculture, c. 1100–1300	54
3	Sinking Land and Disastrous Floods: Water Management, c. 1300–1550	79
4	Surviving in Times of Adversity: Agriculture in the Late Middle Ages, c. 1300–1550	112
5	The Second Reclamation: Urban Capital in the Countryside, c. 1500–1700	137
6	Risen from the Waves: Agriculture after the Second Reclamation, c. 1550–1700	166
7	State and Steam: Water Management, c. 1700–1880	192
8	The Apogee of Lowland Farming, c. 1700–1880	221
9	Nature Tamed? Water Management, c. 1880–1980	245
10	The Indian Summer of Lowland Farming, c. 1880–1980	260
	Conclusion	281
	Epilogue: The North Sea Lowlands after 1980	290
	Bibliography	295
	Index	331

ILLUSTRATIONS

Figures

1 *Terp* of Hogebeintum, Friesland, 2022. Photograph by the author. 29
2 The Holme Post, c. 1928. Collection Huntingdonshire Archives. 37
3 Peat landscape with drainage mill near Vlist, Zuid-Holland. Photo G. J. Dukker. Collection Rijksdienst voor het Cultureel Erfgoed. 98
4 Borsselepolder, Zuid-Beveland, Zeeland, 1973. Photo L. M. Tangel. Collection Rijksdienst voor het Cultureel Erfgoed. 145
5 Harrowing and sowing of rapeseed on Isabellapolder near Aardenburg, Zeeland. Collection ZB Beeldbank Zeeland. 171
6 Cruquius steam pumping station near Haarlem, Noord-Holland, 1974. Photo L. M. Tangel. Collection Rijksdienst voor het Cultureel Erfgoed. 197
7 Farm at Finsterwolde, Groningen, 1976. Photo P. van Galen. Collection Rijksdienst voor het Cultureel Erfgoed. 238
8 Completion of the Afsluitdijk, 28 May 1932. Photo L. Verbost. © Rijkswaterstaat. 249
9 Revolutionised farming: wheat harvest with a Massey-Harris combine harvester in Kats on the island of Noord-Beveland, Zeeland, 1956. Photo F. Coolman. Special Collections, Wageningen University & Research – Library. 277
10 Dark clouds gathering above *Hallig* Nordstrandischmoor, Nordfriesland. Photo Dirk Meier. 292

Maps

1 The North Sea Lowlands — xv
2 The east coast of England — xvi
3 The Wadden coast — xvii
4 The Low Countries around 1350 — xviii
5 Central Holland — xix

Illustrations

| 6 | The Flemish coastal plain and Zeeland | xx |
| 7 | The Zuider Zee works and Delta works | xxi |

Tables

1	Percentage distribution of landownership on the Oude Yevenewatering, 1388 and 1550	114
2	Owner-occupiers and leaseholders in four districts in Friesland, 1511	127
3	Distribution of farms by holding size in four Frisian districts, 1511	128
4	Land reclamation in the Dutch and German lowlands, 1500–1699	141
5	Percentage distribution of landownership in two areas on the Flemish coastal plain, 1665/1670	168
6	Percentage of land owned by town dwellers and urban institutions in three villages in Delfland, 1561/2–1635, 1654 and 1663	173
7	Herd size of cattle owners in Delfland, 1672	175
8	Herd size of the tenants of the Thorney estate, c. 1750	232
9	Wheat yields in five areas in the North Sea Lowlands, 1615–1862 (in hl/ha)	242

PREFACE

It might seem preordained that I, born and raised on a marshland farm in Zeeland Flanders, and descendant of a family that had farmed in the reclaimed marshes since 1667, would end up writing a history of farming in the North Sea Lowlands. However, my career as a historian has been a succession of – mostly lucky – coincidences and for decades I never imagined I would ever write a book like this. Yet, over the years two themes kept popping up in my research, agriculture and drainage, and by 2015 I realised they might be combined in a book. The inspiration came to me during a workshop in Ascona organised by Gérard Béaur. I am grateful to Gérard for that opportunity. The good food and the great view of Lago Maggiore may have contributed to the inspiration.

Once I started working it became apparent that although I knew a lot already, I needed to know even more, and so a period of nine years of voracious reading and not always enthusiastic writing ensued. I could never have accomplished that alone and many people are to thank for the fact that I managed to bring it to a good end. Two people need to be mentioned especially. Firstly, Ewout Frankema, who encouraged me to write this book, read much of the manuscript, and let me work on it for all these years. Secondly, Petra van Dam, who patiently read all my texts, kept me on the right track and urged me to go on when I felt like giving up. Several other people enabled me to profit from their expertise by reading and commenting on (parts of) the manuscript: Paul Brusse, Peter Henderikx, Wijnand Mijnhardt, Elise van Nederveen Meerkerk, Anton Schuurman, Milja van Tielhof, Harm Zwarts, and last but not least my old friend Jan de Putter. Mans Schepers, Hans Mol and Han Nijdam helped fill in some of the gaps in my knowledge of medieval history, and I now fully agree with Mans that archaeologists and historians need to cooperate if we want to better understand medieval agriculture. Others helped in the development of my manuscript into a readable book: Annemieke Verhoek, who made the maps, Elaine McIntyre, who improved my English, Alex Schwartz, who assisted me with the index, and Caroline Palmer at Boydell & Brewer.

Writing is a lonely occupation, but I was lucky to do much of the writing in the pleasant environment of the Economic and Environmental History Group of Wageningen University. Although my colleagues research very

Preface

different periods and parts of the world, it is a solace to be surrounded by people who are also experiencing similar struggles with sources and writing issues, and to be able to complain with them about those problems during lunch. Although it is impractical to mention everyone by name here, I make an exception for our efficient colleagues at the secretariat, Sandra Vermeulen and Barbara Schierbeek, who make our lives at the university so much easier.

Wageningen, June 2024

GLOSSARY

Ambacht	Rural district
Beherdischheit	Perpetual lease of a farm at a fixed rent (Ostfriesland)
Beklemrecht	Perpetual lease of a farm at a fixed rent (Groningen)
Boezem	System of lakes, ponds and canals for storage and transport of excess water
Commission of Sewers	Water authority
Cope	Concession for wetland reclamation
Deichacht	Authority for dike maintenance
Dijkgraaf	1. Judge in cases of water maintenance; 2. Chairman of the board of a water authority
Dijkstoel	Water authority (Guelders)
Gemeenmaking	Communal maintenance of water infrastructure financed by rates levied from landowners/farmers
Hallig	Island without sea defences (Nordfriesland)
Heemraad	1. Jury member; 2. Member of the board of a water authority
Heemraadschap	Water authority
Hochland	Fertile stream ridge close to a river (Germany)
Hollerkolonie	Settlement of Hollanders along the Lower Weser or Lower Elbe rivers
Internal Drainage Board	Local water authority

Glossary

Kabeldeichung	Maintenance in kind of parts of a dike or canal by individual landowners or farmers
Kommuniondeichung	Communal maintenance of water infrastructure financed by rates levied from landowners/farmers
Kwelder	Salt marsh
Landgemeinde	Rural commune
Octrooi	Patent for wetland reclamation
Oktroy	Patent for wetland reclamation
Pfanddeichung	Maintenance in kind of parts of a dike or canal by individual landowners or farmers
Schauung	Inspection of maintenance in kind
Schouw	Inspection of maintenance in kind
Sielacht	Authority for sluice maintenance
Sietland	Low, badly draining part of a river plain (Germany)
Terp	Dwelling mound (Friesland)
Verhoefslaging	Maintenance in kind of parts of a dike or canal by individual landowners or farmers
Warft	Dwelling mound (Nordfriesland)
Watering	Water authority (Flanders, Zeeland)
Werf or werve	Dwelling mound (Flanders, Zeeland)
Wierde	Dwelling mound (Groningen)
Wurt	Dwelling mound (Lower Saxony)
Zijlvest	Authority for sluice maintenance (Groningen, Friesland)

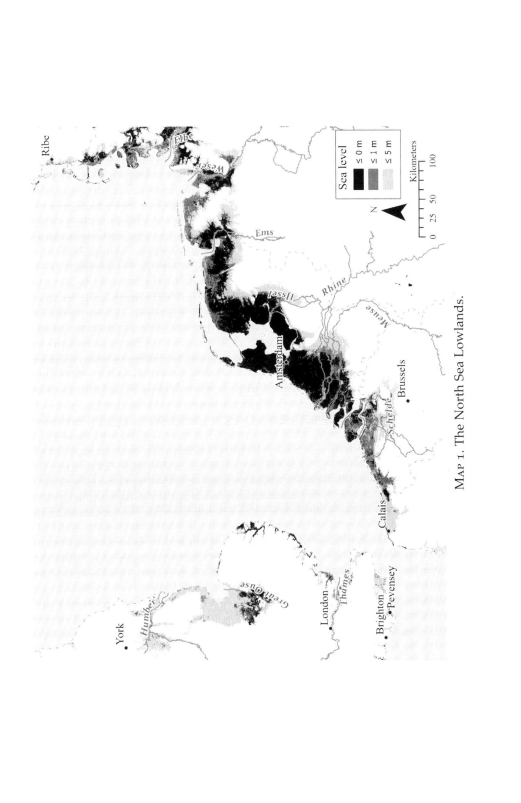

MAP 1. The North Sea Lowlands.

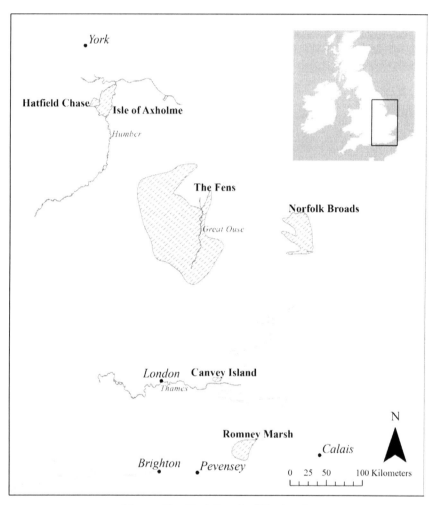

Map 2. The East Coast of England.

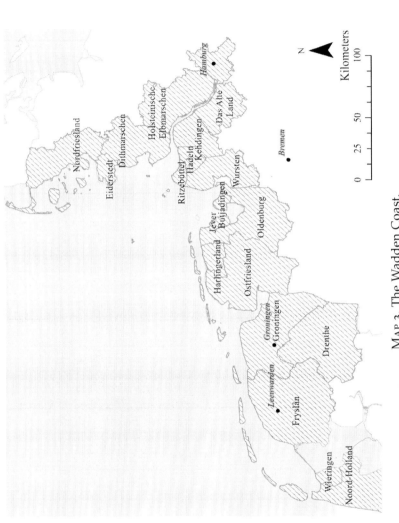

MAP 3. The Wadden Coast.

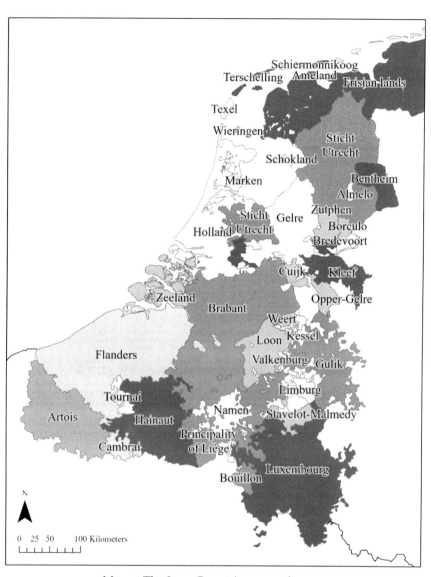

MAP 4. The Low Countries around 1350.

Map 5. Central Holland.

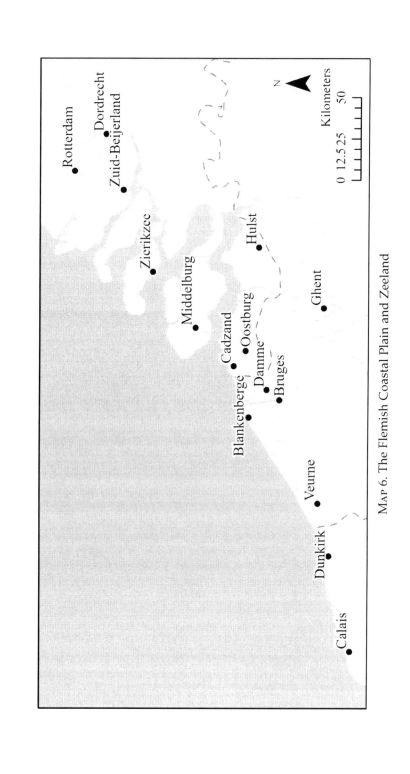

MAP 6. The Flemish Coastal Plain and Zeeland

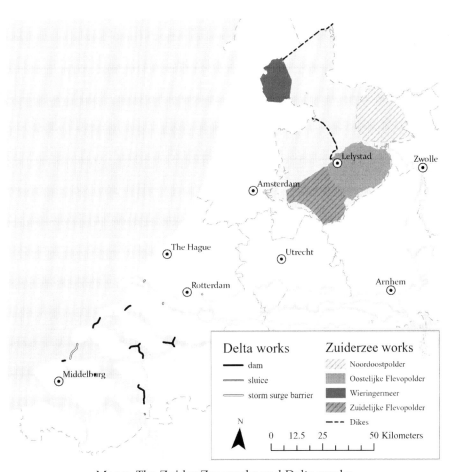

MAP 7. The Zuider Zee works and Delta works

Introduction

The countries bordering the southern North Sea belong to the most agriculturally productive states in the world. The Netherlands in particular is an agricultural giant. Due to its very high productivity, this country, with its small agricultural acreage, is the world's third largest exporter of agricultural products after the United States and Brazil.[1] Although many people are now rightly concerned about intensive farming's environmental downsides, this remains a remarkable achievement. What makes it even more remarkable is that it began in the soggy, inhospitable wetlands along the North Sea coast. Scholars such as Jan de Vries and Jan Bieleman have shown that already in the seventeenth century the coastal provinces of the Netherlands, with their relatively large commercial and productive farms, were worlds apart from the inland provinces and much of continental Europe with its small-scale subsistence farming.[2]

This book aims to explain the advanced nature of farming in the Dutch coastal zone by looking at it from a comparative perspective and over a long period, from c. 900 to 2000. The coastal provinces of the Netherlands were part of a zone stretching from Calais in northern France to Ribe in southern Denmark, composed of reclaimed marshes, fens and floodplains situated around or even below sea level. On the opposite shore of the North Sea similar regions can be found in eastern England from the Pevensey Levels in East Sussex to Yorkshire. I will call them here the North Sea Lowlands.[3] All societies in these reclaimed wetlands needed to adapt

---

[1] E. Thoen and T. Soens, 'Contextualizing 1500 Years of Agricultural Productivity and Land Use in the North Sea Area', in E. Thoen and T. Soens, eds., *Struggling with the Environment: Land Use and Productivity. Rural History in North-Western Europe, 500–2000* (Turnhout, 2015), pp. 455–7; F. Viviano, 'How the Netherlands Feeds the World', *National Geographic* (September 2017), pp. 58–81; J. D. van der Ploeg, *Gesloten vanwege stikstof: Achtergronden, uitwegen en lessen* (Gorredijk, 2023), p. 181.

[2] J. de Vries, *The Dutch Rural Economy in the Golden Age* (New Haven and London, 1974); J. Bieleman, *Five Centuries of Farming: A Short History of Dutch Agriculture* (Wageningen, 2010), pp. 37–40.

[3] G. Bankoff, 'The "English Lowlands" and the North Sea Basin System: A History of Shared Risk', *Environment and History* 19 (2013), pp. 4–5. Bankoff calls the lowland areas of England the 'English Lowlands'. I have extended

to the watery environment by creating physical infrastructure, regulation and organisations. They did so in different ways and through different developments. By comparing these regions we can determine how the way in which their inhabitants dealt with water influenced economic development. Recently, the high agricultural productivity of the Dutch coastal zone has been identified as one of the causes of the development that ultimately led to the Industrial Revolution.[4] This book has the more modest aim of providing insight through comparative research as to how agricultural development was influenced by what Sheilagh Ogilvie identified as the two major constraints on farmers' decision-making: the natural environment and the human environment of institutional rules determining behaviour in rural societies.[5]

A Challenging Landscape

In their natural state, wetlands are ecosystems that are very rich in biodiversity and resources. But they are far from easy to transform into intensively farmed land, and therefore any transformation into farmland often takes place over some time. Contrary to what David Ricardo assumed, humans do not first reclaim the most fertile soils.[6] They start with light, well-draining soils that are the simplest to cultivate. Early medieval Europeans preferred to live on thin, sandy soils that drained easily and could be worked with scratch ploughs.[7] People were attracted by the fertile alluvial soils of the wetlands which promised good pasture and bumper harvests, but they first needed appropriate technology and organisational forms to enable them to tame the water and live safely. Without water management systems, settling lowlands was not an option.[8] A Dutch expert summed up the rewards and the price of settling in lowland areas:

> [empoldered land] forms part of the most intensively used, most productive, and most densely populated part of the world's cultivated

---

this to include also the eastern shore of the North Sea, following the example of K. Ritson, *The Shifting Sands of the North Sea Lowlands: Literary and Historical Imaginaries* (London and New York, 2019), p. 5.

4 J. Scott, *How the Old World Ended: The Anglo-Dutch-American Revolution* (New Haven and London), pp. 284–5.
5 S. Ogilvie, 'Choices and Constraints in the Pre-Industrial Countryside', in C. Briggs, P. Kitson and S. J. Thompson, eds., *Population, Welfare and Economic Change in Britain, 1290–1834* (Woodbridge, 2014), p. 277.
6 D. Ricardo, *On the Principles of Political Economy and Taxation* (Georgetown, 1819), pp. 37–8.
7 D. Levine, *At the Dawn of Modernity: Biology, Culture, and Material Life in Europe after the Year 1000* (Berkeley, 2001), p. 153; see also H. C. Carey, *Principles of Social Science* (3 vols., Philadelphia, 1858), vol. 1, pp. 104–7.
8 P. Vidal de La Blache, *Principes de géographie humaine* (Paris, 1922), pp. 50–1.

Introduction

area [but] it is also the most expensive, the most risky, and the most difficult part in terms of reclamation, operation, and maintenance.[9]

Settlers in lowland areas indeed faced daunting problems. Not only did they need to keep river and sea water out of their fields and homes, but they had the perhaps even bigger challenge of draining the fields to make them dry enough for agriculture. And they had to drain an awful lot of water because many lowland regions operate as large sinks, where the water from enormous upland areas reaches the sea. The Netherlands, for example, are, for a large part, composed of the deltas of the Scheldt, Meuse, and Rhine rivers that carry enormous amounts of water to the sea from a vast catchment area in Germany, France and Switzerland. The same is true for the coastal zone of north-west Germany, where the Ems, Weser and Elbe rivers reach the sea, and on a smaller scale for the English Fens. Meeting these challenges required large investments in dikes, drains and sluices as well as the creation of organisations to maintain these works. And then there were some additional problems such as a chronic shortage of wood for fuel and building material in these mostly treeless areas. It is therefore not that surprising that it took some time before pioneers were willing to settle in such environments. What is much more surprising is that in the end agriculture proved to be so successful in these areas. Of course, their mostly alluvial soils were potentially very fertile, but enormous exertions were required to change deltas and river plains into the prosperous agricultural regions they eventually became.

It is useful to elaborate here on the importance of drainage. Historiography, in the Netherlands in particular, tends to concentrate on the heroic struggle of the inhabitants of the coastal areas against flooding by the sea and major rivers. More than a century ago, however, Raoul Blanchard remarked that the major problem was not flood protection, but drainage.[10] Drainage does not just mean conducting river water to the sea, but also field drainage. Field drainage is crucial because the mostly alluvial soils of the Lowlands tend to get easily waterlogged.[11] This risk occurs especially in winter, when there is an excess of precipitation over evaporation. In waterlogged soils, the root system of crops will lose its ability to respire after 48 hours. There will also be a loss of soil nitrate, which is essential for plant growth, and the soil will have reduced ability

---

[9] A. Volker, quoted in H. J. Walker, 'The Coastal Zone', in B. L. Turner et al., eds., *The Earth as Transformed by Human Action: Global and Regional Changes over the Past 300 Years* (Cambridge, 1990), p. 283.

[10] R. Blanchard, *La Flandre: étude géographique de la plaine Flamande en France, Belgique et Hollande* (Dunkirk, 1906), p. 265.

[11] This discussion is based on H. Cook, 'Soil and Water Management in the English Landscape: Principles and Purposes', in H. Cook and T. Williamson, eds., *Water Management in the English Landscape: Field, Marsh and Meadow* (Edinburgh, 1999), pp. 15–22.

to warm up in spring, thereby inhibiting germination and early plant growth. Ensuring efficient field drainage is a problem that farmers in the North Sea Lowlands have continuously struggled with over the past millennium. Although floods could cause enormous damage and much human suffering, their long-term effects were limited, as Tim Soens has recently demonstrated.[12]

Once structures such as seawalls, sluices or canals had been created, landowners and farmers had to continue to invest in their maintenance. But that was not all. Transformation into farmland can have unintended consequences, especially in lowland regions where people must deal with the tremendous force of flood water. Whereas drainage can lead to the lowering of the surface level of the land, construction of dikes to protect the land can in the long run reduce the storage capacity of sea inlets or riverbeds and thus also increase the risk of flooding. Therefore, the inhabitants of lowland areas were dealing with an environment that was continuously changing, to a large degree as a result of their own actions. This interaction between agriculture and the lowland environment influences the long-term development of both. For example, land use patterns that emerged in the Middle Ages as consequences of the way in which water was managed then can in some cases still be observed today. The vast grassland landscape of the 'Green Heart' of Holland is the result of the water management system that was introduced there during the late Middle Ages, which made only pastoral farming possible. Of course, the demands of agriculture also put their stamp on water management and the environment. Probably the most impressive example of this is the transformation of water management in the seventeenth-century English fenlands, when promoters of drainage wanted to create land where arable crops could be grown. Such interactions with the environment of course occur in all forms of agriculture, but what makes the lowland areas different from other parts of north-western Europe is the tremendous force of water and the perpetual commitment – investments in infrastructure and designing of institutional arrangements – required to control it.

The continuous process of action and reaction between humans and their watery environment is the basis for the rather bleak view several historians hold of the history of water management. On water management systems in China, Mark Elvin remarked: 'Once a community is committed to a system of this sort, it has no easy option – barring a new technological fix – but to allocate labor and resources to maintaining it'.[13] There is more than a grain of truth in this view. Physical infrastructure,

---

12 T. Soens, 'Resilient Societies, Vulnerable People: Coping with North Sea Floods before 1800', *Past and Present* 241 (2018), p. 160.
13 M. Elvin, *The Retreat of the Elephants: An Environmental History of China* (New Haven and London, 2004), p. 155.

such as dikes, canals and sluices, is constructed to facilitate more intensive agricultural use of the land. That land will feed more people and many of them will settle in the former wetlands. Consequently, a return to the pre-reclamation situation is hardly feasible because in that situation the land could feed considerably fewer people. The only alternative is then to continue maintaining the existing system and to spend increasing amounts of labour and capital on it, just to uphold the status quo. So for China Elvin describes water management systems as situations of technological lock-in. This means that a technological system, through positive feedback – making changing course extremely expensive – can become locked in a trajectory even though alternative technologies are available.[14]

It cannot be denied that also in the North Sea Lowlands decisions taken in a long-forgotten past are still haunting farmers, administrators and hydraulic experts today. The continuing land subsidence in peat areas with which the Dutch and English are wrestling nowadays, for example, is the result of the decision to drain the peat areas a thousand years ago. For this reason William TeBrake concluded that the idea of technological lock-in is applicable to Dutch water management as well. TeBrake also added, however, that Dutch water management was successful to the degree that it made continued occupation of the lowland areas possible, and that it had a positive effect on the long-term economic development of the Low Countries.[15] Of course, agriculture in reclaimed wetlands needed to be productive simply to be able to finance the construction and upkeep of the required physical infrastructure.

Water management is not just a matter of creating physical infrastructure. Any infrastructure must be maintained continuously, requiring an institutional framework, as such maintenance is a collective effort. Maintaining a water management infrastructure of flood defences, canals and sluices is a task necessitating the effort of local or regional communities, because individuals do not possess the required resources. Institutions – regulations and the organisations that design them and monitor their implementation – are required to coordinate the efforts of those who have to physically perform the maintenance, usually landowners and/or farmers. Until recently this institutional infrastructure and its relationship with agriculture and rural society and economy has

---

[14] R. Martin and P. Sunley, 'Path Dependence and Regional Economic Evolution', *Journal of Economic Geography* 6 (2006), p. 400; P. Pierson, *Politics in Time: History, Institutions, and Social Analysis* (Princeton and Oxford, 2004), pp. 20–2. The classical example of lock-in is the QWERTY keyboard; see P. A. David, 'Clio and the Economics of QWERTY', *American Economic Review* 75 (1985), pp. 332–7.

[15] W. H. TeBrake, 'Taming the Waterwolf: Hydraulic Engineering and Water Management in the Netherlands during the Middle Ages', *Technology and Culture* 43 (2002), pp. 498–9.

received less attention from historians than the physical infrastructure.[16] This study aims to fill this lacuna. The institutional infrastructure shows a bewildering variety, both in space and in time. Different solutions were found for the same problems. In some German areas institutions were introduced in the twentieth century which in Flanders were already known in the Middle Ages. This makes these lowlands an ideal 'laboratory' within which to study the formation and development of institutional arrangements and their effects on agriculture.[17]

The Institutional Environment

In 1573 the Grebbe Sluice, situated in the central Netherlands between the towns of Rhenen and Wageningen, was in a bad state of repair. Since this was an important sluice through which a large area drained its excess water into the river Rhine, experts proposed the construction of a new sluice. The town of Wageningen, which was liable for a quarter of the maintenance costs of the sluice as representative of its burghers who owned land in the area, refused to contribute to the building of a new one. A less expensive plan, to repair the existing sluice, was also rejected by the magistrates of Wageningen. After fruitless negotiations the other two parties responsible for maintenance of the sluice decided to go ahead and repair it and try to recoup Wageningen's share later. Probably they never succeeded in getting their money back. In the 1580s again repairs were required and again Wageningen refused to cooperate.[18]

Economists will not be surprised at the behaviour of the Wageningen magistrates. Public or collective goods are goods that, when provided to a member of a group, will by necessity be shared by all members of that group. At the national level defence and the maintenance of order are public goods. All citizens of the state profit from these services, regardless of whether they pay tax or not.[19] The goods provided by water management systems are of the same nature. A seawall protects all land situated behind it, whether the owners of that land contribute to its maintenance or not. And a sluice drains all land adjoining it, again regardless of whether the owners pay rates for the sluice's maintenance

---

[16] P. J. E. M. van Dam, P. J. van Cruyningen and M. van Tielhof, 'A Global Comparison of Pre-Modern Institutions for Water Management', *Environment en History* 23 (2017), pp. 336–7; an exception is S. Ciriacono, *Building on Water: Venice, Holland and the Construction of the European Landscape in Early Modern Times* (New York and Oxford, 2006), pp. 32–9 and 87–90.

[17] B. J. P. van Bavel, 'History as a Laboratory to Better Understand the Formation of Institutions', *Journal of Institutional Economics* 11 (2015), p. 70.

[18] T. Stol, *De veenkolonie Veenendaal: Turfwinning en waterstaat in het zuiden van de Gelderse Vallei 1546–1643* (Zutphen, 1992), p. 135.

[19] M. Olson, *The Logic of Collective Action: Public Goods and the Theory of Groups* (2nd edn., Cambridge and London, 1971), pp. 14–15.

Introduction

or not. Because they provide public goods, water management systems are vulnerable to free-rider behaviour of landowners or farmers who want a safe place to live and dry land to cultivate but without paying for it. So long as these systems provide their services to a small group, the free-rider problem will be manageable. A group of a dozen or so farmers who have reclaimed a marsh area will be able to ensure that all members contribute to maintenance of the common drains through informal coordination. But even in such small communities there needs to be at least some informal agreements as to who is to maintain which part of the drain or dike, and the quality of the maintenance work that must be performed. The larger the community, the more complicated such arrangements will become. The community is then in need of institutions, often defined as the rules of the game or as constraints on economic behaviour which humans impose upon themselves.[20]

Institutional arrangements also require organisations. There are many types of organisations: they can be political (political parties, city councils), economic (firms, family farms), religious (churches), etcetera.[21] In the field of water management, organisations are needed because they design rules and monitor their implementation. Most of these rules concern maintenance: who is to be liable for maintenance, how is it to be performed – through labour or by paying rates – and what quality standards are to be met? Another important rule concerns the design of rules: who has the right to do this? Who has access to the general meetings or can become member of the board of an organisation that manages a water management system? Nowadays, the state – an organisation of organisations – is by far the most important player in the field of water management. But during the period of the great reclamations at the beginning of the second millennium, when the first enduring water management systems were created in north-western Europe, states in the modern sense did not yet exist. Initially, control over water management was left to the general administrators of villages or rural districts. Later, from the twelfth century onwards, beginning at the regional level, a new type of organisation emerged, called *watering*, *heemraadschap*, *polder* or later *waterschap* in Dutch, and *Deichacht* or *Deichverband* in Germany, Commission of Sewers in England. In this book they will be referred to as water authorities. During the Middle Ages, most water management remained the responsibility of the general administration, especially at

---

[20] D. C. North, *Institutions, Institutional Change and Economic Performance* (Cambridge, 1990), pp. 3–5; D. C. North, J. J. Wallis and B. R. Weingast, *Violence and Social Orders: A Conceptual Framework for Interpreting Recorded Human History* (Cambridge, 2009), pp. 15–16.
[21] G. M. Hodgson, 'Introduction to the Douglass C. North Memorial Issue', *Journal of Institutional Economics* 13 (2017), pp. 15–16; North, *Institutions*, p. 5.

the local level, but in the early modern period water authorities became predominant.

The water authority was a hybrid type of organisation; it was both public and private. It was an association of landowners that was also invested with political and judicial power – aiming at the promotion of the economic interests of its members but equipped with forms of public authority. An important difference with other private associations was that membership was not voluntary: all owners of land within the territory of the water authority were *qualitate qua* members.[22] As late as the middle of the nineteenth century, in the Netherlands there was discussion about the legal nature of water authorities. Some lawyers still considered them to be private associations of landowners.[23] Water authorities designed and enforced rules for the maintenance of water management systems. One of the most important of their tasks was the prevention of free riding. They met with different degrees of success, depending, amongst others, on recognition by the emerging states or principalities.

Free-riding behaviour of people liable for maintenance of water infrastructure occurred not only in the North Sea area, but for instance also in northern Italy, where large landowners often tried to avoid paying rates for maintenance of infrastructure.[24] An extreme example from the North Sea area is what happened on the island of Nordstrand in 1516. Inhabitants of the island chased the dike judges away when they were trying to verify the quality of the maintenance work. The representative of the king, however, condemned the perpetrators and confirmed the authority of the dike judges.[25] This shows that water authorities needed at least minimal recognition by a higher authority, such as the state, to be able to enforce their rules. Elinor Ostrom demonstrated this in a very different context. The irrigation communities in the arid area of Valencia could implement their decisions on conflicts about water because they were recognised by the kingdom of Aragon.[26] So the attitude of the state or its predecessors such as medieval overlords and territorial princes must be considered when studying the development of water authorities and their rights. In parts of medieval Friesland water authorities received

---

[22] M. van Tielhof, *Consensus en conflict: Waterbeheer in de Nederlanden 1200–1800* (Hilversum, 2021), p. 29.

[23] H. van der Linden, 'Geschiedenis van het waterschap als instituut van waterstaatsbestuur', in B. de Goede et al., eds., *Het waterschap: Recht en werking* (Deventer, 1982), pp. 32–3; M. Schroor, *Wotter: Waterstaat en waterschappen in de provincie Groningen, 1850–1995* (Groningen, 1995), p. 66.

[24] Ciriacono, *Building on Water*, p. 88.

[25] A. Panten, 'Deiche und Sturmfluten in der geschichtlichen Darstellung Nordfrieslands', in T. Steensen, ed., *Deichbau und Sturmfluten in den Frieslanden* (Bredstedt, 1992), p. 16.

[26] E. Ostrom, *Governing the Commons: The Evolution of Institutions for Collective Action* (Cambridge, 1990), pp. 81, 101.

Introduction

hardly any recognition, whereas in the county of Holland they were granted extensive rights, in some cases even extending to the right to impose capital punishment.[27] This variation in recognition by the state influenced the development of institutions and the quality of water management, and indirectly also agricultural development.

Although free riding was a serious problem, it should not be exaggerated. Polders were generally not crowded with people who were deliberately cheating their neighbours. The people living in the lowlands were aware that maintenance of dikes and sluices was an issue of life and death, and the prevailing social norm was that everybody in the community needed to contribute to maintenance in order to ensure survival.[28] It should be realised that institutions are not just constraints, but also instruments to make things possible.[29] In the case of the North Sea Lowlands, social norms, regulations and water authorities enabled humans to live and work safely in reclaimed wetlands continuously exposed to the danger of flooding. Of course, absentee landowners living high and dry in the uplands might not share these norms. A further danger was economic adversity causing landlords and farmers to spend less time and money on such maintenance. They may have intended to do more work once the situation had improved, and in the meantime, maybe their neighbour could do a bit more, but the neighbour was probably also in trouble and trying to cut costs. When, for instance, the other parties involved inquired in 1573 as to why the inhabitants of Wageningen refused to contribute to repairing the Grebbe Sluice, it turned out that this small town was in financial trouble because it had been forced to take in and provision a garrison. Such forced behaviour could ultimately result in disaster, not because of decisions by deliberately calculating individuals, but as a consequence of the messiness of daily life. For such situations rules and organisations were designed to guarantee the quality of maintenance.

That not many persons liable for maintenance shirked their obligations was also due to the fact that trust existed between the people within a water authority. According to Robert Putnam, trust is social capital: social networks and the norms of reciprocity associated with them.[30] Research

---

[27] H. C. Darby, *The Medieval Fenland* (Cambridge, 1940), pp. 147–63; M. van Tielhof and P. J. E. M. van Dam, *Waterstaat in stedenland: Het Hoogheemraadschap van Rijnland voor 1857* (Utrecht, 2006), pp. 48–9.

[28] Bankoff, 'The "English Lowlands"', p. 13; F. Mauelshagen, 'Disaster and Political Culture in Germany since 1500', in C. Mauch and C. Pfister, eds., *Natural Disasters, Cultural Responses: Case Studies toward a Global Environmental History* (Lanham, 2009), p. 52.

[29] J. R. Searle, 'What Is an Institution?', *Journal of Institutional Economics* 1 (2005), p. 17; D. N. McCloskey, *Bourgeois Dignity: Why Economics Can't Explain the Modern World* (Chicago, 2010), p. 304.

[30] R. D. Putnam and K. A. Goss, 'Introduction', in R. D. Putnam, ed., *Democracy in Flux: The Evolution of Social Capital in Contemporary Society* (Oxford, 2002),

on Italy has demonstrated that in the north of that country, where a 'civic tradition' of cooperation in associations has existed since the Middle Ages, trust and solidarity tended to prevail. People there had the experience that when they helped someone else or gave something up for the public interest that altruistic deed would be reciprocated. In southern Italy, however, such a civic tradition was absent, and people could not cooperate to improve their communities.[31] In the North Sea Lowlands too, trust existed between landowners cooperating within water authorities. However, the social capital of water authorities is inward-looking social capital; water authorities represent the economic interests of the landowners within their territory.[32] Trust may exist within these organisations, but the expectations of the reliability of outsiders may still be very low. As a result, when conflicts over maintenance arose, it was often between organisations that were jointly responsible for maintenance of certain infrastructure, such as between the town of Wageningen and the other two parties over the maintenance of the Grebbe Sluice. This may explain why water authorities first emerged at the regional level, because it was at that level where most coordination problems existed due to a lack of trust.

Norms, beliefs and trust are important factors in explaining institutional development, but here we will be mostly concerned with the formal regulation of water management and the organisations designing them (keeping in mind those norms, beliefs and trust). Such institutional arrangements are usually perceived as solutions to economic problems. They are expected to improve efficiency of the sector for which they were designed and to enhance the overall performance of the economy, under certain conditions. The regulations designed and enforced by craft guilds, for example, are considered by many historians as stimulating economic performance by providing their members with cheap credit, by enforcing quality standards and fixed prices, and by protecting guild members from exploitation by more powerful entrepreneurs.[33] When the conditions change, such institutions are assumed to be replaced by new rules which are more attuned to the demands of the new situation. Market forces will eventually replace inefficient institutions with efficient ones.[34] Some years ago, Sheilagh Ogilvie argued against what she called the 'efficiency'

---

p. 3; M. Prak and J. L. van Zanden, *Pioneers of Capitalism: The Netherlands 1000–1800* (Princeton, 2023), pp. 5, 9–10.

[31] R. D. Putnam, R. Leonardi and R. Nanetti, *Making Democracy Work: Civic Traditions in Modern Italy* (Princeton, 1993), pp. 124–30, 172–82. Whether trust survived unscathed in northern Italy until the present is dubious.

[32] Putnam and Goss, 'Introduction', p. 11.

[33] S. R. Epstein, 'Craft Guilds in the Pre-Modern Economy: A Discussion', *Economic History Review* 61 (2008), p. 155.

[34] D. C. North, *Structure and Change in Economic History* (New York, 1981), p. 7.

Introduction

view of institutional arrangements. She pointed to the unequal distributive effects of institutions. According to Ogilvie, institutions not only influence the size of the economic pie – through enhancing efficiency – but also the distribution of the slices of that pie – through apportioning output. This can cause conflict, especially if elites try to influence the distribution of output in such a way that they profit most at the cost of poorer groups. Bas van Bavel has also observed that in the medieval Low Countries dominant social groups manipulated institutional arrangements for their own interests.[35]

The distributive effects which Ogilvie and Van Bavel called attention to can also be observed in water management, although more indirectly, through the apportioning of maintenance costs. There are indeed cases from the Middle Ages and the early modern period demonstrating that elites managed to shift the burden of maintenance at least partly to poorer groups. In Butjadingen, a territory in north-west Germany, for instance, in the seventeenth century about a quarter of the land was exempt from obligations to maintain the seawalls. The owners of this land, mostly clergy and wealthy landlords, left the burden of maintenance to be shouldered by the peasantry. When in 1681 the Danish Crown tried to introduce a more equitable distribution of the costs by abolishing the existing exemptions, local elites managed to sabotage these measures and forced the state to withdraw them four years later. The result was that flooding of the area could only be prevented by massive subsidies from the Danish Crown and by taxing the peasantry to the utmost. The taxpayers suffered, the community suffered, and the elites prospered.[36] Ogilvie's 'conflict view' is a very relevant approach for water management institutions, because water authorities were usually led by oligarchies of wealthy landlords and large farmers. As the example of Butjadingen shows, such elites indeed tended to promote their own interests above those of the community. This is especially important because the development of institutional arrangements is path dependent. They change gradually and incrementally. Institutional change is constrained by the legacy of the past, and one of the strongest constraints is the attitude of elites defending the status quo, which is in their interest, as Ogilvie correctly assumes. In his later work, Douglass North also paid more attention to the fact that institutional arrangements could be far from efficient, and

---

[35] S. Ogilvie, 'Whatever Is, Is Right? Economic Institutions in Pre-Industrial Europe', *Economic History Review* 60 (2007), pp. 662–3; B. J. P. van Bavel, *Manors and Markets: Economy and Society in the Low Countries, 500–1600* (Oxford, 2010), pp. 404–5; Van Bavel, 'History as a Laboratory', pp. 84–5.

[36] W. Norden, *Eine Bevölkerung in der Krise: Historisch-demographische Untersuchungen zur Biographie einer norddeutschen Küstenregion (Butjadingen 1600–1850)* (Hildesheim, 1984), pp. 222–3. Butjadingen was part of the county of Oldenburg which was ruled by the Danish Crown from 1667 to 1773.

that powerful groups in society could prevent the introduction of more efficient alternatives in order to serve their own interests.[37]

Whatever the causes, distributive effects or lack of social capital, institutional change was difficult to achieve. Nevertheless, in most of the North Sea area, water management institutions overall seem to have been efficient in the sense that they helped to prevent major disasters most of the time, which was the best that could be hoped for before the twentieth century. Moreover, institutional change was not impossible. Institutional arrangements were adapted to changing circumstances, here earlier, there later.[38] Questions to be answered in this book are: who promoted or obstructed institutional change and for what reasons? And what effects did institutional change or the lack thereof have on the development of agriculture? Because of the importance of institutions for the functioning of water management systems, and because of the copious archival sources from the thirteenth century onwards, studying the history of water management systems can provide an important contribution to the discussion on the influence of institutions on economic development.

Agrarian History from a Comparative Perspective

This book requires a comparative perspective. Comparison enables us to ascertain what distinguished the development of an area, and to get a grip on the determinants of the course it took. In economic history the units of comparison are often states. For this research, however, the state is not a useful unit of comparison, if only because the present-day states of Belgium and Germany have only existed since the nineteenth century. What is more important is that well into the twentieth century agriculture showed an enormous regional diversity, as did the organisation of water management. So for this period, the most useful unit of comparison is the region.[39] The state is only a relevant unit of comparison for the subject of agricultural policy and national water management policy from the middle of the nineteenth century onwards.

The comparative method can be applied by comparing the diverse regions of the North Sea Lowlands. Whilst facing the same environmental problems they often opted for very different solutions. All these lowlands are part of what Greg Bankoff has termed the North Sea Basin system, an area around the southern North Sea comprising eastern England,

---

[37] D. C. North, *Understanding the Process of Economic Change* (Princeton and Oxford, 2005); North, Wallis and Weingast, *Violence and Social Orders*, p. 17.

[38] M. van Tielhof, 'Forced Solidarity: Maintenance of Coastal Defences along the North Sea Coast in the Early Modern Period', *Environment and History* 21 (2015), pp. 319–50.

[39] Van Bavel, *Manors and Markets*, p. 396; Van Bavel, 'History as a Laboratory', p. 87; R. L. Hopcroft, *Regions, Institutions, and Agrarian Change in European History* (Ann Arbor, 1999), pp. 1–2.

Introduction

most of the coastal Low Countries, north-western Germany, and small parts of northern France and western Denmark, 'continually shaped and reshaped by the processes of storm, flood and erosion'.[40] All in all, the lowlands – here defined as areas close to or even below sea level that require water management systems for any form of agriculture that is more intensive than the gathering of hay, reed or sedge – studied in this book cover a land area of some 35–40,000 km^2, of which about half are situated in the Netherlands, a fifth in England, one sixth in Germany, and less than one tenth in Belgium.[41] Small lowland areas are located in northern France and southern Denmark. Within this basin system specific landscapes emerged that showed many similarities, but also variations caused by differences in soil conditions, social relations, economy, and state formation.

Regional research is a venerable tradition in Dutch agrarian historiography that has its origins in the 1930s and became prominent from the 1950s onwards through the work of the 'Wageningen School'. At first only single regions were studied, but from the 1990s scholars from outside the 'Wageningen School', such as Erik Thoen and Bas van Bavel, added a new element to the study of agricultural regions in the past by comparing several regions and the connections and interactions between them.[42] How to define such regions? Both Joan Thirsk and Erik Thoen have dealt with that question extensively. Thoen calls them 'social agrosystems', Thirsk calls them agricultural regions. They agree that such regions, or production systems as Thoens terms them, are not just defined by the physical environment. Social, economic and political factors, such as property relations and political power, are relevant too; in Thoen's eyes they are even more important than the environment.[43]

The distribution of landownership is an important factor for comparison, not only because land was by far the most important production factor in rural societies, but also because the water authorities were controlled by

[40] Bankoff, 'The "English Lowlands"', p. 4.
[41] G. P. van de Ven, ed., *Man-Made Lowlands: History of Water Management and Land Reclamation in the Neteherlands* (Utrecht, 2004), p. 17 (Netherlands); www.vvpw.be, retrieved 24 June 2016 (Belgium); K. Dierßen, 'Ökosysteme der Nordseemarschen', in L. Fischer, ed., *Kulturlandschaft Nordseemarschen* (Bredstedt and Westerhever, 1997), p. 16 (Germany); www.ada.org.uk, retrieved 24 June 2016 (England).
[42] P. J. van Cruyningen, 'A Cat with Nine Lives: Rural History in the Netherlands after 1900', *TSEG/The Low Countries Journal of Social and Economic History* 11/2, pp. 133, 136–7, 146–7.
[43] J. Thirsk, *England's Agricultural Regions and Agrarian History, 1500–1750* (Basingstoke and London, 1987), pp. 11–19; E. Thoen, 'Social Agrosystems as an Economic Concept to Explain Regional Differences: An Essay Taking the Former County of Flanders as an Example', in B. J. P. van Bavel and P. Hoppenbrouwers, eds., *Land Holding and Land Transfer in the North Sea Area (Late Middle Ages – 19th Century)* (Turnhout, 2004), pp. 47–66.

landowners. They could decide to change the water management system or to prevent change. According to some scholars an unequal distribution of landownership can have pernicious consequences. Tim Soens has demonstrated that in the Flemish coastal plain a shift in landownership from smallholders to large landlords resulted in lower investment in flood protection in the late Middle Ages.[44] Another relevant issue is whether there existed differences in attitudes towards water management between various groups of landowners, such as the church, the nobility, farmers and town dwellers. Research on the south-west of the Netherlands in the early modern era has shown that urbanites were more prepared to invest in flood defences than were noblemen or ecclesiastical landowners.[45]

Political power is an important factor because of the influence of the process of state formation on the development of institutional arrangements regarding water management. So comparison of regions within different states is also required. On the western shore of the southern North Sea only one state existed: England. On the eastern shore the situation was much more complicated. There the principalities and republics that preceded the modern states of Belgium, Germany and the Netherlands must be studied. There were many of these, and their borders shifted. Here areas have largely been selected from the largest principalities that emerged during the Middle Ages and that survived until the Napoleonic period.

Thus far the approach outlined here is similar to that of historians such as Erik Thoen, Bas van Bavel and Tim Soens, who use a comparative approach and attribute much importance to issues of property and power, which is not surprising since they are influenced by the work of neo-Marxist Robert Brenner, who perceives 'social property relations' as the motor of change in medieval and early modern agrarian societies.[46] The Brennerian approach has been of immense value to the rural historiography of the Low Countries, but by focusing so strongly on property and power, the Brennerian historians have, in my view, paid insufficient

---

[44] T. Soens, 'Floods and Money: Funding Drainage and Flood Control in Coastal Flanders from the Thirteenth to the Sixteenth Centuries', *Continuity and Change* 26 (2011), pp. 348–9.

[45] P. J. van Cruyningen, 'From Disaster to Sustainability: Floods, Changing Property Relations and Water Management in the South-Western Netherlands', *Continuity and Change* 29 (2014), pp. 257–8.

[46] Brenner, 'Agrarian Class Structure and Economic Development in Pre-Industrial Europe', in T. H. Aston and C. H. E. Philpin, eds., *The Brenner Debate: Agrarian Class Structure and Economic Development in Pre-Industrial Europe* (Cambridge, 1985), pp. 10–63, idem, 'The Low Countries in the Transition to Capitalism', in P. Hoppenbrouwers and J. L. van Zanden, eds., *Peasants into Farmers? The Transformation of Rural Economy and Society in the Low Countries (Middle Ages – Nineteenth Century) in Light of the Brenner Debate* (Turnhout, 2001), pp. 275–338.

Introduction

attention to natural factors such as soil and the availability or excess of water. They tend to play down the importance of environmental variables for the development of social agrosystems.[47]

However, as Jan Bieleman has remarked, 'farming ultimately deals with managing biological processes',[48] meaning that factors such as climate, soil type and fertility, and the availability or excess of water are indeed important elements of the social agrosystem. They pose constraints but also offer opportunities to humans settling in an area, especially in a challenging environment such as a reclaimed wetland. This book follows Bruce Campbell's advice and aims to bring nature back in as a historical protagonist.[49] That does not mean a reversion to environmental determinism, because people can and do make choices within the boundaries posed by the constraints of the wetland environment. This book will show that the result was a bewildering variety of agricultural practices and water management institutions. All those variations, however, had their origins in responses to the specific demands of the reclaimed wetlands.

Culture is not mentioned as an aspect of social agrosystems or agricultural regions, probably because agrarian historians prefer tangible and measurable variables. Living in reclaimed wetlands and having to deal with the perpetual risk of flooding leads to specific cultural responses, however, which influenced behaviour. Petra van Dam calls this an 'amphibious culture'.[50] These responses could take very different forms, from a 'fatalistic' acceptance of disasters as a frequent life experience – often interpreted as divine retribution for sins – to a strong belief in technology as the solution.[51] These divergent attitudes can both be observed in the North Sea Lowlands and have influenced decisions on investment in flood defences and drainage systems.

*

This book covers the period from the tenth century, when the large-scale reclamations of the wetlands began, until the end of the twentieth century. This millennium is divided chronologically into five periods,

---

[47] Thoen, 'Social Agrosystems', pp. 48–9.
[48] J. Bieleman, 'Farming System Research as a Guideline in Agricultural History', in B. J. P. van Bavel and E. Thoen, eds., *Land Productivity and Agro-Systems in the North Sea Area Middle Ages – 20th Century* (Turnhout, 1999), p. 235.
[49] B. M. S. Campbell, 'Nature as an Historical Protagonist: Environment and Society in Pre-Industrial England', *Economic History Review* 62 (2010), pp. 281–314.
[50] P. J. E. M. van Dam, 'An Amphibious Culture: Coping with Floods in the Netherlands', in P. Coates, D. Moon, and P. Warde, eds., *Local Places, Global Processes* (Oxford, 2016), pp. 78–82.
[51] Bankoff, 'The "English Lowlands"', p. 5, Bankoff, 'Cultures of Disaster, Cultures of Coping: Hazard as a Frequent Life Experience in the Philippines', in Mauch and Pfister, eds., *Natural Disasters*, pp. 268–9.

covering the periods c. 900 to 1300, 1300 to 1550, 1550 to 1700, 1700 to 1880, and 1880 to 1980, roughly following the upward and downward trends of the lowland economy. An epilogue will discuss the period around the turn of the millenium, during which it has become clear that lowland farming and even the physical existence of these lowlands themselves are in jeopardy.

The book is mostly based on the existing historiography on lowland agriculture and water management in the Netherlands, Belgium, England and Germany. The literature will be discussed in the following chapters, but some general remarks should be made here, because the availability of literature has been one of the factors that has determined the selection of regions to be studied in this book. There is substantial literature on the history of agriculture in all four countries, but in Germany and England historiography on lowland agriculture is more limited than in the Dutch-Flemish area. German and English historians have concentrated more on upland agriculture, which is understandable considering that the lowlands cover only two per cent of the German land area and less than ten per cent of that of England. Of the English lowlands, agriculture in South Lincolnshire has been studied most intensively thanks to the work of H. E. Hallam and Joan Thirsk.[52] In Germany, there is hardly any literature on the agrarian history of the lowlands in the Middle Ages and early modern period.[53] Much work on the German lowlands has been published in the yearbooks and publication series of regional historical societies; the regional or local focus of that work means that it sometimes lacks a view of the broader context. There is also a tendency in German historiography to focus more on the social and cultural aspects of rural history, instead of on the more economic approach that is applied in this book.[54]

The geographical distribution of literature on the history of water management is even more unbalanced. Historiography on this subject in the Netherlands and Germany is immense, to the extent that in the Netherlands and north-western Germany the 'fight against the waterwolf' is part of national and regional identity, with regional historical societies, lawyers, geographers, hydraulic engineers, and a few historians publishing extensively on water management history. Here too, there often is a strong

---

[52] H. E. Hallam, *Settlement and Society: A Study of the Early Agrarian History of South Lincolnshire* (Cambridge, 1965); J. Thirsk, *English Peasant Farming: The Agrarian History of Lincolnshire from Tudor to Recent Times* (2nd edn., London and New York, 1981).

[53] O. S. Knottnerus, 'Agrarverfassung und Landschaftsgestaltung in den Nordseemarschen', in L. Fischer, ed., *Kulturlandschaft Nordseemarschen*, pp. 88, 102–5.

[54] P. Blickle, 'German Agrarian History during the Second Half of the Twentieth Century', in E. Thoen and L. Van Molle, eds., *Rural History in the North Sea Area: An Overview of Recent Research (Middle Ages – Twentieth Century)* (Turnhout, 2006), pp. 149, 161–3.

Introduction

regional or local focus, and authors mostly concentrate on the legal and technical aspects of water management. The environmental and economic aspects only started to receive attention from the 1990s onwards.[55] The Flemish historiography of water management is considerably more modest than that of Germany and the Netherlands, but thanks to the work of Tim Soens our knowledge has been enormously increased in recent years.[56] English historiography on water management is much less extensive and shows even more imbalances than that of the continental countries. Following the 1940 classic by H. C. Darby on the draining of the Fens,[57] many books and papers have been published on the drainage projects of the seventeenth century, mostly focusing on the riots that accompanied the implementation of these projects. The emergence and functioning of water management systems in the periods before and after the reign of the Stuarts have received much less attention.

Due to these imbalances the existing historiography shows several lacunae, which makes consequent systematic comparison between all regions in all periods impossible. This means that this book will be mostly focused on the Low Countries and that comparison will be made with English and German regions when and where possible. Prior to introducing the main research areas, it will be beneficial to give the reader an impression of where the regions are situated: their characteristics will be discussed in the following chapters. On the eastern shore of the North Sea, from south-west to north-east, there is firstly the Flemish coastal plain, stretching from Sangatte west of Calais up to Antwerp. To the north of Flanders the archipelago of Zeeland and southern Holland, in the delta of Scheldt and Meuse, is situated. North of this lie the vast peaty plains of mainland Holland and Utrecht. In the centre of the Netherlands the lowlands reach deep inland into the floodplains of the rivers Rhine, Waal, Meuse and IJssel. This is the Dutch central river area. A smaller lowland area is situated to the east of the Zuider Zee in Guelders and Overijssel. The Wadden coast stretches from the Dutch province of Friesland eastward and northward to south-western Jutland in Denmark. It is divided into many smaller regions (Friesland, Groningen, Ostfriesland, Jever, Butjadingen, Wursten, Hadeln, Kehdingen, Altes Land, Dithmarschen, Nordfriesland), which in the past were rather isolated from each other. On the western shore of the southern North Sea there is no contiguous lowland zone. The main lowland regions here that will be discussed are Romney Marsh in Kent, the marshes in the Thames estuary, the vast peaty plains in Cambridgeshire and southern Lincolnshire, and Hatfield Chase and the Isle of Axholme on the border of Lincolnshire and Yorkshire.

During the heady years of the 'new history' in the decades following the Second World War, when many historians believed that it would

---

[55] See Van Cruyningen, 'A Cat', pp. 147–8.
[56] See bibliography for relevant works.
[57] H. C. Darby, *The Draining of the Fens* (Cambridge, 1940).

be possible to formulate generalised laws to explain historical change, economic historian M. M. Postan warned – very wisely – that this was exagerratedly optimistic. According to Postan, historians cannot produce general laws because they write accounts of single combinations of circumstances. It is 'in the revelations of the general in the particular that the contribution of the historical method to social science will be found'.[58] That is what this book is trying to do. It provides a survey of the history of farming on the North Sea coast – the particular – to answer more general questions concerning the influence of environment and institutions on agricultural development. It will show that institutions were crucial for the development of agriculture, but that the institutions themselves developed in continuous interaction with environmental changes, which in their turn were to a large degree the result of interference by humans.

---

[58] M. M. Postan, *Fact and Relevance: Essays on Historical Method* (Cambridge, 1971), pp. 31–2; for the disillusionment that set in during the 1970s see L. Stone, 'The Revival of the Narrative: Reflections on a New Old History', *Past and Present* 85 (1979), pp. 3–24.

1

The Great Reclamation: Transformation of the Wetlands, c. 900–1300

In Europe, the centuries around the beginning of the second millennium are sometimes termed the period of the Great Reclamation. During these years vast tracts of woodlands and wetlands were reclaimed.[1] This chapter will show how, and under which circumstances, the Great Reclamation, when hundreds of thousands of hectares of wetlands were transformed into farmland, took place in the North Sea Lowlands. These circumstances were of great importance because they determined the economic and social relations that emerged in the reclaimed areas, and they in their turn influenced the organisation of water management and agricultural development in the long run.

A discussion of the physical environment and settlement patterns on the eve of the Great Reclamation will be followed by a section on the Great Reclamation itself, showing how it came about in different landscapes and under varied institutional arrangements. This second section will be followed by two sections focusing on the long-term consequences of the medieval wetland reclamations. The first is on the unintended consequences of the reclamations, such as the lowering of the surface level, which are haunting farmers and engineers to the present day. The second is on the political and economic freedom of the inhabitants of the wetlands. People in reclaimed wetlands tended to enjoy more freedom than inhabitants of older settled areas, but how much freedom did they actually enjoy and what consequences could this have had for agricultural development? But we will begin by addressing the demographic, economic, social and political developments in the North Sea area during the high Middle Ages.

The North Sea Area during the Great Reclamation

The Great Reclamation took place in the context of demographic and economic expansion. Estimates indicate that the European population

---

[1] M. Bloch, *Les caractères originaux de l'histoire rurale Française* (2nd edn., Paris, 1952, reprint 1999), pp. 56–7.

increased from 30–40 million around AD 1000 to 70–80 million by 1300. This may look impressive, but it means that population increased by no more than 0.25 per cent per year. In fact, population growth had already started in the seventh century, but with interruptions caused by plague and Viking raids. From the late tenth century, population growth became sustained and was accompanied by commercialisation and urbanisation.[2] This change in the structure of economy and society was of more importance than the increase in population numbers as it created opportunities for interregional trade and specialisation. The North Sea Lowlands shared in this development. The population of what is now the territory of the Netherlands and Belgium is estimated to have doubled between the early seventh and early tenth centuries and then to have quintupled from c. 0.4 million around AD 900 to two million around 1300. The increase of the rural population was one of the reasons why the reclamation of the inhospitable wetlands along the southern North Sea coast began in the high Middle Ages.[3]

Urbanisation can also be a cause of land reclamation, as an increasing urban population requires more food, as well as more raw materials for urban industry, such as wool, flax and dyestuffs. During the first phase of the Great Reclamation, however, this hardly played a part in the North Sea Lowlands. There were very few urban centres and those that did exist were situated in the far south, in the Meuse valley. Even in Flanders, the earliest province in the coastal part of the Low Countries to urbanise, major urban centres only began to emerge in the second half of the eleventh century.[4] In the northern Low Countries urbanisation began even later. In Holland and Zeeland, around 1200, only some five per cent of the population may have lived in towns of more than 5000 inhabitants. Then rapid growth set in, and by around 1300 the percentage of the population living in cities had increased to 13 per cent.[5] In England, by 1300, London, situated on the edge of the Lowlands, became a major city with some 80,000 inhabitants.[6] Raising urban demand may have stimulated land reclamation from the twelfth century onwards, but the initial thrust towards reclamation had come from the increasing rural population.

---

[2] W. P. Blockmans and P. Hoppenbrouwers, *Introduction to Medieval Europe, 300–1550* (London and New York, 2007), p. 107.

[3] Van Bavel, *Manors and Markets*, p. 36; B. Ayers, *The German Ocean: Medieval Europe around the North Sea* (Sheffield, 2016), pp. 8–9; H. van der Linden, 'Het platteland in het Noordwesten met nadruk op de occupatie circa 1000–1300', in *Algemene Geschiedenis der Nederlanden*, vol. 2 (Haarlem, 1982), pp. 53–4.

[4] D. Nicholas, 'Of Poverty and Primacy: Demand, Liquidity, and the Flemish Economic Miracle', *American Historical Review* 96 (1991), p. 18.

[5] Van Bavel, *Manors and Markets*, pp. 36, 280–1.

[6] C. Dyer, *Making a Living in the Middle Ages: The People of Britain 850–1520* (New Haven and London, 2002), p. 190.

Raising production to feed an increasing population can be achieved in two ways: by raising the productivity of the existing agricultural land or by increasing the amount of land under cultivation. Increasing land productivity was not easy in the Middle Ages, because technological change was slow. Often, the only way to increase yields was to cultivate the land more intensively by applying more labour. The alternative was to reclaim areas such as fens and marshes. Around AD 1000 the rural population preferred to reclaim the vast tracts of relatively untouched land that still existed in Europe. In the North Sea Lowlands these consisted mainly of marshes and fens. For the wet, heavy soils of these areas, ploughs were required that could turn the topsoil. This was the so-called mouldboard plough. This plough consisted of a horizontal beam to which were connected a coulter, a knife that cut the topsoil vertically, and a share, a knife that cut horizontally. Behind these two elements the curved mouldboard was constructed, which turned the slice that the two knives had cut from the soil. By turning the topsoil, weeds were killed and mineral nutrients were brought to the surface from the subsoil, making cultivation of wetlands feasible. The pattern of ridge and furrow that was created in this way also improved drainage of the field. Through the furrows, water could flow towards the drainage ditches.[7] Most parts of these ploughs were wooden, but share and coulter were made of iron. Many mouldboard ploughs were wheeled, but some lighter types only had a wooden 'foot' to keep the plough stable. Historians have assumed that this plough was introduced in the Carolingian period and that this was one of the changes that made reclamation of the Lowlands feasible.[8] Recent archaeological research has raised doubts, however. It appears that in the Frisian lands the mouldboard plough was already known at the beginning of the first millennium AD.[9]

The sustained population growth and land reclamation which occurred from the tenth century onwards may at least in part have been caused by a change in the climate.[10] During the European Medieval Climate Anomaly, from the tenth to the late thirteenth century, mean annual temperatures were 0.5 °C higher than in the twentieth century and the summers were hotter. The 900–1050 period in particular was hot.[11] Data like this has

[7] R. C. Hoffmann, *An Environmental History of Medieval Europe* (Cambridge, 2014), pp. 122–4; Blockmans and Hoppenbrouwers, *Introduction*, pp. 107–10.
[8] L. White, *Medieval Technology and Social Change* (Oxford, 1962), pp. 41–2, 52–3.
[9] J. Nicolay and H. Huisman, 'Ploughing the Salt Marsh: Cultivated Horizons and their Relation to the Chronology and Techniques of Ploughing', in J. Nicolay and M. Schepers, eds., *Embracing the Salt Marsh: Foraging and Food Preparation in the Dutch–German Coastal Area up to AD 1600* (Eelde, 2022), pp. 71–2.
[10] B. M. S. Campbell, *The Great Transition: Climate, Disease and Society in the Late-Medieval World* (Cambridge, 2016), p. 50.
[11] C. Pfister and H. Wanner, *Klima und Gesellschaft in Europa: Die letzten Tausend Jahre* (Bern, 2021), p.167; Hoffmann, *An Environmental History*, pp. 320–2;

to be used carefully, as long-term averages can hide strong short-term temporal swings in temperatures. Moreover, temperature is not the only factor to be taken into account; precipitation and extreme events such as storms are also relevant. Moreover, there are regional variations to consider, and climate changes can have divergent impacts on different environments.[12] A drop of 0.5 °C, for instance, will have a more damaging impact in northern than in more southern latitudes, because it shortens an already short growing season. One should therefore be very careful when using climate data in the interpretation of agricultural development. In the Netherlands, for instance, there is some evidence that the first half of the tenth century was exceptionally dry, which may have stimulated reclamation of the peat areas.[13] However, evacuation of excess water from the peat areas in Holland and Friesland did become easier due to floods that created sea inlets. As the peat lands could now drain towards these inlets, this may explain drier conditions in the peat areas.[14]

In older research, expansion of population and settlement of the coastlands was often explained by reduced influence of the sea, a regression phase. Conversely, when sea levels rose – a transgression phase – it was assumed that land had to be given up. From the 1970s onwards this rather environmentally deterministic thinking became increasingly criticised.[15] Nowadays, most researchers will agree that economic, demographic and technological developments were more relevant for expansion or contraction of settlement than sea level rise. Transgressions and regressions did exist, but not all tidal basins responded in the same way. Occupation of the coastal zone was hardly influenced by sea level rise,

---

W. Behringer, *Kulturgeschichte des Klimas: Von der Eiszeit bis zur globalen Erwärmung* (Munich, 2007), pp. 103–5; A. G. Jongmans et al., *Landschappen van Nederland: Geologie, bodem en landgebruik* (Wageningen, 2013), p. 56.

[12] J.-P. Devroey and A. Nissen, 'Early Middle Ages, 500–1000', in Thoen and Soens, eds., *Struggling with the Environment*, pp. 21–2.

[13] H. A. Heidinga, 'Indications of Severe Drought during the Tenth Century AD from an Inland Dune Area in the Central Netherlands', *Geologie en Mijnbouw* 63 (1984), pp. 243–7; J. M. Bos, B. van Geel and J. P. Pals, 'Waterland – Environmental and Economic Changes in a Dutch Bog Area, 1000 AD – 2000 AD, in H. H. Birks et al., eds., *The Cultural Landscape – Past, Present and Future* (Cambridge, 1988), p. 323, C. de Bont, *Amsterdamse boeren: Een historische geografie van het gebied tussen de duinen en het Gooi in de middeleeuwen* (Hilversum, 2014), p. 81.

[14] Van Tielhof and Van Dam, *Waterstaat*, pp. 24–5, 30–1; H. Schoorl, *Zeshonderd jaar water en land: Bijdrage tot de historische geo- en hydrografie van de Kop van Noord-Holland 1150–1750* (Groningen, 1973), pp. 11–13.

[15] See, for instance, the contributions in A. Verhulst and M. K. E. Gottschalk, *Transgressies en occupatiegeschiedenis in de kustgebieden van Nederland en België* (Gent, 1980).

which was – at 5 centimetres on average per century – modest during the last 2000 years.[16]

Population growth forced people to look for opportunities in the vast wetlands, made possible by changing hydrological circumstances, climate change and technology, and such developments were also stimulated by urbanisation after 1100. Did the state also play a part in the Great Reclamation? To answer that question we need to know what kind of 'states' existed in the research area around the turn of the millennium. States in the meaning of national states, coercion-wielding organisations governing multiple regions and their cities by means of centralised, differentiated, and autonomous bureaucratic structures, as defined by Charles Tilly, did not yet exist around the year 1000. The development towards this type of state only began in this period.[17] The only country in the southern North Sea area that already had a centralised royal bureaucracy in the eleventh century was England, but even this state was vulnerable and could not wield the amount of power modern states now can rely on.[18] On the opposite shore of the North Sea, following the slow disintegration of the Carolingian empire during the ninth century, the coastal areas were divided between the kingdom of France and the Holy Roman Empire. The power of king and emperor was limited, however, and after a while local power holders managed to create semi-autonomous units which we can call territorial principalities. The first and for a long time most powerful territorial prince in the research area was the count of Flanders, whose territory had already emerged in the late ninth century. Flanders was the only principality on the eastern shore of the southern North Sea that was part of France. In the Holy Roman Empire the counts of Holland and Oldenburg and those of Guelders and Brabant became powerful overlords from the eleventh to twelfth centuries onwards. To these worldly lords must be added several prince bishops. In the northwest corner of the empire the most powerful of these were the bishop of Utrecht and the archbishop of Bremen. In the southern part of the Jutland peninsula the king of Denmark struggled for power with the count of Holstein, whilst Frisian immigrants enjoyed some autonomy.[19]

Territorial principalities did not emerge everywhere along the eastern shore of the North Sea. In many places along the Wadden coast from

---

[16] P. C. Vos, *Origin of the Coastal Landscape: Long-Term Landscape Evolution of the Netherlands during the Holocene, Described and Visualized in National, Regional and Local Palaeogeographical Maps* (Groningen, 2015), pp. 8–11; K.-E. Behre, *Landschaftsgeschichte Norddeutschlands: Umwelt und Siedlung von der Steinzeit bis zur Gegenwart* (Neumünster, 2008), pp. 28–32, 76–7, 324.

[17] C. Tilly, *Coercion, Capital, and European States, AD 990–1992* (Cambridge, MA and Oxford, 1992), pp. 1–5.

[18] Blockmans and Hoppenbrouwers, *Introduction*, pp. 170–1.

[19] F. Müller and O. Fischer, *Das Wasserwesen an der schleswig-holsteinischen Nordseeküste* (3 vols., Berlin, 1917–57), vol. 3/1, pp. 114–18.

Friesland to Dithmarschen central authority disintegrated. Here some 40 to 50 *Landgemeinden* (rural communes) emerged. These were small territories, some not larger than a few villages, and they were 'free' areas, meaning that they recognised no overlord, and their society was based on the free ownership of land. Since most of the *Landgemeinden*, such as Oostergo, Westergo, Fivelgo and Rüstringen, were Frisian, the expression 'Frisian freedom' was often used, but some of these territories, such as Dithmarschen, Drenthe and Stedingen, were actually Saxon. They are sometimes called 'farmers' republics' because the heads of the free landowning families were legally equal and all had a say in the government of the area.[20] In reality, the *Landgemeinden* were oligarchies, in which the wealthiest families held the most power. Recent research even claims that the leading families already perceived themselves as nobles in the high Middle Ages.[21] Although prone to feuding, in peaceful times these small republics were prosperous communities.[22] Until c. 1300 they functioned well and were even able to ward off attacks by territorial lords, but the question is whether they would be able to solve coordination problems in water management.

The territorial principalities developed into small proto-states with an embryonic central bureaucracy. What was especially important for the development of water management was that the territorial lords had managed to lay their hands on the royal rights to the wilderness, either by grant from the king or emperor or through usurpation. By claiming ownership of the wilderness of fens, marshes and the foreshore of already occupied land they could reclaim these lands themselves or, more usually, grant parts of the wilderness for reclamation to individuals or groups. In this way, they could stimulate the reclamation of fens and marshes. This happened especially in the Low Countries, where the counts of Flanders and Holland and the bishop of Utrecht had gained control of these rights at an early stage, before the turn of the millennium.[23]

---

[20] W. Ehbrecht, 'Gemeinschaft, Land und Bund im Friesland des 12. bis 14. Jahrhunderts', in H. van Lengen, ed., *Die Friesische Freiheit des Mittelalters – Leben und Legende* (Aurich, 2003), pp. 136, 158; H. van Lengen, 'Bauernfreiheit und Häuptlingsherrlichkeit', in K.-E. Behre and H. van Lengen, eds., *Ostfriesland: Geschichte und Gestalt einer Kulturlandschaft* (Aurich, 1996), pp. 113–22; H. Nijdam, 'Preface to the Edition and Translation of the Old Frisian Tekst', in H. Nijdam, J. Hallebeek and H. de Jong, eds., *Frisian Land Law: A Critical Edition of the Freeska Landriucht* (Leiden and Boston, MA, 2023), p. 9.

[21] G. de Langen and J. A. Mol, *Friese edelen, hun kapitaal en boerderijen in de vijftiende en zestiende eeuw: De casus Rienck Hemmema* (Amsterdam, 2022), p. 188.

[22] O. S. Knottnerus, 'De Waddenzeeregio, een uniek cultuurlandschap', in D. van Marrewijk and A. Haartsen, eds., *Waddenland: Het landschap en cultureel erfgoed in de Waddenzeeregio* (Groningen and Leeuwarden, 2001), pp. 49–50.

[23] H. van der Linden, *De cope: Bijdrage tot de geschiedenis van de openlegging der Hollands-Utrechtse laagvlakte* (Assen, 1955), pp. 81–4; C. Dekker, *Het Kromme*

Landscape and Settlement at the End of the First Millennium

Before discussing the Great Reclamation, it is useful to take a look at landscape and settlement in the southern North Sea Lowlands at the end of the first millennium. What kind of landscapes and soils can be observed here during the reign of the Ottonian emperors? Which areas were already settled and what were the means of subsistence of the people living there?

The geography of the various lands on the eastern shore of the southern North Sea from Calais to southern Jutland around AD 900 shows remarkable similarities. Some 4500 years ago, the coast of the western part of the Low Countries was protected from the sea by a range of dunes. The low, flat area behind those dunes changed into a humid freshwater environment fed by rivers. In the stagnant water plants lived and died. On dying, the plants sank to the bottom, and after a while the remains of partly decayed dead plants formed a substrate on which new plants could grow. The process was then repeated, and in this way peat formed, a soil type that consists of 80 per cent or more water and at most 20 per cent decayed plant particles.[24] Peat formation also occurred in depressions further inland. Once peat starts growing, it can rise to several metres above sea level and cover more and more of the surrounding area. Around AD 900, much of the coastal area was covered by metres-high raised peat bogs. These were extremely wet environments, where the water table almost reached surface level. They were difficult to penetrate and almost impossible to cultivate unless the water table could be lowered considerably.

The fertility of peat soils varies. Most fertile are the low peat areas that are fed by nutrient-rich river water or ground water (eutrophic peat). When peat bogs continue to grow, however, they will eventually reach such a height that ground water or river water can no longer reach the highest layers. From then on, peat is formed that is only fed by rainwater and is poor in nutrients (oligotrophic peat). Many of the raised peat bogs that were predominant in most of the Netherlands were too high for the ground water to reach the top, so they were dependent solely on rainwater. Where the coastal dunes were breached, especially in the south-western delta, the land was regularly flooded, causing a layer of marine clay to be deposited on top of the peat. Much of the south-west was characterised by these clay-on-peat soils. The Flemish coastal plain,

---

Rijngebied in de Middeleeuwen: Een institutioneel-geografische studie (Zutphen, 1983), p. 164.

[24] Hoffmann, *An Environmental History*, p. 136; Ayers, *The German Ocean*, p. 9; Jongmans et al., *Landschappen*, pp. 544–5; I. D. Rotherham, *Peatlands: Ecology, Conservation and Heritage* (London and New York, 2020), pp. 7–8.

stretching from the area of Dunkirk to present-day Zeeland Flanders, was a relatively narrow band of salt marshes. These also were often clay-on-peat soils. Behind this coastal band an area of peat bogs was situated, beginning north-west of Ghent and reaching well into western Brabant.[25]

Other soil types in the Netherlands were river and marine clay. These soils can be very fertile, but a lot depends on their structure. Light clays, composed of relatively large soil particles, are well drained and can be easily tilled. Heavier clays, of small soil particles, are often impermeable, and they tend to get waterlogged, but then become rock hard during dry periods. The difference between these two types of clay can be observed in another part of the Low Countries, the central river area. Here, sediment was deposited by the rivers Rhine and Meuse. The largest and heaviest particles were deposited close to the river, where they formed relatively high levees and stream ridges with well-drained light clays. The smaller and lighter particles were deposited further on, where they resulted in heavy impermeable clays that were badly drained and very difficult to cultivate. In later times most of them were agriculturally useless back swamps.[26]

In the north, on the Wadden coast, the landscape was somewhat different. There was no continuous range of dunes there, but instead a range of barrier islands, so the sea had more influence on the formation of the landscape. Behind the barrier islands marine clay was deposited and a broad zone of salt marshes (*kwelders*) was formed. As in the river clay area, the heaviest particles were deposited first, resulting in relatively high ridges of friable, well-drained clay. Behind these ridges, further from the sea, during the third to sixth centuries AD, heavy clays were deposited.[27] These were badly drained and difficult to plough. Even more landward vast peat areas could be observed reaching well into Drenthe, where they bordered on Pleistocene sands. This landscape with its succession of clay, peat and sandy soils stretched all along the southern North Sea coast from the present-day province of Friesland to the estuaries of Elbe and Weser.[28] In Carolingian times there was no clear border here between land and sea, just endless mudflats and salt marshes. This only changed with the construction of seawalls in the high Middle Ages. The western coast of Schleswig-Holstein now belongs to the Wadden Sea area, but in the days of Charlemagne this was a peat area behind a barrier of dunes and salt marshes. This changed in the Middle Ages due to catastrophic

---

[25] A. Verhulst, *Landschap en landbouw in middeleeuws Vlaanderen* (Brussels, 1995), pp. 77–81; K. A. H. W. Leenders, *Verdwenen venen: Een onderzoek naar de exploitatie van thans verdwenen venen in het gebied tussen Antwerpen, Turnhout, Geertruidenberg en Willemstad (1250–1750)* (Tilburg, 2009), p. 307.
[26] Jongmans et al., *Landschappen*, pp. 381–3.
[27] Jongmans et al., *Landschappen*, pp. 812–13.
[28] Behre, *Landschaftgeschichte*, pp. 33–4.

floods that destroyed the barrier and much of the peat lands behind it (see also p. 86).[29]

Almost all of these Dutch and German shores of the southern North Sea can be termed lowland. On the opposite shore the situation is somewhat different. England has more steep coasts, so the coastal lowlands there are less extensive. Nevertheless, there are some important lowlands on the North Sea coast, such as Romney Marsh in Kent, the Norfolk Broads, the Thames and Humber estuaries, the marshes along the Lincolnshire coast, and especially the fenlands around the Wash. These fenlands show remarkable similarities with those of the Holland–Utrecht plain. Both cover a large area of 3000 km^2 or more, and through both the excess water of a very large catchment area must be carried to the sea. In the case of the Holland–Utrecht plain the catchment area even stretches as far as Switzerland. Both experienced difficulty evacuating the enormous amounts of water that reached them from the uplands because they were locked off from the sea, in the Dutch case because of a continuous dune ridge, in the English case because of a relatively high ridge – about four metres above sea level – of 'silt fens' consisting of light to medium clays and comparable with the *kwelders* of the eastern shore.[30]

In the Carolingian period, settlement in the area that was to become the Netherlands remained mostly limited to the coastal barriers, the riverbanks and the sandy uplands in the east and south.[31] The Old Dunes that protected the lowlands in the ninth century, were densely populated and cultivated. The Young Dunes that formed from the tenth century to the east and on top of the Old Dunes remained a wilderness.[32] Although the sandy soils were not very fertile, peasants preferred them because they were well-drained and could be easily cultivated with simple implements. For settlers, good natural drainage and easy cultivation were decisive.

For the same reasons, the light clay soils of the higher stream ridges in the central river area were among the first in the lower parts of the country to be reclaimed, already in the Merovingian period. At some places there

[29] D. Meier, 'From Nature to Culture: Landscape and Settlement History of the North Sea Coast of Schleswig-Holstein, Germany', in E. Thoen et al., eds., *Landscapes or Seascapes? The History of the Coastal Environment in the North Sea Area Reconsidered* (Turnhout, 2013), pp. 93, 96–100; Müller and Fischer, *Das Wasserwesen*, vol. 3/2, pp. 17–19.

[30] H. Godwin, *Fenland: Its Ancient Past and Uncertain Future* (Cambridge, 1978), p. 137.

[31] G. J. Borger, 'Draining – Digging – Dretching: The Creation of a New Landscape in the Peat Areas of the Low Countries', in J. T. A. Verhoeven, ed., *Fens and Bogs in the Netherlands: Vegetation, History, Nutrient Dynamics and Conservation* (Dordrecht, 1992), p. 137.

[32] F. Beekman, *De Kop van Schouwen onder het zand: Duizend jaar duinvorming en duingebruik op een Zeeuws eiland* (Utrecht, 2007), p. 57; Van de Ven, *Man-Made Lowlands*, pp. 39–41.

was even continuous occupation from the Roman period onwards.[33] On the stream ridges arable agriculture was practised, and the low back-swamps were used for gathering of fuel and – if dry enough – for grazing cattle. In Holland and Zeeland settlement was mostly limited to the dunes and the sandy soils immediately behind them, and in Holland also to the clay banks along rivers such as the Oude Rijn. The immense peat and clay-on-peat areas in these provinces were almost completely uninhabited. Along the northern coast, people had settled on artificial mounds, called *terpen* or *wierden*, on the clay ridges. This *terpen*-area stretched further along the German Wadden coast, where the mounds were called *Warften* or *Wurten*. Such artificial mounds also existed on the Flemish coastal plain and in Zeeland, where they were called *werven*.[34] On the opposite shore of the North Sea, settlement in the lowlands was mostly concentrated on the higher silt ridges and on the islands of bedrock situated in the fen areas.[35]

Despite the forbidding aspect of the muddy marsh areas, they appear to have been quite densely populated, especially in the north.[36] This has led earlier historians, such as Bernard Slicher van Bath, to the view that since cereals could hardly be grown in this environment, the Frisians must have been stock farmers and traders, who purchased bread grains by exchanging animal products and cloth – the famous *pallia Fresonica* – with other regions in the North Sea area and the Baltic.[37] However, the outcome of archaeological research shows that this view is no longer tenable. The difference between Frisia and the inland areas was smaller than assumed. Cereals such as barley, rye and emmer wheat, and fibre plants such as flax and hemp, were grown on the slopes of the *terpen*.[38] In the period shortly before the start of the Great Reclamation, from c. 850 AD, the layout of the *terpen* was changed: buildings were moved to the slopes and the top was now used for crop cultivation, particularly

---

[33] P. A. Henderikx, 'The Lower Delta of the Rhine and the Maas: Landscape and Habitation from the Roman Period until c. 1000', *Berichten van de Rijksdienst voor het Oudheidkundig Bodemonderzoek* 36 (1986), pp. 483, 490.

[34] Tys, 'Medieval embankment', pp. 207–9; Soens, Tys and Thoen, 'Landscape transformation', p. 136; J. A. J. Vervloet, *Inleiding tot e historische geografie van de Nederlandse cultuurlandschappen* (Wageningen, 1984), p. 87.

[35] S. Rippon, *The Transformation of Coastal Wetlands: Exploitation and Management of Marshland Landscapes in North West Europe during Roman and Medieval Periods* (Oxford, 2000), pp. 169, 172.

[36] B. H. Slicher van Bath, 'The Economic and Social Conditions in the Frisian Districts from 900 to 1550', *A. A. G. Bijdragen* 13 (1965), p. 100.

[37] Slicher van Bath, 'The Economic and Social Conditions', pp. 103–6; Van Bavel and Hoffmann still adhered to this view in 2010 and 2014 (Van Bavel, *Manors and Markets*, pp. 212–18; Hoffmann, *An Environmental History*, pp. 71–5), but Vervloet already rejected it in 1984 (Vervloet, *Inleiding*, p. 64).

[38] M. Schepers, 'Variatie in cultuurplanten in Friese terpen', in J. Nicolay and G. de Langen, eds., *Friese terpen in doorsnede: Landschap, bewoning en exploitatie* (Groningen, 2023), p. 320.

FIGURE 1. *Terp* of Hogebeintum, Friesland, 2022. Photograph by the author. Like all remaining dwelling mounds this is only the mutilated remainder of a much larger structure, on which farms, arable fields and sometimes a church were situated. Much of the fertile soil was removed during the nineteenth century to be used as fertiliser.

of productive winter grains. In this way, arable production could be increased. Moreover, experiments have demonstrated that the higher silt ridges in the *terpen* area were not only suitable for grazing, but also for growing salt-tolerant summer crops like barley and flax, made feasible by digging ditches and erecting low embankments. Therefore the region was considerably more self-sufficient in crop production than previously assumed.[39] Cattle and sheep breeding were important elements of the farming system of the Frisian coastlands, but basically the *terp* dwellers operated mixed farms.[40]

[39] J. Nicolay and G. de Langen, 'Synthese: De steilkanten onderling vergeleken', in Nicolay and De Langen, eds., *Friese terpen in doorsnede: Landschap, bewoning en exploitatie* (Groningen, 2023), p. 435; G. de Langen, *Middeleeuws Friesland: De economische ontwikkeling van het gewest Oostergo in de vroege en volle Middeleeuwen* (Groningen, 1992), p. 62; W. van Zeist, 'Milieu, akkerbouw en handel van middeleeuws Leeuwarden', in M. Bierma et al., *Terpen en wierden in het Fries-Groningse kustgebied* (Groningen, 1988), pp. 136–7; Nicolay and Huisman, 'Ploughing', p. 59.

[40] W. Prummel and H. C. Küchelmann, 'The Use of Animals on the Dutch and German Wadden Coast, 600 BC – AD 1500', in Nicolay and Schepers, eds., *Embracing the Salt Marsh*, pp. 114, 119; M. Schepers, C. Smit, and J. F.

For English and German marshlands there is also evidence of the cultivation of salt-loving crops in un-embanked areas.[41] In the coastal plain of Flanders and Zeeland hundreds of sheep farms were situated on *werven*. While the sheep grazed on the salt marshes, the farmers added to their own diet with fishing and fowling.[42] Although the coastal dwellers produced more food than was assumed in older historiography, they probably still needed to import some cereals. They could exchange wool, cloth and meat for grain, and that exchange was feasible because they could be easily reached by ship. Therefore trade was important for these coastal communities, but that does not mean that all Frisians were traders. They may even have profited from passive trade, whereby trade was mostly conducted by international traders, not by the *terp* dwellers themselves. Like the hobbits, most Frisians stayed at home to grow food and eat it.[43]

Another product of the coastal marshes was salt, indispensable for the preservation of food like meat, fish and dairy products. Salt production existed in all coastal marshes along the southern North Sea, in the Low Countries, England and Germany. It could be extracted from sand or mud, but particularly from peat saturated with sea water. From one cubic metre of wet peat 24 kg of salt could be produced, as experiments have demonstrated.[44] By its nature, this activity mostly took place in marshes or fens that were not embanked. There, the working of the tides ensured that the soil was permeated with salt. Sods were cut from the peat layers

---

Scheepers, 'Akkeren op de kwelder', in A. Nieuwhof and A. Buursma, eds., *Van Drenthe tot aan 't Wad: Over landschap, archeologie en geschiedenis van Noord-Nederland. Essays ter ere van Egge Knol* (Groningen, 2023), p. 50; M. Schepers and K.-E. Behre, 'A Meta-Analysis of the Presence of Crop Plants in the Dutch and German Terpen Area between 700 BC and AD 1600', *Vegetation History and Archaeobotany* 32 (2023), p. 316.

[41] S. Rippon, 'Human Impact on the Coastal Wetlands of Britain in the Medieval Period', in Thoen et al., eds., *Landscapes or Seascapes?*, p. 338; S. Krabath, 'Mittelalterlicher Deich- und Landesausbau im Nordwestdeutschen', in N. Fischer, ed., *Zwischen Wattenmeer und Marschenland: Deiche und Deichforschung an der Nordseeküste* (Stade, 2021), pp. 77–8.

[42] D. Tys, 'The Medieval Embankment of Coastal Flanders in Context', in Thoen et al., eds., *Landscapes or Seascapes?*, pp. 207–10.

[43] R. van Schaïk, 'Een samenleving in verandering: De periode van de elfde en twaalfde eeuw', in M. G. J. Duijvendak et al., *Geschiedenis van Groningen: Prehistorie – Middeleeuwen* (Zwolle, 2008), pp. 136–8; G. de Langen and J. A. Mol, 'The Distribution of Farmland on the Medieval and Prehistoric Salt Marshes of the Northern Netherlands: A Retrogressive Model of the (Pre-)Frisian Farm, Based on Historical Sources from the Early Modern Period', in Nicolay and Schepers, eds., *Embracing the Salt Marsh*, p. 49.

[44] D. Meier, 'Man and Environment in the Marsh Area of Schleswig-Holstein from Roman until Late Medieval Times', *Quaternary International* 112 (2004), p. 67.

and salt was processed from them. Salt trade became an important source of income in several regions, such as Nordfriesland.[45] The fuel required for the boiling process was also provided from the peat layers, the sods of which could be used as excellent fuel.

The wetlands of the southern North Sea area in Carolingian and Ottonian times were not wastelands. Some of the coastal marshes were relatively densely populated and produced salt, wool, meat, dairy products and cloth, and even crops such as barley and flax. All of this was possible without protecting the land from floods by way of seawalls. The peat areas were less accessible due to their waterlogged nature and were exploited less intensively. They provided fish, waterfowl and wood for the inhabitants of the treeless marshes, and for the Holland fen areas there is evidence for transhumance, the grazing of cattle during the dry summer months.[46]

The Great Reclamation

Around AD 1000 there was not much real virgin land remaining in western Europe. As demonstrated in the previous section, even the wetlands had been touched by humankind. The Great Reclamation mainly meant an intensification of human land use. For wetlands, landscape archaeologist Stephen Rippon has discerned three stages of intensity of use. First, there is exploitation of the resources present in the wetlands, by grazing animals, cutting reed, fishing or catching wildfowl. This occurred long before the Great Reclamation and left the wetland environment mostly unchanged. Only the most remote corners of the vast fenlands of Holland and England may have been free from such human interference. The second stage is modification, when humans try to improve productivity of the land by digging ditches or constructing low dikes, but the landscape remains vulnerable to floods. In the third stage the landscape is transformed: high dikes are built to keep the water of sea and rivers out, and more sophisticated drainage systems are designed to drain excess water.[47] The wetland is transformed into an agrarian landscape. What could be observed during the Great Reclamation was mostly transformation, but there were also areas where people limited their interference

---

[45] Rippon, *The Transformation*, pp. 42–6; T. Williamson, *The Norfolk Broads: A Landscape History* (Manchester, 1997), pp. 45–7; I. G. Simmons, *Fen and Sea: The Landscapes of South-East Lincolnshire* (Oxford, 2022), p. 20; Tys, 'The Medieval Embankment', p. 209; Behre, *Landschaftsgeschichte*, pp. 278–83; Henderikx, 'Economische geschiedenis', in P. Brusse and P. A. Henderikx, eds., *Geschiedenis van Zeeland: Prehistorie–1550* (Zwolle, 2012), pp. 133–4; D. Meier, *Die Halligen in Vergangenheit und Gegenwart* (Heide, 2020), pp. 27–9.

[46] De Bont, *Amsterdamse boeren*, p. 37.

[47] Rippon, *The Transformation*, pp. 1–2.

to modification, and others where exploitation remained the preferred way to use the resources of the wetlands. Since increasing population pressure occurred everywhere in the area under study in the high Middle Ages, it is useful to keep an eye on these exceptions, because they show there were alternatives to transformation of wetlands. Comparison of these areas with those that were transformed can provide insight into the social, economic or environmental reasons as to why people decided to either transform wetlands or just modify them.

In the following, the progress of the Great Reclamation will be described for the three major landscape types in the area under study: river plains, coastal marshes and peat areas. In the central river area of the Netherlands, change was more gradual than in the marshes and fens. Parts of the fertile, well-drained clays of the higher stream ridges had already been reclaimed earlier, especially in the eastern part of the river area. Transformation started here in the ninth century with piecemeal reclamation of more land on these high ridges and reclamation of lower stream ridges in the west of the area. From the tenth to thirteenth centuries the back-swamps were also taken into cultivation.[48] At first sight the reclamation of these heavy impermeable clays seems surprising, because in the post-medieval period they at best occasionally yielded some bad quality hay. During the Middle Ages, however, their quality very probably had deteriorated, due to the unintended consequences of reclamation which led to increasing drainage problems. In the tenth century drainage was still relatively easy, so reclaiming was feasible. It is unlikely that these heavy clays were used as arable land, but they could provide farmers with pasture and meadows.

Around AD 800, when only the highest stream ridges were inhabited and cultivated, flood protection was not a high priority in the river area.[49] As more and more of the lower land was taken into cultivation the need to protect this land from river floods increased. Small streams were dammed up, and here and there low banks were constructed. During the high Middle Ages these banks were heightened and connected, forming continuous embankments. In this way, around the beginning of the fourteenth century dike rings protected most land along the main rivers from floods all year round.[50]

In the coastal marsh areas the construction of winter dikes began around AD 1000 in the Low Countries, England and north-west Germany.[51]

---

[48] B. J. P. van Bavel, *Transitie en continuïteit: De bezitsverhoudingen en de platteland-seconomie in het westelijke gedeelte van het Gelderse rivierengebied, ca. 1300 – ca. 1570* (Hilversum, 1999), p. 63.

[49] Dekker, *Het Kromme Rijngebied*, p. 586.

[50] Van de Ven, *Man-Made Lowlands*, p. 92; G. J. Borger and W. Ligtendag, 'The Role of Water in the Development of the Netherlands – a Historical Perspective', *Journal of Coastal Conservation* 4 (1998), p. 111.

[51] De Langen, *Middeleeuws Friesland*, pp. 32–3, 37; Ayers, *The German Ocean*, p. 10; K. A. Rienks and G. L. Walther, *Binnendiken en slieperdiken in Fryslân* (Bolswert, 1954), p. 32.

Reclamation proceeded here from the coast landward. The inhabitants of the existing settlements in the dunes and the highest clay ridges close to the sea first reclaimed the fertile lands close to their villages or *terpen*, and then worked their way inland. In a first stage, sometimes ring dikes were created on the higher ridges, creating oval enclosures of up to 80 hectares to increase safety of crop cultivation.[52] High winter dikes had not yet been constructed. In marsh areas like Zeeland, people continued to live on low dwelling mounds. Here and there small dikes and dams were built to extend the area under cultivation. From the twelfth century onwards higher winter dikes were constructed to protect larger areas from floods. Population pressure may have stimulated this, not only because more land was required, but also because there were now sufficient labourers to build such long and relatively heavy dikes.[53]

Reclamation of the fenland areas in Holland, Utrecht and Friesland gained momentum in the tenth century. It started in northern Holland and Friesland and reached the Holland–Utrecht plain somewhat later, in the eleventh century.[54] In Friesland, where the reclamations had started in the ninth century from the west and north, the south-east of the province was reached around the middle of the eleventh century.[55] In the Holland district of Rijnland most available land was already under cultivation in the thirteenth century, and by 1300 the frontier in the Dutch fenlands was closed.[56] There were different ways of reclaiming peat areas, depending on whether a raised peat bog, a peat ridge or a peat plain was reclaimed. In all three cases the basis was the cutting of a system of straight, parallel ditches to drain the area.[57] The result was an ordered landscape, characterised by the long straight lines of the drainage ditches stretching kilometres deep into the peat.

In England too, large parts of the fens and marshes were reclaimed. Wetland reclamation had already begun here in the ninth century.[58] In southern Lincolnshire alone more than 25,000 hectares of land were

---

52 Tys, 'The Medieval Embankment', p. 215; Rippon, 'Human impact', p. 339; De Langen, *Middeleeuws Friesland*, p. 62; Krabath, 'Mittelalterlicher Deich- und Landesausbau', p. 79.
53 Henderikx, 'Land, bewoning, sociale structuren', in Brusse and Henderikx, eds., *Geschiedenis van Zeeland*, p. 92.
54 Bos et al., 'Waterland', p. 323; De Langen, *Middeleeuws Friesland*, p. 92; W. H. TeBrake, *Medieval Frontier: Culture and Ecology in Rijnland* (College Station, 1985), pp. 191–2.
55 J. A. Mol, 'De middeleeuwse veenontginningen in Noordwest-Overijssel en Zuid-Friesland', *Jaarboek voor Middeleeuwse Geschiedenis* 14 (2011), p. 98.
56 TeBrake, *Medieval Frontier*, p. 202.
57 De Bont, *Amsterdamse boeren*, pp. 50–9.
58 R. Van de Noort, *North Sea Archaeologies: A Maritime Biography, 10,000 BC – AD 1500* (Oxford, 2011), p. 118.

taken into cultivation between the tenth and the late thirteenth centuries.[59] There were conspicuous differences from the reclamations in the Low Countries, however. In southern Lincolnshire most reclamation took place in the marshes seaward of the existing settlements. These marshes had originally mostly been used for grazing cattle. Following the construction of seawalls, part of the pasture was converted into arable land. Part of the fens landward of the settlements was also reclaimed, but here reclamation was more limited, and large tracts of the fens remained in use for grazing.[60] This is a rational way of land use for mixed farming enterprises: the light clays on the seaside could be easily drained and ploughed and thus were very suitable for arable agriculture. The landward fens were much more difficult to drain and plough, so they could better remain pastureland. This option was not open to the pioneers in the Holland–Utrecht plain, because, apart from the narrow bands of clay along the rivers and of sandy soils behind the dunes there was only peat, the only type of soil offering any possibility of expansion. Another explanation for the more piecemeal character of fenland reclamation in England is that, although the fens were owned by manorial lords, the adjoining village communities managed them as commons.[61] These communities perceived the fens as an important resource for the mixed farms – especially valuable pasture – and tried to preserve this resource.[62] In the principalities of the Low Countries, however, the rights to the 'wilderness' of the fens had been claimed by counts and prince-bishops, and it was in their interest to have as much land reclaimed as possible for the purpose of raising their income.

In England, it was not only the peat areas but also parts of the marshes that remained untransformed. Marshes in southern Essex such as Canvey Island were not embanked. In their un-embanked state these marshes were very suitable for grazing sheep, the products of which could be profitably sold at the nearby London market. With a population of about 80,000 around the year 1300, London was at that time the biggest city in the area under study. Landlords considered that the profits from sheep farming outweighed the risks of flooding, especially since during high floods the sheep could find refuge on artificial mounds. They saw

---

59 H. E. Hallam, ed., *The Agrarian History of England and Wales. Volume II 1042–1350* (Cambridge, 1988), p. 151.
60 Hallam, *Settlement and Society*, pp. 148, 172; Hallam, *The Agrarian History*, p. 151.
61 A. J. L. Winchester, *Common Land in Britain: A History from the Middle Ages to the Present Day* (Woodbridge, 2022), pp. 26, 37.
62 H. Renes, 'The Fenlands of England and the Netherlands: Some Thoughts on their Different Histories', in T. Unwin and T. Spek, eds., *European Landscapes: from Mountain to Sea* (Tallinn, 2003), p. 105; Hallam, *Settlement and Society*, pp. 171–2.

no reason to engage in costly and risky reclamation projects while the marshes remained profitable in their current, lightly modified, state.[63]

Quantifying the extent of the Great Reclamation is not easy, but for the Netherlands a rough estimate can be made, with the oldest measurements for the cadastre from the 1820s as a point of departure. At that time, the land area of the lowland part of the Netherlands can be estimated at about 1.4 million hectares.[64] The net outcome of land loss and land reclamation from the late Middle Ages to the early nineteenth century must be subtracted from this. It can be estimated at about 0.4 million hectares (see also pp. 141–2). From the remaining one million hectares the area that was already settled at the beginning of the tenth century must then be subtracted. That area was limited to riverbanks, high stream ridges, dunes, *terpen*, and maybe some high salt marshes. If we estimate this generously at 0.2 to 0.3 million hectares, that would imply that some 0.7 to 0.8 million hectares of land were transformed from the tenth to the thirteenth centuries. This is only a very rough estimate, but it does convey a sense of the enormous size of the reclamations in this period. Since most of this land was reclaimed after AD 1000, it can be estimated that over the following three centuries about a quarter of a million hectares may have been reclaimed per century. This is the same scale that was reached in the twentieth century with the gargantuan drainage projects in the Zuider Zee area. Those later projects were implemented with modern technology such as draglines, whereas the men and women reclaiming the medieval fens and marshes only had spades and baskets at their disposal.

The Great Reclamation of the lowlands was an impressive achievement, but must be nuanced. In the first place, during the eleventh to thirteenth centuries a lot of land was submerged, especially in Noord-Holland and the Zuider Zee area.[65] Secondly, as noted above, much of the land that was reclaimed was not wasteland. It was often already used for grazing, fuel gathering, fishing, fowling and sometimes even for growing crops.[66] To use Rippon's terminology, the Great Reclamation meant a change from exploitation or modification to transformation of the wetlands.[67] Before, there were fluent boundaries between areas that were used in different degrees of intensity. After, a strict boundary was drawn between intensively cultivated land within the dikes, and the land that had remained outside the dikes.

---

[63] Dyer, *Making a Living*, p. 190; Rippon, 'Human Impact', pp. 346–7.
[64] *Grootte der gronden tijdens de invoering van het kadaster* ('s-Gravenhage, 1875).
[65] Y. T. van Popta, *When the Shore Becomes the Sea: New Maritime Archaeological Insights on the Dynamic Development of the Northeastern Zuyder Zee Region (AD 1100–1400)* (Eelde, 2021), pp. 25–6; Schoorl, *Zeshonderd jaar*, p. 14.
[66] E. Thoen and T. Soens, 'The Low Countries, 1000–1750', in Thoen and Soens, eds., *Struggling with the Environment*, p. 229.
[67] Rippon, *The Transformation*, pp. 1–2.

Both in the Low Countries and in England vast tracts of wetlands were reclaimed between AD 1000 and 1300, but as a result of a combination of geographical (the ubiquity of peat soils in Holland and Utrecht), institutional (the existence of commons in England) and economic factors (the profitability of sheep farming on un-embanked marshes in the Thames estuary), in England large parts of the marshes and of the peat fens were only exploited or at most modified. From an environmental viewpoint, and with the benefit of hindsight, the attitude of English landlords and farmers towards land reclamation can be considered fortuitous, if only because they avoided the unintended landscape changes that confronted, and confront, the population of the reclaimed wetlands on the eastern shore of the North Sea.

The Unintended Consequences of Reclamation

In the previous section reference was made to the unintended consequences of wetland reclamation. Draining and embanking these lands led to the start of natural processes that in several cases haunt the inhabitants of the reclaimed wetlands to the present day, and in the past led to many (near-)disasters. Preventing such disasters or repairing the damage they caused has occupied the population of the wetlands for ten centuries and has influenced the development of agriculture, if only because of the high expenditure involved. So, it is useful to provide an explanation of these harmful processes because their consequences will regularly recur in the following chapters. There were two main processes: lowering of the surface level due to drainage, and silting-up of rivers and loss of storage capacity due to the construction of embankments.

Subsidence of the land due to compaction and oxidation is mostly a problem of peat soils. When clay soils are drained, some land subsidence will occur during the first years after drainage, but after that initial compaction the surface level of clay soils will not change much.[68] Peat is much more susceptible to surface lowering. As previously mentioned, peat is composed of up to 20 per cent decayed plant particles and 80 per cent or more water, making draining imperative if one wants to cultivate peat soils. The draining of that large amount of water from the soil will inevitably lead to compaction and thus lowering of the surface level. Harmful as it may be, compaction will not destroy the substance of the peat layers. Much more dangerous are the accompanying processes of dehydration and oxidation. When cultivated peat land is not under plant cover, the wind can lift part of the dry plant particles and thus remove part of the topsoil. Also, when peat is drained, the pores on the topsoil are no longer filled with water, but with air. This causes the plant particles

---

[68] K. C. Numan, *Burghorn van kwelder tot polder: De eerste bedijking in Hollands Noorderkwartier, 1457–1461* (Heerhugowaard and Wormer, 2020), pp. 42–3.

Figure 2. The Holme Post, c. 1928. Collection Huntingdonshire Archives. When set up in 1851, the top of the post was at surface level, by 1928 eleven feet (c. 3.3 metres) of soil had disappeared.

in the soil to oxidise and change into carbon dioxide. In the short run, this has a positive effect, because oxidation makes nutrients available for plant growth. In the long run, however, the soil literally goes up into thin air and peat layers of six or seven metres can disappear completely. Of these processes oxidation is the most harmful; it causes some 85 per cent of the shrinkage of peat soils.[69]

The sometimes dramatic consequences of peat drainage are poignantly illustrated by the Holme Post. This cast iron post, allegedly from the dismantled Crystal Palace, was driven into the soil in 1851 when Whittlesey Mere (south of Peterborough) was drained. In that year, the post's top was at surface level; 80 years later it was almost 3.5 metres above that level.[70] The Holme Post is an extreme case, but everywhere where peat soils are drained, considerable shrinkage of the soil occurs. The speed of the process depends on the way the land is used. Arable agriculture is most harmful. As the land is ploughed and left fallow for some of the year, oxidation and wind erosion can do their work and destroy part of the topsoil. This can lead to shrinkage of 2.5 centimetres per year. When the land is permanently under grass and not drained deep, shrinkage can be limited to 0.5 centimetres per annum.[71] In that case compaction continues, but oxidation is halted and wind erosion prevented because the land is permanently covered by vegetation. Therefore, in fenlands, stock raising and dairy farming are preferable to arable farming or horticulture, but, even when under grass drained peat soils will shrink up to 0.5 metre per century. The most damaging effect of shrinking peat for agriculture is that the lower the surface level becomes, the more difficult it will be to evacuate excess water and keep the land dry enough for agriculture. In the first stage it will become impossible to drain the land in autumn and winter, and cereal cultivation will end. As the land lowers further, it will get so wet as to prevent cattle from grazing there in all but the driest periods of the year. In a final stage, the level of the land will be lower than that of the rivers into which it needs to drain.

---

[69] Godwin, *Fenland*, p. 133; T. Edelman, *Bijdrage tot de historische geografie van de Nederlandse kuststrook* ('s-Gravenhage, 1974), pp. 44–5; G. J. Borger, *De Veenhoop: Een historisch-geografisch onderzoek naar het verdwijnen van het veendek in een deel van West-Friesland* (Amsterdam, 1975), pp. 196–7; S. van Asselen, *Peat Compaction in Deltas: Implications for Holocene Delta Evolution* (Utrecht, 2010), p. 30; P. J. E. M. van Dam, 'Sinking Peat Bogs: Environmental Change in Holland, 1350–1550', *Environmental History* 6 (2001), p. 34.

[70] G. Fowler, 'Shrinkage of the Peat-Covered Fenlands', *Geographical Journal* 81 (1933), p. 149.

[71] G. Fowler, 'Old River-Beds in the Fenlands', *Geographical Journal* 79 (1932), p. 210; C. Schothorst, 'Drainage and Behaviour of Peat Soils', in H. de Bakker and M. W. van den Berg, eds., *Peat Lands Lying below Sea Level in the Western Part of the Netherlands, their Geology, Reclamation, Soils, Management and Land Use* (Wageningen, 1982), pp. 154–5; Edelman, *Bijdrage*, p. 71.

In that case drainage and agriculture become impossible and the land must be abandoned.

In some areas the problems were compounded by relief inversion, especially in the islands between the main rivers in Holland. When these were reclaimed, the central peat bogs were higher than the stream ridges along the rivers. Due to drainage, however, after some centuries the level of the peat bogs became lower than that of the clay ridges, which had hardly shrunk. As a result regions such as the Alblasserwaard became saucer-like with high edges from which the water could hardly escape.[72] Some settlements had to be given up due to peat shrinkage. In most cases, however, technological solutions were found by way of drainage mills and later steam and diesel pumps. In the long run, however, these solutions compounded the problems. The more water that was pumped away, the quicker the soil shrank. From the eleventh to the twenty-first centuries agriculturists in the fenlands have wrestled with being caught in this vicious cycle from the moment the fens were reclaimed.

In some regions the processes of shrinkage and oxidation were slowed down because farmers kept the land under grass and limited pumping to spring and summer when the cattle were in the fields (see p. 120). This was the least destructive way to exploit peat soils, and it raises the question of whether people realised that drainage caused lowering of the surface level and knew about the processes that caused it. If they did, they could have used that knowledge to slow down subsidence of peat lands. The available evidence, however, shows that for most of the period under study people were unaware of these processes. As late as 1729 the English engineer Thomas Badeslade showed he was not aware of soil subsidence in the fens and its causes. Both in England and the Dutch Republic the idea that drainage caused lowering of the surface only seems to have emerged during the eighteenth century.[73] As far as we know, the first author who explicitly stated that the surface level of peat soils was lowered due to drainage was the West-Frisian lawyer Zacharias L'Epie in 1734. He correctly depicted this lowering as a continuous process from the moment West-Friesland was enclosed by dikes in the thirteenth century, and estimated the lowering since then until 1731 to be about

---

[72] J. C. Besteman, 'Van Assendelft naar Amsterdam: Occupatie en ontginning van de Noordhollandse veengebieden in de Middeleeuwen', in D. E. H. de Boer, E. H. P. Cordfunke and H. Sarfatij, eds., *Holland en het water in de Middeleeuwen: Strijd tegen het water en beheersing van het gebruik van water* (Hilversum, 1997), p. 29; De Bont, *Amsterdamse boeren*, pp. 42–3; M. van Tielhof, 'Betrokken bij de waterstaat: Boeren, burgers en overheden ten zuiden van het IJ tot 1800', in E. Beukers, ed., *Hollanders en het water: Twintig eeuwen strijd en profijt* (Hilversum, 2007), p. 66.

[73] H. C. Darby, *The Draining of the Fens* (2nd edn., Cambridge, 1956, reprint 2011), pp. 122–3.

two metres. L'Epie also had an inkling of the causes of this process. He correctly assumed that a lot of the drained soil was composed of plant matter. He further assumed this substance tended to 'rot' down in the course of time, causing the soil to shrink.[74] Fifteen years later, the engineer Cornelis Velsen voiced comparable views of the shrinkage and wastage of peat soils.[75] In England, agronomist Arthur Young also showed he was aware of the relationship between drainage and peat shrinkage when he wrote in 1805: 'the interior of the district [the Fens] has subsided by the various efforts made to drain it'.[76]

The publications of L'Epie and Velsen appear to have had little effect on the theory and practice of water management. They did not prevent, for instance, the introduction of year-round steam pumping in the nineteenth century, which considerably accelerated the processes of shrinkage and wastage (see pp. 198, 204). One of the first scientists who really understood what was happening in the fens was the geologist Sidney B. J. Skertchly, who wrote in 1877: 'from the moment the drainage works commenced the land began to subside, owing to the abstraction of water from the porous soil'.[77] In 1933, Major Gordon Fowler reported on the lowering of the surface level in the Great Level, which he correctly attributed to drainage.[78] The Dutch were slower to grasp what was going on in their fens. During the 1940s Joost Hudig, professor of agricultural chemistry and fertilisation at the agricultural college (*Landbouwhogeschool*) at Wageningen, and his student Johannes Duyverman reconstructed the soil subsidence process and its causes in the Holland–Utrecht plain for the period from the Great Reclamation onwards. They were able to demonstrate that due to shrinkage and oxidation the surface level of peat areas was lowered by some 2.5 metres over eight or nine centuries.[79]

The paragraphs above demonstrate that at least until the nineteenth century the processes of shrinkage and oxidation were not really understood. When farmers in fenlands did adopt less destructive ways of farming and draining, they were not based on any environmental knowledge. Often, their behaviour was simply destructive. The most damaging agricultural method that was applied in peat areas was 'paring and burning'. This involved the vegetation and topsoil of peat land being

---

74 Z. L'Epie, *Onderzoek over de oude en tegenwoordige natuurlyke gesteltheyt van Holland en West-Vriesland* (Amsterdam, 1734), pp. 67–8, 79–80.
75 W. van der Ham, *De Grote Waard: Geschiedenis van een Hollands landschap* (Rotterdam, 2003), p. 26
76 Quoted in Darby, *The Draining*, p. 173.
77 S. B. J. Skertchly, *The Geology of the Fenland* (London, 1877), p. 154.
78 Fowler, 'Shrinkage', pp. 149–50.
79 J. Hudig and J. J. Duyverman, 'De cultuur der zogenaamde laagveengronden en hun moeilijkheden', *Mededelingen van de Nederlandsche Heidemaatschappij* 7 (1949), p. 3; J. J. Duyverman, *De landbouwscheikundige basis van het streekplan: Het centrale veengebied Utrecht en Zuid-Holland* (Wageningen, 1948), pp. 158–66.

burnt and the ashes mixed with the soil to improve fertility. In oligotrophic peat in particular this worked well because the ash reduced the natural acidity of this nutrient-poor soil. Paring and burning resulted in good crop yields for up to six or seven years. On the continent paring and burning was only practised on the high inland moors, not in the lowlands, but it occurred frequently in the English fenlands. The environmental price paid for better yields was high. The lowering of the surface level sped up because the topsoil was burnt, and the fires produced huge clouds of white, acrid smelling smoke that could reach as far as Poland.[80]

Even though dikes protect sea or river marshes from floods, they too have unintended consequences. When marshes are not embanked, floods will spread over a large area. In doing so they will spend much of their force, resulting in the water level on the submerged marsh remaining low because of the large area being flooded. Consequently, damage will remain limited. In contrast, when a dike is constructed water will gather in front of that dike. When the dike is breached, that water will burst through with tremendous force and the level of water will be considerably higher because of the smaller area that is now flooded.[81] Moreover, construction of embankments tends to stimulate people to live in flood-prone areas. So long as marshes are not embanked, humans tend to settle on the edges of adjacent uplands or on dwelling mounds in the marshes. Dikes provide a feeling of safety so people will be more easily encouraged to settle in embanked marshes. All these factors combined lead to a strong increase in the amount of damage and of human and animal victims when floods breach dikes. There will be fewer flood events but their impact will be greater. To prevent this, higher and more strongly constructed embankments can be built, but that will increase expenditure.

Dikes force the sea to deposit its sediment in large heaps in front of them, thus causing rapid accretion of new land.[82] This has usually been considered a beneficial by-effect of dike construction because it offers the opportunity of embanking more very fertile land. After the first dikes had been built in Zeeland, for example, large areas in front of those dikes were added to the cultivated area by the construction of new embankments. Although this was of course beneficial to lords and communities, in the long run it also had negative effects. The more land that was embanked along the tidal rivers, the smaller the storage capacity of these rivers became. This was especially detrimental when storm surges forced huge amounts of water into the river mouths. Due to new embankments this mass of water now has less area over which to spread so its level will become higher, thus increasing the risk of flooding to the cultivated land behind the dikes. This risk was especially high on the landward side of the funnel-shaped

---

[80] Darby, *The Draining*, pp. 165–6; Behre, *Landschaftsgeschichte*, pp. 223–4.
[81] De Langen, *Middeleeuws Friesland*, p. 59.
[82] Hallam, *Settlement and Society*, pp. 123–4.

sea inlets.[83] Again, the solution could be sought in higher and better dikes, resulting in higher costs for construction and maintenance.

Enclosing riverbeds between high dikes also had negative long-run effects. When a river was not enclosed between dikes, the sediment it transported was distributed over almost all the land situated along the river. After the construction of embankments it was only deposited in the area between those embankments.[84] This caused the riverbed to silt up, whereas the level of the land protected by the dikes stayed the same or even became lower due to compaction. After a while, the level of the river could become higher than that of the land, causing increasing risk of flooding and impeding the drainage of the cultivated land. The problems were compounded by seepage of river water to the cultivated land, making the lower arable land too cold and wet for cultivation.[85] And again, technical solutions were expensive.

Looking at this catalogue of problems with the benefit of hindsight, we might question the wisdom of the humans who reclaimed these wetlands. Some of their decisions were indeed highly questionable, but we should realise they did not have the sophisticated technical knowledge we now possess. Nevertheless, they were surprisingly successful in maintaining their agricultural activities in the reclaimed wetlands. The dynamic and challenging environment of these wetlands should not only be perceived as a cause of problems but also as a stimulus for improving both water management systems and agricultural performance. The inhabitants of the lowlands were forced to look for ways to adapt to their dynamic environment or to find technological innovations in water management in order to survive. Adaptation often meant accepting a wetter environment and less intensive forms of farming – particularly cattle raising instead of arable farming.

New technology, such as drainage mills or later steam pumps, permitted continuation of intensive farming but also required higher investments in the implementation of that technology. Since the landowners had to finance this, the land needed to produce high yields, so landowners and farmers were forced to enhance agricultural performance. By doing so, highly productive agriculture could emerge, but they also ran the risk of depleting the soil. Impressively, in most marshes and fenlands along the southern North Sea soil depletion was avoided, and agriculture remained exceptionally productive through a skilful combination of adaptation and innovation. This 'forced innovation', as Peter Hoppenbrouwers has

---

[83] P. Vos and F. D. Zeiler, 'Overstromingsgeschiedenis van Zuidwest-Nederland, interactie tussen natuurlijke en antropogene processen', *Grondboor en Hamer* 62 (2008), p. 91.
[84] Borger and Ligtendag, 'The Role', p. 111.
[85] Van de Ven, *Man-Made Lowlands*, p. 165.

termed it, both in hydraulics and in agriculture, may be the secret of the success of agriculture in the lowlands.[86]

Wetland Reclamation and Freedom

In the 1950s, two scholars, the American medievalist Bryce Lyon and the Dutch legal historian Hendrik van der Linden, claimed that the way in which the wetlands were reclaimed resulted in a society in which free farmers could freely dispose of their land. Lyon did this in a 1957 paper on the coastal plain of Flanders, Van der Linden in his 1955 dissertation on the Holland–Utrecht plain.[87] Of these two authors, it was Van der Linden who exerted an immense influence on Dutch historiography. If Lyon and Van der Linden are correct in assuming that the reclamation of the wetlands resulted in farmers who were personally free, with no labour or other obligations to manorial lords and able to freely dispose of their land, such as selling it or leaving it to their children, then this would mean that a fundamental difference existed with the uplands. It would mean that in the lowlands of Holland and Flanders, and maybe also elsewhere in the North Sea Lowlands, institutions prevailed that presented agriculturists with opportunities to improve their farming operations, which the unfree tenants of the uplands did not have.

Van der Linden's and Lyon's publications were among the last contributions to a debate on the 'freedom' of the lowlands on both shores of the southern North Sea that began in the 1930s. It is no coincidence that this debate occurred in a period in which freedom and democracy were seriously threatened by totalitarian ideologies such as Nazism and communism. Staunch defenders of liberal democracy like the Swiss historian Adolf Gasser and the Dutch historian Bernard Slicher van Bath hoped that by research into the origins of freedom and democracy they could better defend them from the totalitarian threat.[88] What makes this debate relevant in the context of this book is that several participants, such as Slicher van Bath and the German historian Hermann Aubin, noticed a relationship between the reclamation of the lowlands, the organisation of water management and political, personal and economic freedom.[89] Van

---

[86] P. Hoppenbrouwers, 'Van waterland tot stedenland: De Hollandse economie ca. 975–ca. 1570', in T. de Nijs and E. Beukers, *Geschiedenis van Holland tot 1572* (Hilversum, 2002), p. 112.

[87] B. Lyon, 'Medieval Real Estate Development and Freedom', *American Historical Review* 63 (1957), pp. 52–3; Van der Linden, *De cope*, pp. 160–82, 324–5.

[88] A. Gasser, *Geschichte der Volksfreiheit und der Demokratie* (Aarau, 1939), pp. 222–4; B. H. Slicher van Bath, *Boerenvrijheid* (Groningen and Batavia, 1948), p. 23.

[89] B. H. Slicher van Bath, 'Problemen rond de Friese middeleeuwse geschiedenis', in Slicher van Bath, *Herschreven historie: Schetsen en studiën op het gebied der*

der Linden and Lyon provided an empirical basis to the rather vague notions of these earlier scholars.

Before discussing Lyon's and Van der Linden's views, it is useful to look at the way in which society in the regions surrounding the lowlands was organised, to get an idea of the meaning of freedom in the Middle Ages. The heartland of the manorial system was to the south, in northern France, southern Belgium and the Rhineland. In its 'classic' form this meant that most of the agricultural land belonged to large estates – manors – divided into two main parts. The first part was split into holdings that were given out to tenants. The second part, the demesne, was exploited by the lord of the manor himself. The demesne could measure hundreds of hectares. The work on the demesne was performed by the tenants, who owed the lord labour duties of sometimes three days per week in return for the use of their land holdings. They also had to deliver products such as eggs, poultry or pigs to their lord. Most of these tenants were unfree serfs who were legally tied to the manor and could not leave it without the lord's permission. Yet, not all estates took the 'classic' form; on many estates tenants did not perform labour duties but instead paid dues in kind or money. At the end of the first millennium the manorial system expanded to the north into Saxony and to the west into England.[90] In the North Sea Lowlands as well the manorial system was not unknown. It was also well developed in the early reclaimed areas of the Dutch river area in Guelders and Utrecht. Large domains were also known in the Frisian *terpen* area along the Wadden Sea coast and in the Saxon areas of the north-west German marshes, although in the Frisian areas free farmers were predominant. In the Frisian lands those large domains did not take the form of bipartite manors and serfdom was abandoned comparatively early, in the eleventh century.[91] In Holland, manors existed in the *geest* areas behind the dunes, on the stream ridges along the rivers. Manors and servile peasants also existed in Zeeland – where they had already disappeared by the twelfth century – but they were largely absent from the Flemish coastal plain.[92] There were many variations of the manorial

---

middeleeuwse geschiedenis (Leiden, 1949), pp. 279–80; H. Aubin, 'Von den Ursachen der Freiheit der Seelande an der Nordsee', *Nachrichten der Akademie der Wissenschaften zu Göttingen I. philologisch-historische Klasse* (1953), p. 44. Hermann Aubin was far from a staunch democrat. During the Third Reich he was quite regime friendly.

[90] C. Wickham, 'Social Relations, Property and Power around the North Sea, 500–1000', in B. van Bavel and R. Hoyle, eds., *Social Relations: Property and Power. Rural History in Northwestern Europe* (Turnhout, 2010), pp. 29–34; Van Bavel, *Manors and Markets*, p. 75; A. Verhulst, *Précis d'histoire rurale de la Belgique* (Brussels, 1990), pp. 29–35; Prak and Van Zanden, *Pioneers*, p. 32.

[91] De Langen and Mol, 'The Distribution', pp. 47–8.

[92] Van Bavel, *Manors and Markets*, p. 78; Dekker, *Het Kromme Rijngebied*, pp. 37, 269–70; F. Swart, *Zur friesischen Agrargeschichte* (Leipzig, 1910), pp. 171–2;

system, but as Bas van Bavel remarks, characteristic of all variations was 'non-economic coercion, i.e. the exploitative, non-contractual relation between the lord and the occupants of the land'.[93] In the newly reclaimed wetlands, according to Lyon and Van der Linden this non-economic coercion was absent.

Since Van der Linden had most influence, this section will mostly concentrate on his work, with some references to Lyon's article. After his 1955 PhD dissertation, Van der Linden published for 30 years on the reclamation of the fenlands. I will base this section mostly on his dissertation, his extremely influential chapter in the general history of the Low Countries from 1982, and a paper from the same year in which he formulated his views in the most radical way.[94] His point of departure is that the count of Holland and the bishop of Utrecht possessed the rights to the wilderness that had previously belonged to the king.[95] They used those rights to stimulate the reclamation of the vast fenlands of the Holland–Utrecht plain, expecting increasing income from the new cultivated land and its inhabitants. They did this by concluding *cope* contracts with individuals or groups wishing to reclaim fenland. The word *cope* means purchase, and the *copers* purchased a plot of fenland from the count to ensure the right to reclaim it. The *copers* were not always the individuals who reclaimed the land. They could also be *locatores*, promoters, often noblemen, who laid out the area that was to be reclaimed and then attracted farmers to do the heavy work in return for a holding in the new land. The farmers who reclaimed the land and settled there had to pay a *denier* (a penny) yearly to the count or bishop as a symbolic recognition of the overlord's authority. The pioneer farmers became full owners of their farms, which they could freely sell or bequeath to their heirs. There was no personal bond between farmer and count, only the public law relationship between prince and subject.[96] Place names in the Holland–

A. Janse, *Ridderschap in Holland: Portret van een adellijke elite in de late Middeleeuwen* (Hilversum, 2001), pp. 139–40; Henderikx, 'Land, bewoning', p. 103; Thoen, 'Social agrosystems', p. 56; C. Boschma-Aarnoudse, *Boer en poorter: Waterland en de Zeevang in de middeleeuwen* (Heerhugowaard and Wormer, 2022), pp. 49–50; A. van Steensel, 'Noord-Beveland: de geschiedenis van een middeleeuws eiland', in *Archief: Mededelingen van het Koninklijk Zeeuws Genootschap der Wetenschappen* (2023), p. 60.

[93] Van Bavel, *Manors and Markets*, p. 75.
[94] Van der Linden, *De cope*; Van der Linden, 'Het platteland'; Van der Linden, 'History of the Reclamation of the Western Fenlands and of the Organizatons to Keep them Drained', in De Bakker and Van den Berg, eds., *Peat Lands*, pp. 42–73.
[95] Van der Linden, *De cope*, pp. 81–2, 103.
[96] Van der Linden, *De cope*, pp. 110–12, 160, 172, 184, 325, 329; Van der Linden, 'History', pp. 60–1.

Utrecht fenland, such as Nieuwkoop and Boskoop, still refer to the *cope* contracts that formed the legal basis for these settlements.

Van der Linden suggested that the Holland counts may have been inspired by Flemish examples and introduced these arrangements in Holland at the end of the tenth century.[97] A chronicle describing events around the town of Vlaardingen in 1018 appears to confirm this. It states that Count Dirk III had given out plots of land to settlers to reclaim and had obliged them to pay a yearly recognition.[98] However, when the count undertook reclamations in the peat lands in the area of Delft in southern Holland around the turn of the millennium the work was done by serfs in a manorial context.[99] Still, the Flemish connection should not be completely disregarded. Lyon showed in his paper that a charter of the Flemish Count Boudewijn V in 1067 granted the same kind of free landownership to pioneer farmers in the Flemish coastal plain as Van der Linden observed in the *cope* agreements.[100] Erik Thoen observed that the Flemish count managed to prevent the emergence of a layer of local lords and that most farmers in the coastal plain were completely free.[101] Others have suggested that the bishop of Utrecht and the rich cathedral chapters of that city may have spawned the *cope*, but the evidence is lacking for this.[102] Whatever the origin, in his 1982 article Van der Linden concluded on a jubilant note that from the Holland–Utrecht plain free peasant ownership of land was diffused to many areas along the North Sea Coast, such as the Saxon areas in north-west Germany, Romney Marsh, Holland in Lincolnshire, and the Cambridgeshire fens.[103]

From 1982 onwards, Van der Linden's views of the reclamation of the Holland–Utrecht fens were paradigmatic in Dutch historiography, but criticisms arose too.[104] For instance, in the south-west of Utrecht where the manorial system was strongly developed, eleventh-century reclamations were undertaken by manorial lords and the manorial system was extended into the newly reclaimed land. No free peasant landownership was introduced here.[105] This criticism does not really undermine Van

---

[97] Van der Linden, *De cope*, pp. 352–3.
[98] H. van Rij, ed., *Alpertus van Metz. Gebeurtenissen van deze tijd. Een fragment over bisschop Diederik I van Metz. De mirakelen van de heilige Walburg in Tiel* (Hilversum, 1999), p. 77.
[99] C. de Bont, *Delft's Water: Two Thousand Years of Habitation and Water Management in and around Delft* (Delft and Zutphen, 2000), pp. 37, 43–4.
[100] Lyon, 'Medieval Real Estate', pp. 52–3.
[101] Thoen, 'Social Agrosystems', p. 56.
[102] Hoppenbrouwers, 'Waterland tot stedenland', p. 107.
[103] Van der Linden, 'History', pp. 61–2.
[104] For instance, B. Augustyn, *Zeespiegelrijzing, transgressiefasen en stormvloeden in maritiem Vlaanderen tot het einde van de XVIe eeuw: Een landschappelijke, ecologische en klimatologische studie in historisch perspektief* (Brussels, 1992), pp. 393–415.
[105] Dekker, *Het Kromme Rijngebied*, pp. 269–73.

der Linden's thesis, because these small-scale reclamations had little to do with the reclamation of the vast fenlands he was focusing on. More damaging were the criticisms in the 1993 dissertation of Buitelaar, who concluded that large-scale fenland reclamations along the Vecht river in the bishopric of Utrecht had been undertaken in a manorial context and that as late as the thirteenth century there were still servile tenants in this area. Buitelaar accepts, however, that in the county of Holland *cope* agreements resulted in free landownership.[106]

In Holland *cope* contracts have only been preserved from 1233 onwards. That is not surprising; very few documents of any kind from the eleventh and twelfth centuries have been preserved. Luckily, the *cope* agreement was exported to north-west Germany, and there older contracts have survived. In 1063 the archbishop of Bremen was granted the royal rights to the wilderness in his diocese. He attracted settlers from Holland to reclaim land along the lower Weser and lower Elbe rivers. Contracts for reclamation in these areas have been preserved from 1113, 1142, 1149, 1171, 1181 and 1201. Together, they provide a clear insight into the conditions of reclamation and the rights of farmers in the new land, and they demonstrate that these rights were 'imported' from Holland.[107] Van der Linden had already used these documents to support his thesis, and the work of German scholars has proved that he was correct in doing so.

The contracts from the archbishopric of Bremen show that, as in Holland, the settlers were only obliged to pay a symbolic recognition to the archbishop. In the *Hollerkolonien*, as the settlements of the Hollanders were called, *Hollerrecht* (Hollander law) was applied, which was quite different from the law that was valid in the adjoining older Saxon communities. The most conspicuous element of this *Hollerrecht* was that the settlers could dispose freely of their farms by leasing, selling or mortgaging and their offspring could inherit the holding, just as in the Holland fenlands.[108] From this we can conclude that *cope* agreements resulting in a peasantry that could freely dispose of its holdings already existed in the eleventh century and that these agreements and thus free peasantry were exported to north-west Germany in the early twelfth century. Therefore, farmers in the reclaimed wetlands of Holland, Flanders and the archbishopric of Bremen were economically free. That does not necessarily mean that

---

[106] A. L. P. Buitelaar, *De Stichtse ministerialiteit en de ontginningen in de Utrechtse Vechtstreek* (Hilversum, 1993), pp. 139, 149, 153, 296.

[107] L. Deike, *Die Entstehung der Grundherrschaft in den Hollerkolonien an der Niederweser* (Bremen, 1959), pp. 59–62; H. A. Pieken, *Die Osterstader Marsch: Werden und Wandel einer Kulturlandschaft* (Bremen, 1991), p. 234.

[108] Deike, *Die Entstehung*, p. 62; K.-J. Lorenzen-Schmidt, 'Reiche Bauern: Agrarproduzenten der holsteinischen Elbmarschen als Betriebsführer und Konsumenten', in D. Freist and F. Schmekel, eds., *Hinter dem Horizont. Band 2: Projektion und Distinktion ländlicher Oberschichten im europäischen Vergleich, 17.-19. Jahrhundert* (Münster, 2013), p. 31.

they were also personally free. The German *cope* contracts clearly show that serfs migrating to the newly reclaimed land did not lose their servile status. The archbishop of Bremen – and very probably also the count of Holland – did reveal a preference for personally free farmers, and the Bremen contracts contain several clauses to prevent serfs from settling in the new land. They especially did not want serfs of other lords as settlers, because when a serf died, part of his inheritance – usually the best cow or best draught animal – reverted to his lord, which meant that wealth was extracted from the new settlement.[109] So probably most, but not all, settlers were personally free. Both free and unfree settlers, however, could freely dispose of their property and in the course of time merged into a group of legally equal, free farmers. In the late Middle Ages no serfs remained in the *Hollerkolonien* along the Lower Weser and Lower Elbe.[110]

The previous paragraph has shown that freedom of farmers did indeed spread to Saxon north Germany, as Van der Linden stated. However, in some German areas freedom was short-lived. In the *Hollerkolonien* around the city of Bremen farmers had already lost the right to free disposal of their holdings in the thirteenth century. This was mostly caused by the defeat of the farmers from Stedingen – one of the rural republics – in their war against the surrounding principalities. To crush the initially very successful Stedingen army, a crusade had been proclaimed, and in 1234 the rebel army was destroyed and the possessions of the rebels, who had been declared heretics, were confiscated. Although most of the *Hollerkolonien* were not part of Stedingen, many farmers from these settlements appear to have fought on the Stedingen side and lost the property of their holdings. Another cause of expropriation may have been that farmers ran into debt and had to give up their holdings. Reclaiming land was expensive, profits from farming were insecure and borrowed money could not always be repaid on time.[111] The complete disappearance of peasant ownership, however, indicates that the confiscation after the War of the Stedinger was the most important cause of its demise in this area. Farmers did not revert to servile status after 1234. They remained personally free, but they became tenants and had to pay high rents – a third of the harvest – to noble and ecclesiastical landlords.[112]

In the *Hollerkolonien* along the left bank of the lower Elbe to the north of Hamburg free peasant ownership survived for much longer. Here the

---

[109] Deike, *Die Entstehung*, pp. 60–1.
[110] A. E. Hofmeister, *Besiedlung und Verfassung der Stader Elbmarschen im Mittelalter* (2 vols., Hildesheim, 1979–81), vol. 2, pp. 94–7.
[111] Deike, *Die Entstehung*, pp. 101–8; A. E. Hofmeister, *Seehausen und Hasenbüren im Mittelalter: Bauer und Herrschaft im Bremer Vieland* (Bremen, 1987), pp. 68, 83, 98.
[112] Hofmeister, *Seehausen und Hasenbüren*, p. 174; Pieken, *Die Osterstader Marsch*, p. 177.

low *Sietland*, the wet clays situated behind the already cultivated stream ridges close to the river, the *Hochland*, was settled from the 1130s onwards by colonists from Holland. Count Rudolf II of Stade and his successors granted the settlers the same conditions their compatriots in the Weser area had received. Farmers in the new settlements had free disposal of their property, whereas in the older Saxon settlements in the fourteenth century there still were *Liten*, unfree farmers who were tied to the land. In due time, peasant landownership was eroded in the *Hollerkolonien*, and church, noblemen and urbanites acquired land in the area. Still, as late as the early seventeenth century, in the Alte Land, where most *Hollerkolonien* on the left bank of the Elbe were situated, farmers owned some two thirds of the agricultural land.[113]

The case of the lower Weser area shows that free peasant landownership based on *cope* agreements did not guarantee freedom in the long run. It only succeeded in Holland, Utrecht and the Lower Elbe marshes. However, although the *cope* model may not have been as successful as Van der Linden claimed, there is no denying that in all North Sea Lowlands farmers enjoyed more freedom than their counterparts in the uplands. It seems the *cope* contract was not the only major factor at work. The landscape seems to offer a better explanation for the freedom of the population of the lowlands. The manorial system primarily aimed at large-scale production of bread grains. Required for this was a large area of land that was suitable for arable farming, and a relatively high population density as arable farming is labour intensive. In the eighth and ninth centuries, when manorialism emerged, arable land was scarce and population sparse in the wetlands, so no fertile soil conditions existed here for the rise of the manorial system.[114] Even after drainage, the landscape mostly remained too wet for large-scale arable farming (see pp. 72–4). The only exception is the Dutch river area, which had been settled before the Great Reclamation and where large arable fields could be created on the stream ridges.

Did farmer freedom also spread to Romney Marsh and the fenlands of Lincolnshire and Cambridgeshire as Van der Linden claimed? There is no evidence that supports this view. Indeed, farmers in the English lowlands generally seem to have enjoyed more freedom than their counterparts in the uplands. In South Lincolnshire, for instance, manorialism was poorly developed and more people had free status than elsewhere in England.[115] However, that was not the result of any spread of Dutch influence. In the first place, much of the lowlands on the English east coast were part of the Danelaw, where the manorial system was less developed and thus

---

113 Hofmeister, *Besiedlung und Verfassung*, vol. 2, pp. 19, 99–101, 167–8, 205.
114 Janse, *Ridderschap*, p. 140; Van Bavel, *Manors and Markets*, p. 81; De Vries, *The Dutch Rural Economy*, pp. 24–5.
115 Hallam, *Settlement and Society*, p. 31; Simmons, *Fen and Sea*, p. 5.

farmers experienced more freedom. Secondly, a greater degree of freedom was customary in all recently reclaimed and settled areas in Europe, from the English Fenlands to German forest reclamatlons and areas occupied by Christian settlers during the *Reconquista* on the Iberian Peninsula.[116] Dutch settlers were not required to effect the spread of freedom.

The wetland landscape stimulated freedom for the rural population in conjunction with the preference of large landlords for reclamation by independent farmers instead of under their own supervision by coerced labour. To lure pioneers to the fens and marshes, advantageous conditions had to be offered, such as free possession of the land. It was not only the counts of Flanders and Holland and the bishop of Utrecht who did this, but also the Danish king when Frisians settled in what is now Nordfriesland.[117] Not only territorial princes but also local lords and abbeys preferred this. In the county of Zeeland, the land of local lords was reclaimed by farmers in exchange for free ownership of the land. The same holds true for the English lowlands, where landlords stimulated reclamation by offering low rents and free status to pioneering peasants.[118] In this way, landlords could increase their revenues without investing much in risky enterprises.

The Great Reclamation offered few opportunities to extend the manorial system to the reclaimed wetlands. In the vast fenland areas of Holland and Friesland, manors were too small – where they existed at all – to serve as centres for reclamation. However, more importantly, the territorial lords were not keen on settling serfs in the new land. The counts of Flanders and Holland and the bishops of Utrecht and Bremen were not interested in controlling labour but in establishing public authority over their territories. Their revenues did not derive from the exploitation of land but mostly from taxation.[119] Where manorialism had existed in the lowlands, it dissolved early on. Around 1300, manors had disappeared from all lowland regions in the Low Countries, even from the river area where they had been most prominent. The same holds true for much of the German Wadden coast, where in the Frisian areas and the *Hollerkolonien* manorialism had always been weak or absent.[120] From that

---

[116] Hallam, *Settlement and Society*, pp. 201, 207; W. Abel, *Geschichte der Deutschen Landwirtschaft vom frühen Mittelalter bis zum 19. Jahrhundert* (3rd edn., Stuttgart, 1978), p. 35; M. Bloch, *Feudal Society* (1939, reprint London and New York, 2014), pp. 53, 197.

[117] D. Meier, H. J. Kühn and G. J. Borger, *Der Küstenatlas: Das schleswig-holsteinische Wattenmeer in Vergangenheit und Gegenwart* (Heide, 2013), p. 44.

[118] E. Miller and J. Hatcher, *Medieval England: Rural Society and Economic Change 1086–1348* (London, 1978), p. 40; Henderikx, 'Economische geschiedenis', p. 129.

[119] Van der Linden, 'Het platteland', pp. 69–78; Van Bavel, *Manors and Markets*, pp. 84–5.

[120] Henderikx, 'Land, bewoning', p. 104; Van Bavel, *Manors and Markets*, pp. 79, 86, 88–9; F. Konersmann and W. Troßbach, 'Agrarverfassung im

time, the farmers in these areas were personally and economically free. The disappearance of the manorial system from the older settlements was at least partly caused by the freedom of the settlers in the new land. For farmers, freedom and the absence of heavy obligations to a lord meant the reclaimed lowlands were more attractive to settle than the manorial areas with their heavy labour duties. This, combined with the wish of the territorial princes to create a population of free subjects within their territories, without rival claims of manorial lords, resulted in the early demise of manors in the lowland principalities on the eastern shore of the North Sea.[121]

On the opposite shore manorialism proved more resilient; in England it only started to decline from the 1350s, after the Black Death.[122] It should be kept in mind, however, that in the English lowlands the manorial system was weakly developed as, for instance, on the Norfolk Broads.[123] In Lincolnshire many farmers were 'sokemen', who had almost the same rights as free men and could sell and grant their land to others. In the new lands reclaimed in South Lincolnshire after AD 1000, farmers possessed even greater freedom. Most farmers here were sokemen, and many of them lived on unmanorialised land that was not connected with a demesne farm and the tenants of which did not have to perform labour services. In the fens of Cambridgeshire too the tenants on newly reclaimed land were free and only paid a money rent.[124] Comparing medieval South Lincolnshire with the Low Countries, H. E. Hallam concluded: 'Both societies were free and open, dynamic in growth, co-operative and autonomous in their institutions, and disinclined to take lordship seriously'.[125] Whereas elsewhere in medieval Europe the rural population was merged into one class of dependent serfs and villeins, the peasantry of the North Sea Lowlands enjoyed relative freedom.[126]

Übergang', in R. Kießling, F. Konersmann and W. Troßbach, eds., *Grundzüge der Agrargeschichte. Band I: Vom Spätmittelalter bis zum Dreißigjährigen Krieg* (Cologne, Weimar and Vienna, 2016), pp. 186–7; O. S. Knottnerus, 'Yeomen and Farmers in the Wadden Sea Coastal Marshes, c. 1500–c. 1900', in Van Bavel and Hoppenbrouwers, eds., *Landholding and Land Transfer*, p. 152.

121 Van der Linden, *De cope*, p. 336; B. J. P. van Bavel, 'The Emergence and Growth of Short-term Leasing in the Netherlands and Other Parts of Northwestern Europe (Eleventh–Seventeenth Centuries): A Chronology and Tentative Investigation into its Causes', in B. J. P. van Bavel and P. R. Schofield, eds., *The Development of Leasehold in Northwestern Europe, ca. 1200–1600* (Turnhout, 2008), p. 202.

122 M. Bailey, *The Decline of Serfdom in Late Medieval England: From Bondage to Freedom* (Woodbridge, 2014), pp. 285–306.

123 Williamson, *The Norfolk Broads*, p. 21.

124 Hallam, *Settlement and Society*, p. 201; R. H. Britnell, *The Commercialisation of English Society 1000–1500* (Cambridge, 1993), pp. 61, 68–9, 147.

125 Hallam, *The Agrarian History*, p. 507.

126 Levine, *At the Dawn*, pp. 208–9.

Conclusion

Around 1300 the frontier in the lowlands was closed. In the Low Countries in particular an extraordinary feat had been accomplished: about three quarters of a million hectares of land had been added to the cultivated area. In these reclaimed wetland areas new societies emerged which were very different from those on the uplands. Lowland societies along the southern North Sea coast had two characteristics in common. Farming was exceptionally free and individualistic, but society was also strongly communally organised by medieval standards. Both these traits had their origins in the wetland landscape and the demands on the individuals who reclaimed and settled that landscape.

Freedom in the reclaimed wetland societies had two motivating factors. Firstly, manorialism did not have a chance to develop fully in these areas. It emerged in the early Middle Ages, at a time when most of the lowlands were still uninhabited. Only on the stream ridges along the rivers could manors emerge. England was the main exception. Here manors did exist, although in the English lowlands the manorial system was considerably weaker than in the uplands. Even after reclamation, the former wetlands were not entirely suitable for manorial agriculture. The manorial economy was characterised by large open fields dedicated to grain cultivation. The wet environment of the lowlands was still mostly suited to animal husbandry. Most of the land was too wet for cereals, especially winter cereals, and the need to cut ditches and drains precluded the creation of large fields.

Secondly, to attract pioneers to the inhospitable wetlands, princes and lords were prepared to offer settlers generous conditions, the most important of which were personal freedom and full ownership of their holdings. Settling in the wetlands did not imply that the pioneers automatically became free; serfs who settled there did not lose their servile status. In most areas, however, after a while any differences in legal status blurred and a community of relatively free and equal farmers emerged. Territorial lords supported this development because they preferred a direct relationship with their subjects, without the interference of local lords. These free farmers could also freely sell and bequeath their holdings or use them as collateral for loans, and they could also freely choose how to cultivate their land without intervention by neighbours or the village community.[127] The existence of these free pioneer areas contributed to the early disappearance of serfdom from the parts of the Low Countries lowlands where it had existed earlier. By 1300, the manorial system and serfdom had all but disappeared. Only in England did serfdom survive into the second half of the fourteenth century.

---

[127] Hopcroft, *Regions, Institutions*, p. 97.

Freedom was the positive legacy of the Great Reclamation of the lowlands. Its negative legacy lies in the unintended consequences of such reclamation. The worst of those was the shrinkage and oxidation of the drained peat lands, causing a continuous lowering of the surface level of the land and increasing difficulties in evacuating excess water from the fields. As will be demonstrated in the next chapters, overcoming these and other difficulties required increasing investments in new infrastructure. This also resulted in 'forced innovation' of agriculture because the income from the land had to be increased so as to finance these investments. Therefore, it could be argued that the unintended consequences also had positive effects.

2

Pioneering in the Wetlands: Water Management and Agriculture, c. 1100–1300

Introduction

Investing in wetland reclamation is not a one-off investment: ditches, drains, banks and sluices must be maintained continuously. Drains and ditches need to be dredged, sluices and dikes require repair. Since most of this maintenance necessitated a collective effort, communities had to design rules to apportion areas of the maintenance tasks to landowners or farmers and to monitor the quality of the maintenance performed. At first the regulation of water management was the responsibility of the general local or regional administration of a village, seigneurie or *Landgemeinde*. Later specialised water authorities emerged that organised and regulated water management within their territories. This development of the organisation of water management is discussed in the first part of this chapter. How agriculture fared in the reclaimed wetlands is addressed in the second part.[1]

Organisation of Water Management in the Middle Ages

Since most reclamations were undertaken by relatively small groups of farmers, it is not surprising that the upkeep of water works was organised at the level of local communities everywhere in the drained wetlands of the southern North Sea area. Parishes, villages, hamlets or seigneuries were the units within which these maintenance duties were performed.[2]

---

[1] This chapter begins around 1100, two centuries later than the previous chapter, owing to the extreme scarcity of sources on water management and agriculture for the period prior to the twelfth century.

[2] S. J. Fockema Andreae, 'Embanking and Drainage Authorities in the Netherlands during the Middle Ages', *Speculum* 27 (1952), p. 158; Van de Ven, *Man-Made Lowlands*, p. 56; T. Soens, *De spade in de dijk? Waterbeheer en rurale samenleving in de Vlaamse kustvlakte (1280–1580)* (Gent, 2009), p. 21; P. A. Henderikx, *Land, water en bewoning: Waterstaat en nederzettingsgeschiedenis in de Zeeuwse en Hollandse delta in de Middeleeuwen* (Hilversum, 2001), p. 189; K.-H. Peters, 'Entwicklung des Deich- und Wasserrechts im Nordseeküstengebiet',

At a very early stage it was the local communities who started to design rules for maintenance. Hardly any of these rules were put into writing, but they can be distilled from the records that become more numerous from the thirteenth century onwards. Especially important for this are the charters and bylaws that codified the unwritten rules which had originated in earlier centuries.[3] In the Frisian lands, the situation was somewhat different. Here, the *Landgemeinden* were responsible for water management, so the units were somewhat larger.[4] Wherever the general local or regional administration was responsible for water management, it was perceived as a public responsibility.[5]

By far the most important issue for which rules had to be designed was the extent of liability for maintenance. Everywhere in the area under study the same kind of system was applied. At first, works were maintained by the landowners whose land adjoined them. So, a farmer whose land abutted a dike was responsible for maintaining the stretch of dike in front of his land, and the same principle was applied for ditches and drains. Even so, this system was not always equitable, as some owners might need to maintain an expensive stretch of dike, while others might only be liable for maintaining a few simple ditches. So, in most places from the twelfth century onwards a new system was introduced in which all landowners in the community were made proportionately liable for maintenance, irrespective of whether or not their land adjoined a dike or a drain.[6] In Dutch this system was called *verhoefslaging* or *verhevening*, in German *Kabeldeichung* or *Pfanddeichung*. Each landowner was liable for maintenance of a stretch of dike or drain (*slag, Kabel*) in proportion to the amount of land owned, meaning the more land one owned the more responsibility ensued for dike or drain maintenance.[7] Although the owner of the land

---

in J. Kramer and H. Rohde, eds., *Historischer Küstenschutz: Deichbau, Inselschutz und Binnenentwässerung an Nord- und Ostsee* (Stuttgart, 1992), p. 187; Darby, *The Medieval Fenland*, pp. 146–7; H. G. Richardson, 'The Early History of Commissions of Sewers', *English Historical Review* 34 (1919), pp. 389–90; E. Feikes, *Die geschichtliche Entwicklung der Deichlast in Nordfriesland* (Stuttgart and Berlin, 1937), p. 69.

[3] Van Tielhof and Van Dam, *Waterstaat*, pp. 46–7; H. E. Hallam, 'The Fen Bylaws of Spalding and Pinchbeck', *Lincolnshire Architectural and Archaelogical Society Reports and Papers* 10 (1963), pp. 40–56.

[4] Peters, 'Entwicklung', pp. 187–90; J. P. Winsemius, *De historische ontwikkeling van het waterstaatsrecht in Friesland* (Franeker, 1947), p. 34.

[5] P. A. Henderikx, 'De zorg voor afwatering en dijken op Walcheren voor circa 1400', in P. A. Henderikx et al., eds., *Duizend jaar Walcheren: Over gelanden, heren en geschot, over binnen- en buitenbeheer* (Middelburg, 1996), p. 25.

[6] Henderikx, *Land, water en bewoning*, p. 157; Van de Ven, *Man-Made Lowlands*, pp. 57–8.

[7] Van Tielhof, *Consensus en conflict*, pp. 77–8.

was liable for such maintenance, when a farm was leased out, this liability was usually transferred to the tenant farmer.

Normally the liability for dike maintenance was based on ownership of land protected by the dike. There was an interesting exception to this rule, however. In Dithmarschen, farmers with grazing rights on the foreshore were liable for dike maintenance. This is not as illogical as it may sound. The fertile marshes outside the dikes provided these farmers with sufficient income for dike maintenance. Moreover, this system guaranteed that the foreshore, which protected the dikes from the worst onslaught of the waves, was kept in good repair. However, where erosion nibbled away at the foreshore, eventually the burden of maintenance had to be shifted to the land behind the dikes.[8]

Each farmer fulfilled his duty through his own labour, or family labour, and delivery of materials such as earth, wood or straw. It was tempting for individual farmers to apply as little work as possible or use inferior materials, especially when times were adverse for farming. So, the second most important issue for which rules were required was quality control. This was done through inspections of the standard of maintenance of infrastructure: *schouw* in Dutch or *Schauung* in German. How this quality control was performed is described in what are probably some of the oldest recorded regulations for water management in the North Sea area, which originated in the eleventh to thirteenth centuries in what is now the Dutch province of Friesland. The clauses concerning water management are assumed to date from c. 1160 to c. 1270.[9] These rules prescribed that yearly on St. Benedict's Day (21 March) dikes and sluices had to be in good order, evidencing that those responsible for maintenance had repaired any damage done during the previous stormy winter season. At that day, the *schout* (sheriff) and a group of sworn witnesses inspected the dikes and sluices. The jurors were asked to state whether maintenance had been performed sufficiently. If they concluded it was not, the sheriff asked a judge (*asega*) to pronounce a verdict. As St. Benedict's Day is quite early after the winter season, this was an appropriate occasion on which to determine what damage the winter storms had caused and what the landowners still needed to do to repair that damage. At later dates the inspection was repeated to check if the repairs had taken place. In some regions up to six inspections took place every year.[10] If landowners

[8] Müller and Fischer, *Das Wasserwesen*, vol. 3/5, p. 50.
[9] W. J. Buma and W. Ebel, eds., *Westerlauwerssches Recht. Jus municipale frisonum* (Göttingen, 1977), pp. 17, 21; Nijdam, 'Preface', pp. 12–15, 53. For a dissenting opinion on the dating of these texts: J. R. G. Schuur, 'De plaatsing van de Schoutenrechten in hun historische context', *It Beaken* 76 (2014), pp. 1–33.
[10] E. Siebert, 'Entwicklug des Deichwesens vom Mittelalter bis zur Gegenwart', in J. Ohling, ed., *Ostfriesland im Schutze des Deiches*, vol. 2 (Pewsum, 1969), p. 145.

remained negligent, they could be obliged to pay a fine. In extreme cases, if a landowner continued to refuse to perform sufficient maintenance, his house could be demolished and his land confiscated.[11]

The above rules were valid in a part of the Frisian lands, but the same kind of arrangements existed elsewhere in the lowlands. The main difference between Frisia and the territorial principalities was that the sheriff did not need the intervention of a judge to execute the verdict of the jurors. After a while the jurors, chosen from prominent members of the community, were replaced by elected aldermen. Inspections were, however, held everywhere, and the aldermen pronounced a verdict on the quality of maintenance which was executed by the sheriff. Usually, a second inspection took place some weeks after the first to give landowners a second chance to meet their obligations. Negligent landowners were fined, and maintenance was performed at their cost.[12] In some areas the sheriff and aldermen retired to an inn and spent their time there eating and drinking at the cost of the negligent landowner until maintenance was performed correctly.[13]

The yearly inspection by the local sheriff and jurors or aldermen was a general feature of water management in the Low Countries and north-west Germany. Yet less is known about the way in which the quality of maintenance was checked on the other side of the North Sea. Water management was not in the hands of the manor courts in the English fenlands, but in the hands of the village communities.[14] This means there was no judge such as the *schout* or *Schulze* with the authority to execute the verdicts of the jurors. Darby mentions that in the Lincolnshire village of Wainfleet a yearly inspection of the ditches took place on St. Andrew's Day (30 November) and that defaults were presented to the earl of Lincoln, who could sentence the negligent landowners to pay a 16 shilling fine.[15] The fact that water management in the English fenlands was not part of the jurisdiction of the local court must have meant that it was more difficult to sanction deficient maintenance than in the Low Countries or Germany, since there was no judge involved with the authority to punish offenders. In the marshes of the Thames estuary, however, the maintenance of the sea defences was organised by the manors. The lord of the manor appointed officials who oversaw repair and presented negligent tenants for punishment by the manorial court.[16] Here the

---

[11] Buma and Ebel, *Westerlauwerssches Recht*, pp. 81–3, 217–23.
[12] A. A. Beekman, *Het dijk- en waterschapsrecht in Nederland vóór 1795* (2 vols., 's-Gravenhage 1905-7), p. 1437; Winsemius, *De historische ontwikkeling*, pp. 83–6.
[13] Beekman, *Het dijk- en waterschapsrecht*, pp. 888–97.
[14] Hallam, *Settlement and Society*, p. 167.
[15] Darby, *The Medieval Fenland*, p. 148.
[16] J. A. Galloway, 'Storm Flooding, Coastal Defence and Land Use around the Thames Estuary and Tidal River c. 1250–1450', *Journal of Medieval History* 35 (2009), p. 178.

situation resembles that in the Low Countries and effective sanctioning was possible.

Some constructions, such as sluices or bridges, could not easily be divided into small parts for maintenance. They also demanded more technical expertise than the simple shifting of earth that comprised most maintenance work of dikes and drains. In such cases, often a group of villages was made liable for maintenance. These groups generally reached agreements with a large landlord such as a nobleman, an abbey or a monastery, or with a city. Those large landlords had sufficient funds and knowledge to keep these more complex structures in a good state. It could even be profitable for them, because sluices were good places for fishing and the fishing rights could be leased out. Moreover, trade took place at sluices because ships often unloaded their goods there and levies could be imposed on that trade. This was an incentive for not only abbeys and cities, but also noblemen, in the Low Countries from Flanders to Friesland to take on the maintenance obligations for sluices.[17] The drawback of this kind of organisation of maintenance was of course that the institutions or persons entrusted with it were tempted to cut costs in order to earn as much as possible from the sluice. In the areas where upkeep of sluices was left to abbeys or noblemen for the greater period of time, such as Friesland, this indeed seems to have occurred in the later Middle Ages.[18] In spite of this drawback, entrusting one party with maintenance was preferable to a situation in which liability was not clearly determined. This was the case with the Grand Sluice at Boston, which had fallen into complete disrepair by 1315, about 150 years after its construction. It was then up to a jury to decide who had to pay for repair.[19]

It needs to be stressed that in the high Middle Ages the local authority only had a judicial, not an administrative task. Local authorities had neither policy nor budget for water management. Maintenance of dikes and drains was a responsibility of the owners and tenants of the farms. The only power the *schout* and his company had was to judge whether those liable for maintenance had done their duty according to unwritten – and later codified – rules. The people accompanying the *schout* formed a jury – which is why they were often called jurors: *iuratores*, *gezworenen* or

---

[17] Soens, *De spade*, p. 19; Fockema Andreae, 'Embanking and Drainage Authorities', p. 162; Winsemius, *De historische ontwikkeling*, pp. 27, 102; J. A. Mol, 'Mittelalterliche Klöster und Deichbau im westerlauwersschen Friesland', in Steensen, ed., *Deichbau und Sturmfluten*, pp. 53–4.

[18] S. J. Fockema Andreae, 'Dijken of wijken', in J. J. Kalma et al., *Geschiedenis van Friesland* (Drachten, 1968), pp. 194–5.

[19] Hallam, *The Agrarian History*, p. 501.

Geschworene – that reached a verdict. The *schout*, as representative of the count or lord, had the authority to execute that verdict.[20]

The Emergence of Water Authorities

The system outlined above functioned well for a long time. Communities were small, their members had regular face-to-face contact and tendencies to free-riding behaviour could be easily controlled. At the local level this method of organising water management often survived until the nineteenth century and the organisation of water management remained in the hands of the local general administration. However, from the twelfth century onwards problems at the regional level emerged which could not be solved easily.

These problems were felt first and foremost in drainage affairs. As more land was drained and reclaimed and parts of the surface area were lowered as a result of that drainage, it became increasingly clear that the local authorities were not capable of constructing and maintaining the systems required to keep the land dry. Drainage of large peat areas, islands in estuaries, or river plains required large infrastructural works such as canals, dams or large sluices, and coordination to make these works function optimally. Such works had to be maintained by groups of communities.[21] These communities formed informal groups that reached agreements on the maintenance duties of each member. These 'maintenance groups' were formed from the bottom-up, which turned out to be their major weakness. Since they were voluntary units of equals, there were serious risks of conflict and free-rider behaviour. This was the cause of the dilapidated state of the Grand Sluice of Boston mentioned above. The organisations had no legal authority to compel members to fulfil their maintenance duties. After some generations, such voluntary associations tended to disintegrate.[22] The emerging states could prevent this by imposing measures. The stages of the process of state formation determined what kind of solution was preferred. In the principalities of the Low Countries the 'maintenance groups' were granted charters giving them authority to force free riders to cooperate on penalty of confiscation of land. In England, where state formation had progressed further, royal ad hoc commissions were appointed to decide on solutions of maintenance conflicts, and they could sentence the parties to comply.

[20] Van der Linden, 'Geschiedenis', p. 10.
[21] C. Dekker, *Zuid-Beveland: De historische geografie en instellingen van een Zeeuws eiland in de Middeleeuwen* (Assen, 1971), p. 512.
[22] J. J. J. M. Beenakker, *Van Rentersluze tot strijkmolen: De waterstaatsgeschiedenis en landschapsontwikkeling van de Schager- en Niedorperkoggen tot 1653* (Alphen aan den Rijn, 1988), pp. 15–17; Winsemius, *De historische ontwikkeling*, pp. 87, 198; Darby, *The Medieval Fenland*, pp. 151–4.

The problems were most difficult to solve in the farmers' republics along the Wadden Coast because they lacked a powerful central authority. To demonstrate these divergent developments, the principalities of Holland, Zeeland and Flanders, the rural republics of Friesland and the Groninger Ommelanden, and the kingdom of England will be studied here.

Within the principalities of the Low Countries, specialised water management organisations were formed that functioned separately from the general public administration. These organisations were called *wateringen* in Flanders and Zeeland, *heemraadschappen* in Holland and Utrecht, or *dijkstoelen* in Guelders. In this book they will be known as water authorities. These were granted charters by a count, duke or bishop who provided them with judicial and administrative authority, but only in the field of water management. In this way a new, specialised layer of public administration emerged, enmeshed in the structure of the emerging states, but with a large degree of autonomy.

In the Rijnland district of Holland, there is evidence that local communities were already cooperating in the maintenance of canals, dams and bridges in the late twelfth century.[23] In 1226 the *scrutatores* of this organisation were mentioned, very probably board members and the predecessors of the later *heemraden* of Rijnland. It is possible the count already had recognised the authority of the board in 1226, but he had certainly done so almost 30 years later, in 1255. Count Willem II explicitly recognised its authority when he stated that in the future he would never again give permission to build a sluice in the territory of Rijnland without first conferring with the *heemraden*. Finally, in 1286, Count Floris V granted Rijnland a charter in which the rules that had emerged in the previous century were codified.[24] A bottom-up initiative of landowners or local lords in rural communities was first recognised by the count, and then a charter was granted providing it with the authority to inspect works and impose fines on negligent landowners. Membership was no longer voluntary: all communities and owners of land situated within the territory of the water board were now obligatory members of what had once been a voluntary association and could be sanctioned by the water authority. Recognition by the count did not always take as long as in the case of Rijnland. In 1277 the count cooperated with local lords in the Alblasserwaard area on the eastern border of the county in founding a water authority for the region, so this local initiative was immediately incorporated.[25] In the county of Zeeland, which was united in a personal union with Holland, the same kind of developments can be observed. From the late twelfth century, seigneuries on the islands of Walcheren

---

[23] Van der Linden, 'Geschiedenis', p. 12; Van Tielhof and Van Dam, *Waterstaat*, p. 41.
[24] Van Tielhof and Van Dam, *Waterstaat*, pp. 43–6.
[25] Henderikx, *Land, water en bewoning*, pp. 213–24.

and Zuid-Beveland started to cooperate in the evacuation of excess water through organisations called *wateringen*, which were recognised by the count and integrated into the administrative organisation of the county.[26]

In this way, by the first quarter of the fourteenth century the low parts of Holland, Zeeland, Utrecht, Guelders and Overijssel were covered with a network of regional water authorities that had received charters from count or bishop and now had jurisdiction over all landowners in their territory. In the part of Holland north of the IJ most water management remained locally organised, but the count also incorporated some regional dike maintenance authorities.[27] That incorporation benefitted the water authorities is clear, but what did the territorial lords hope to gain from this? In the first place, granting charters provided a source of income, because the water authorities had to pay for them. In 1316, for example, the count of Guelders 'sold' a concession for the construction of a new drainage canal – including regulation of liability for maintenance – to inhabitants of the Tielerwaard area.[28] In the second place, counts and bishops could use the incorporation of water authorities to strengthen their still tenuous control over their territories. The agreement of 1277 in the Alblasserwaard area, for example, served to bring the independent lords in the region under the control of the count of Holland and thus strengthen the count's position in the Holland–Utrecht border area.[29] Finally, in the incorporated water authorities the prince had the right to appoint the *dijkgraaf*. The most important task of the board members of the new authorities was basically the same as that of the local authorities: to annually inspect the state of maintenance of drains, sluices or dikes and pronounce a verdict. If that verdict was that maintenance had been insufficient, the *dijkgraaf*, as a judge invested with the count's authority, could execute the verdict and have maintenance performed at the cost of the negligent landowner. Both regional water boards and the count profited: the water authorities were granted judicial authority and the count got a firm grip on the new organisations by appointing their most important official.[30]

It can be said that the emergence of regional water authorities in much of the Low Countries was a consequence of the rather weak position

---

[26] Dekker, *Zuid-Beveland*, pp. 514–18, 523–5; Henderikx, *Land, water en bewoning*, pp. 144, 150–1.

[27] D. Aten, 'Een afgerond geheel: Waterstaat en waterschappen ten noorden van het IJ tot 1800', in Beukers, ed., *Hollanders en het water*, pp. 29–31.

[28] O. Moorman van Kappen, 'De historische ontwikkeling van het waterschapswezen, dijk- en waterschapsrecht in de Tieler- en Bommelerwaarden tot het begin der negentiende eeuw', in O. Moorman van Kappen, J. Korf and O. W. A. van Verschuer, *Tieler- en Bommelerwaarden 1327–1977: Grepen uit de geschiedenis van 650 jaar waterstaatszorg in Tieler- en Bommelerwaard* (Tiel and Zaltbommel, 1977), pp. 27–8.

[29] Henderikx, *Land, water en bewoning*, pp. 220–1.

[30] Van der Linden, 'Geschiedenis', pp. 15–16.

of central authority in the emerging principalities. Incorporating water authorities strengthened that position. It is expected that where the process of state formation had progressed further, the development of water management organisations will have been different. This was indeed the case in the county of Flanders, the oldest of the principalities in the Low Countries. Whereas in the eleventh century, Holland, Utrecht and Guelders were still no more than rather vaguely delineated geographical entities, the county of Flanders had already taken form, covering an area from the Aa river in the south to the Scheldt estuary in the north. By that time, the county already had a strong administrative structure, based on districts called castellanies (*kasselrijen*).[31] In Flanders too, water management was at first organised at the local level as a part of general public administration, and as in the other principalities, in the thirteenth century specialised authorities for water management emerged, called *wateringen*. Here too, these authorities incorporated a territory that was more suited to the demands of water management. In the castellany of the Franc de Bruges alone, there were over 100 *wateringen* in the sixteenth century, some of which were large, others very small.[32] In Flanders, the water authorities were not a new regional administrative layer as in the more northern principalities, but specialised organisations subject to supervision by the aldermen of the castellanies.[33] As in the other principalities, the Flemish water authorities were recognised by the 'state' and had jurisdiction over all landowners within their territory.

The paragraphs above have demonstrated that overlords and emerging states played an important part in the emergence of water authorities everywhere. That does not mean, however, that these organisations were absent in the rural republics of the Wadden coast. There too, local communities in the thirteenth century joined in regional organisations for the maintenance of infrastructure, which in this area were called *zijlvesten*.[34] *Zijl* is the northern Dutch word for sluice and most *zijlvesten* were founded to maintain these works. In Friesland and Groningen, where an overlord was absent, abbeys and monasteries, by far the most important landlords in these territories, took the lead in forming these maintenance communities.[35] In Groningen, the *zijlvesten* developed into permanent organisations, often with a board

---

[31] W. P. Blockmans, *Metropolen aan de Noordzee: De geschiedenis van Nederland, 1100–1560* (Amsterdam, 2010), p. 134.
[32] Soens, *De spade*, pp. 21–2, 331–6.
[33] S. J. Fockema Andreae, 'L'eau et les hommes de la Flandre maritime', *Tijdschrift voor Rechtsgeschiedenis* 28 (1960), pp. 188–9.
[34] Winsemius, *De historische ontwikkeling*, pp. 47, 53; Schroor, *Wotter*, pp. 30–1; W. Ligtendag, *De Wolden en het water: De landschaps- en waterstaatsontwikkeling in het lage land ten oosten van de stad Groningen van de volle Middeleeuwen tot ca. 1870* (Groningen, 1995), pp. 187–8; W. J. Wieringa, *Het Aduarder zijlvest in het Ommelander waterschapswezen* (Groningen, 1946), pp. 5, 9.
[35] Schroor, *Wotter*, p. 28; Wieringa, *Het Aduarder zijlvest*, p. 13; Fockema Andreae, 'Dijken of wijken', p. 186.

headed by an abbot of a powerful abbey, such as the abbot of Aduard, who was chairman of the Aduarder *zijlvest*. In Friesland, however, the *zijlvesten* did not develop into permanent water authorities. They were communities of landlords and rural districts that were liable for maintenance of sluices. Most of them, however, did not have their own boards, but decisions were taken at meetings of representatives of the cooperating communities.[36]

What happened in the Frisian lands makes clear that intervention by a prince or a state was not a necessary condition for the emergence of water authorities. Local initiative was able to establish these organisations without any interference from above. It remains to be seen, however, whether these voluntary associations would be able to continue without free-riding behaviour and whether they would be able to stop any such behaviour without judicial authority granted by the state. The *zijlvesten* were only really put to the test in the late Middle Ages (see pp. 105–7). An example from the late thirteenth/early fourteenth century, however, shows that maintenance of major infrastructure by a voluntary association could be problematic. West-Friesland (now a part of North-Holland) was protected from the sea by a dike surrounding the whole territory. The villages in the area had voluntarily agreed to each maintain a section of the dike, with the landowners in the villages required to perform the actual maintenance. During the thirteenth century more and more villagers shirked their responsibility and the dike fell into disrepair. In 1320 the count of Holland – who had conquered West-Friesland in 1289 – had to intervene. He reorganised water management and installed a water authority to repair the dike and supervise maintenance in the future.[37] Without the count's intervention, the West-Frisian sea defences might have collapsed.

Of the countries in the southern North Sea area, England had the most developed central bureaucracy, already in place before William the Conqueror seized power in 1066. It is not surprising that in this relatively centralised state the development of the organisation of water management took a different turn from the still embryonic 'states' on the continent. Until the mid-thirteenth century developments still ran parallel. Maintenance was organised on the local level and performed by landowners or their tenants. For the maintenance of larger infrastructure, voluntary regional associations were formed, and as on the continent, in the course of time these organisations were confronted with free-rider behaviour and conflicts. The state of disrepair of the Grand Sluice of Boston, referred to above, was caused by uncertainty about and conflicts over the liability for its maintenance. As in the Low Countries, voluntary organisations did not have the judicial authority to force their members

[36] Wieringa, *Het Aduarder zijlvest*, pp. 15–17; Winsemius, *De historische ontwikkeling*, pp. 98–101, 201.
[37] Beenakker, *Van Rentersluze*, pp. 15–17, 24–5.

to comply with the rules that had been agreed upon in the past. In the Low Countries, this was solved by incorporating these associations; the development in England took a very different course.

Romney Marsh, a marsh area in southern Kent, shows the remarkable similarities between the development in the Low Countries and England until the mid-thirteenth century. A voluntary association for the maintenance of the drainage system and the embankments had emerged here, administered by a board of 24 *iuratores*, very similar to the maintenance groups that developed into the *heemraadschappen* in Holland. Romney Marsh even received a charter in 1252. However, when disputes arose between the *iuratores* five years later, Henry de Bathe was issued a commission of *oyer and terminer*. As a royal commissioner he had to hear and determine any disputes that had arisen.[38] This marks the beginning of the Commissions of Sewers, bodies that played a prominent role in English water management until 1930. The judicial powers that were granted to the water authorities in the Low Countries, were also granted to the Commissions of Sewers in England. When conflicts arose about water management, the Crown appointed such a commission which had full authority to deal with them. The commissioners could impose rates, and could even imprison individuals who did not pay the rates or refused to follow their orders. These orders were based upon the verdicts of juries the commissioners could call together. In the Middle Ages, Commissions of Sewers were not permanent; they were dismissed when the problems were solved. From 1327 to 1366 alone no less than 287 Commissions of Sewers were appointed.[39]

The relatively strong, centralised English state did not permit the rise of an autonomous new layer of regional administration. Instead it kept the judicial authority over disputes in water management to itself and only invested ad hoc commissions with the right to settle individual disputes. The power of the state should not be overestimated, however. The Commissions of Sewers were composed of powerful local landlords who possessed a large degree of autonomy. The main difference from the principalities in the Low Countries was that in England the commissioners were officials appointed by the Crown, while the Dutch *heemraden* were elected by the landowners. *Heemraden* could autonomously pursue their own policies and the prince had few means to influence, let alone coerce, them to change their policies. *De facto*, the commissioners of sewers also had a lot of autonomy, but *de jure* the Crown had the power to coerce them, if only by simply deposing them and replacing them with

---

[38] Richardson, 'The Early History', p. 389.
[39] J. E. Morgan, 'The Micro-Politics of Water Management in Early Modern England: Regulation and Representation in Commissions of Sewers', *Environment and History* 23 (2017), pp. 412–15; Richardson, 'The Early History', pp. 387–8, 392.

more pliant commissioners. This difference was to prove crucial when the Crown decided to implement large-scale draining schemes in the fens in the seventeenth century.[40]

The outcome in England was a somewhat shaky organisational structure. The voluntary associations that had emerged remained without any real judicial authority, whereas the Commissions of Sewers were granted that authority but only case by case. This often meant that the same problems, such as free riding and other conflicts over maintenance, kept resurfacing. In southern Lincolnshire, for instance, a group of townships had united in the thirteenth century to maintain the drainage system and to protect the land from flooding. In the 1340s disputes arose over the liability for maintenance and a 1345 Commission of Sewers needed to solve these conflicts and prescribe new rules. In 1500, however, again a Commission was appointed to determine which communities were liable for maintenance and for how much.[41] Free riding had probably resurfaced after a century and a half, over which time some of the older regulations had inevitably also been forgotten. The same is true for Romney Marsh, where conflicts over liability for maintenance kept recurring, and Commissions of Sewers were appointed in 1361, 1474 and 1478 to settle these conflicts.[42]

Around the beginning of the fourteenth century three ways of organising water management had crystallised in the North Sea Lowlands. In the rural republics of the Wadden Sea coast, *zijlvesten* had emerged that were founded by local communities voluntarily forming groups for the maintenance of important sluices. Mostly, these *zijlvesten* remained voluntary associations of an informal nature, in particular in the present-day province of Friesland. The same kind of organisation emerged in England, the other extreme on the scale of state formation in the area under research. Here, the strong centralised state did not incorporate the voluntary organisations for water management, but instead appointed royal Commissions of Sewers, branches of the central government, to settle disputes. In the emerging principalities of the Low Countries, in Holland, Zeeland, Utrecht and Flanders in particular, the voluntary associations – mostly bottom-up initiatives – were incorporated and became regional or local water authorities. From the fourteenth century onwards, these water management organisations would have to deal with increasing environmental problems due to soil subsidence, decreasing storage capacity of sea inlets, and climate change.

---

[40] E. H. Ash, *The Draining of the Fens: Projectors, Popular Politics, and State Building in Early Modern England* (Baltimore, 2017), pp. 64–6.

[41] A. E. B. Owen, 'The Levy Book of the Sea: The Organization of the Lindsey Sea Defences in 1500', *Lincolnshire Architectural and Archaeological Society Reports and Papers* 9 (1961), pp. 36–41.

[42] S. Dimmock, *The Origin of Capitalism in England, 1400–1600* (Leiden and Boston, MA, 2014), pp. 269–70.

Property Relations

In the previous section the term ownership was used to describe the rights of farmers in Holland and north-west Germany to the land. In the Middle Ages this word did not carry the same meaning as today. Nowadays, the owner of real estate in principle has an exclusive right to use and dispose of it as he or she wishes. In the Middle Ages, several individuals or organisations could have rights to a land holding. In the manorial context, for example, the lord had the right to demand labour services from the tenant who held land from him, but that tenant held the right to cultivate the land. Moreover, the church had the right to demand a tithe – usually one tenth of the harvest – to pay for maintenance of the church and provide an income for the parish priest. Even farmers who were 'free owners' of their land, such as the pioneers in the fenlands in Holland, had to pay tithes. Because of these multiple rights to land, it is better to use the word possession instead of ownership.[43]

In much of the North Sea Lowlands, in particular those reclaimed after AD 1000, peasant possession of land took rather less complicated forms. As demonstrated above (see pp. 45–6), apart from tithes, pioneer farmers in the wetlands of Flanders, Holland, Utrecht and the lower Elbe region only paid a small yearly sum to a count or bishop, and their right of possession closely resembled the present-day concept of ownership. Another 'modern' form of access to land was leasehold; a lease of land for a limited period of usually five to twelve years in return for a yearly rent in money or in kind. The only right the tenant had here was to cultivate the land during the period stipulated. If the tenant wished to continue using the land, he had to conclude a new agreement with the landlord, and on that occasion the lease sum could be adjusted.[44] This form of possession spread throughout several parts of the lowlands at an early date, from the thirteenth century, beginning in eastern England, coastal Flanders, and Holland.[45] Leasehold was an 'economic' way of landholding, not based on submission of a peasant to a lord, but on a market transaction, the criteria of which were the outcome of supply and demand. During the Middle Ages it was to become one of the most important forms of access to land in much of the lowlands.

At first sight it seems to be a good thing to be a free owner or free tenant not subjected to a lord who could demand heavy labour duties, but free status also had drawbacks. Servile tenants were not only subject

---

[43] B. J. P. van Bavel and R. Hoyle, 'Introduction: Social Relations, Property and Power around the North Sea, 500–2000', in Van Bavel and Hoyle, eds., *Social Relations*, p. 12.

[44] Van Bavel, 'The Emergence', p. 180.

[45] Van Bavel, 'The Emergence', pp. 185–9; T. Soens and E. Thoen, 'The Origins of Leasehold in the Former County of Flanders', in Van Bavel and Schofield, eds., *The Development of Leasehold*, pp. 34–9.

to their lords but also received a measure of protection, and during the Middle Ages their rights to their holding became stronger. They could not easily be evicted, and their children could inherit use of the holding.[46] Free owners, on the other hand, could run into debt due to high investments in reclamation or to natural disaster. They might have to mortgage their farm and eventually maybe lose possession to their creditors. The position of free tenants was even more fragile; they could lose their land within a couple of years. For individual farmers, freedom was not an unmitigated blessing. From a broader viewpoint, however, many scholars consider freedom to have a beneficial effect on agricultural development. Exposure to a free market for land forces farmers to work efficiently to be able to acquire lease land or to prevent their holding from becoming encumbered with debts. In the competition for land, it is the most able, most productive and most efficient farmers who will survive by reducing costs and investing in improvements, and in this way the productivity of agriculture in general will improve.[47]

Another way of gaining access to land was through rights of common. Many rural communities in medieval Europe possessed commons: woods, fens and marshes that were not reclaimed and were in the possession of the village community as a whole. They provided the villagers with important resources such as fuel (peat and wood), building materials (timber, reed, sedge) and especially grazing for cattle, pigs or sheep. As mentioned above, in England the commons in the fenlands were not reclaimed because they were – rightly – considered to provide crucial grazing grounds to the mostly pastoral economy. In the principalities on the eastern North Sea shore the situation was different. The princes in Flanders, Holland and Utrecht claimed the rights to the 'wilderness', including marshes and fens. Parts of these vast areas were used as commons to gather fuel or thatch or catch fish and wildfowl, and the drier parts of the fens may also have been used for grazing. During the Great Reclamation the territorial princes, however, granted charters for the reclamation of these areas, and by 1300 virtually all commons had disappeared from the Dutch coastal areas. Only in some older villages, on clay or sandy soils, did some commons remain.[48] In the north-German marshes commons were scarce too; in Ostfriesland they had disappeared by the twelfth century.[49] Only in the fens of the English lowlands did extensive commons survive until well into the eighteenth century.

[46] Miller and Hatcher, *Medieval England*, pp. 130–1, 138.
[47] R. Brenner, 'The Agrarian Roots of European Capitalism', in Aston and Philpin, eds., *The Brenner Debate*, pp. 214–15; Brenner, 'The Low Countries', pp. 278, 314, 320–2; Van Bavel, *Manors and Markets*, pp. 245–6.
[48] S. J. Fockema Andreae, *Warmond, Valkenburg en Oegstgeest* (Dordrecht, 1976), p. 53.
[49] A. Mayhew, *Rural Settlement and Farming in Germany* (London, 1973), p. 178.

Commons have often received a bad press. Nineteenth-century reformers perceived commons as obstacles to progress that had to be privatised in order to reclaim lands that were 'lying waste'. They did not recognise that commons provided valuable resources to village communities. Even more disastrous for the reputation of the commons was Garrett Hardin's 1968 article on the 'tragedy of the commons', in which he described how common grazing lands tended to get depleted because every herdsman in the community was tempted to drive as many animals onto the commons as he could.[50] Later research has demonstrated that Hardin was mistaken. There was no free-for-all on historic commons, only a well-defined group had access, and the number of animals they could graze or the amount of fuel they could gather were limited. Commons were able to manage their resources in a sustainable way.[51] Since the North Sea Lowlands consisted of a western part with extensive commons and an eastern part almost without commons, the area under study provides opportunities to gauge the effects of commons on land management.

Compared with the uplands, property relations in the lowlands had 'modern' traits in the high Middle Ages: relatively free peasant landownership, early rise of leasehold and – in Germany and the Low Countries – absence of commons. As stated above, this was caused by the wetland environment that precluded the formation of manors, combined with the preference of lords for reclamation by independent peasants. It was further caused by the fact that this challenging environment was reclaimed relatively late and was therefore unburdened by an institutional legacy from the past. The societies emerging in reclaimed wetlands and woodlands in upland Europe in the high Middle Ages were frontier societies not burdened by existing social and economic structures, where new arrangements could be introduced.[52] New legal forms of access to land could be developed here. As Bas van Bavel has demonstrated, the specific social structure of a region is formed during the years it is settled and reclaimed. This is what happened in the lowlands, and it provided them with a set of modern institutional arrangements. Whether this was also the cause of the specific economic development of the lowlands in the centuries to come, as Van Bavel assumes, will be tested in the next chapter.[53]

---

[50] G. Hardin, 'The Tragedy of the Commons', *Science* 162 (1968), pp. 1243–8.
[51] T. De Moor, 'Avoiding Tragedies: A Flemish Community and its Commoners under the Pressure of Social and Economic Change during the Eighteenth Century', *Economic History Review* 62 (2009), pp. 2, 17–18.
[52] Lyon, 'Medieval Real Estate'; A. R. Lewis, 'The Closing of the Mediaeval Frontier 1250–1350', *Speculum* 33 (1958), pp. 476–7.
[53] Van Bavel, *Manors and Markets*, pp. 387–9.

Farming in the Lowlands

Documentary sources that provide information on the practice of farming – sown crops, types of cattle, implements, buildings – from the period of the Great Reclamation are very scarce, and the documents that have been preserved mostly date from the second half of the thirteenth century. For the period prior to 1250 we must rely on archaeological finds. This means we can at best get vague impressions of factors such as holding size and cattle density and must accept that we cannot obtain reliable data on crop yields. Tracing agricultural development for the three centuries following AD 1000 is thus extremely difficult, although for the Flemish coastal plain some changes can be observed. It is possible, though, to provide a sketch of the state of agriculture in the twelfth and thirteenth centuries. I will focus here mostly on the Flemish coastal plain, the Holland–Utrecht fenlands and the Lincolnshire fenlands, three regions that have been studied intensively and that represent the two main landscape types in the lowlands: marshes and fens.

For the Flemish coastal plain more data is available than for most other lowlands around the North Sea. There are indications that here technological change occurred in the eleventh to thirteenth centuries, new crops were introduced, and agriculture was commercialised.[54] Here, as in Holland, the count had acquired the rights to the wilderness, but, contrary to the count of Holland, he also possessed large estates in the coastal zone, which were managed on his behalf by officials. These estates were mostly oriented towards animal production, such as a 1600-hectare estate that existed around AD 900 to the east of the IJzer river that could accommodate 10,000 sheep. From such estates the first initiatives for embankments were taken. By improving his possessions in the coastal plain, the Flemish count could strengthen his position in the market for products such as dairy, meat and wool for the emerging urban markets.[55] This did not mean that the Flemish coastal plain became an area of large estates. As stated above, a large part of the plain was reclaimed by farmers, who were granted free status as compensation for their efforts.[56] As a result, by the thirteenth century the Flemish coastal area had become a region characterised by smallholdings and modest family

---

[54] A. Verhulst, 'L'intensification et la commercialisation de l'agriculture dans les Pays Bas Méridionaux au XIIIe siècle', in *La Belgique rurale du moyen-age à nos jours. Mélanges offerts à Jean-Jacques Hoebanx* (Brussels, 1985), pp. 99–100.

[55] Tys, 'The Medieval Embankment', pp. 211–12, 219, 231; E. Thoen, 'The Count, the Countryside and the Economic Development of the Towns in Flanders from the Eleventh Century to the Thirteenth Century: Some Provisional Remarks and Hypotheses', in E. Aerts et al., eds., *Studia Historia Economica: Liber Amicorum Herman Van der Wee* (Leuven, 1993) pp. 261, 265.

[56] Lyon, 'Medieval Real Estate', pp. 52–3.

farms, although large and even very large holdings were not unknown.[57] When 190 farms of rebels from the coastal plain who had participated in the revolt of 1323–8 were confiscated, more than half were smaller than 10 *gemeten* (4.4 hectares), but 22 were larger than 50 *gemeten* and 12 of those even surpassed 100 *gemeten*. The Saint-John's Hospital of Bruges in 1300 even owned a 248-hectare farm in Zuienkerke, to the north of the city.[58]

The replacement of large estates by more modest farm units was accompanied by a shift from predominantly animal husbandry towards more mixed farming. Although stock raising and dairy farming remained important, more crops were sown. The younger polders in particular, reclaimed from the twelfth century onwards, had clay soils that were very well suited for arable farming. The larger holdings in the coastal plain were mostly dedicated to cereal farming. In 1295, on two very large farms in present-day Zeeland Flanders, large amounts of wheat and oats were sown. On one of them 31 hectares of wheat, 23 hectares of oats and 26 hectares of pulses were grown.[59] The presence of large amounts of beans, which can also be observed elsewhere in the Flemish coastal plain, indicates that already a crop rotation system may have been in place. In this rotation, leguminous plants, such as beans, peas and vetches, which can draw nitrogen from the air, replenished the stock of nitrogen in the soil that was being depleted by cereal cultivation. This is confirmed by a lease contract from 1344 of a smaller farm in the same area with 26 hectares of arable. The tenant was to sow 43 per cent of the arable with winter grains (mostly wheat), 24 per cent with summer grains (barley and oats) and 19 per cent with beans, peas and vetches. The remaining 14 per cent was to be left fallow.[60] This means that a considerable amount of the land that would have been fallow in a three-course rotation was now under leguminous crops. On smallholdings and family farms more labour-intensive crops were cultivated. A very important crop on these small farms was madder, the roots of which contain a red dyestuff that can be used to colour textiles. This crop was mentioned in the coastal plain from the late twelfth century onwards. Planting and harvesting the roots were extremely labour-intensive activities, so this was an attractive crop for smallholders, who could apply cheap family labour.[61]

[57] Thoen, 'Social Agrosystems', p. 53.
[58] J. Mertens, *De laat-middeleeuwse landbouweconomie in enkele gemeenten van het Brugse Vrije* (Brussels, 1970), p. 42; L. Vervaet, 'Goederenbeheer in een veranderende samenleving: Het Sint-Janshospitaal van Brugge ca. 1275 – ca. 1575' (unpublished PhD thesis, University of Gent, 2014), p. 489.
[59] E. Thoen, 'Technique agricole, cultures nouvelles et économie rurale en Flandre au Bas Moyen Age', in *Plantes nouvelles en Europe Occidentale au Moyen Age et à l'époque moderne. Flaran* 12 (1990), pp. 60–1.
[60] M. K. E. Gottschalk, *De Vier ambachten en het Land van Saaftinge in de Middeleeuwen* (Assen, 1984), p. 255.
[61] Verhulst, 'L'intensification', p. 99; Thoen, 'Technique agricole', pp. 55–6.

Specialisation can be discerned in the agriculture of the Flemish coastal plain. The remaining large holdings continued to raise stock and practise large-scale cereal cultivation; the numerous smallholders cultivated labour-intensive madder and will probably have had some milk cows and have grown some grain for their own family. All farms in the coastal plain could produce for the large cities such as Bruges, Ghent and Ypres that had emerged here since the second half of the eleventh century. Another proof of the progressive nature of high medieval Flemish agriculture is the fact that horse traction was already predominant here in the twelfth century.[62] It was not only urban markets that stimulated agricultural development in the coastal plain, but also the bourgeoisie and religious and charitable institutions that invested in the countryside.[63] Still, the performance of agriculture in the Flemish coastal plain should not be overestimated. Wheat yields of 10–11 hectolitres per hectare in the fourteenth century were not impressive, especially not when compared with the environs of Lille, where wheat yields of 20 hl/ha or more were prevalent.[64] Admittedly, these are yields from large, atypical farms, as with almost all yield data from the Middle Ages.[65] Therefore, yields on smaller, more intensively farmed holdings may have been higher. Nevertheless, agriculture in the Flemish coastal plain was certainly specialised and market oriented.

In the *cope* reclamations of Holland and Utrecht the land was parcelled out in a very regular way. From the axis of reclamation, often a small peat river, drainage ditches were dug into the peat bogs at regular distances of around a hundred metres. The plot in between two ditches became an agricultural holding and would be 1200–1350 or 2250–2700 metres long. In this way two types of agricultural units were created, smaller ones of 10–15 hectares and larger ones of 21–30 hectares. In the north-German *cope* reclamations the units were larger, about 48 hectares,[66] although this does not say much about the area that was actually farmed. The pioneers started by reclaiming the area closest to the reclamation axis, which often was the most fertile part, because there the peat had been improved by

---

[62] E. Thoen, 'The Birth of the "Flemish Husbandry". Agricultural Technology in Medieval Flanders', in G. Astill and J. Langdon, eds., *Medieval Farming and Technology: The Impact of Agricultural Change in Northwest Europe* (Leiden, 1997), p. 81.

[63] Thoen, 'The Count', pp. 265, 277.

[64] Mertens, *De laat-middeleeuwse landbouweconomie*, p. 78; A. Derville, *L'agriculture du Nord au Moyen Age (Artois, Cambrésis, Flandre wallonne)* (Villeneuve d'Ascq, 1999), pp. 112–13.

[65] J. Hatcher, 'Peasant Productivity and Welfare in the Middle Ages and Beyond', *Past and Present* 262 (2024), pp. 281–314.

[66] Van der Linden, 'Het platteland', pp. 59–60; D. Fliedner, *Die Kulturlandschaft der Hamme-Wümme Niederung: Gestalt und Entwicklung des Siedlungsraumes nördlich von Bremen* (Göttingen, 1970), p. 38.

nutrient-rich deposits from river floods. It was only at a later stage that the land deeper in the peat bogs was reclaimed. Since not all land in the plots was cultivated, and holdings after a while were often divided, because this was an area of partible inheritance in which all children had a right to a share in the holding, farms in the Holland–Utrecht fenlands were modest family farms.

The modest size of the holdings was reflected in the farm buildings. These were built near the axis of reclamation, often on low mounds of clay to prevent them from sinking into the soggy peat. People lived in longhouses, rectangular or boat shaped buildings that provided living space for the farm family, stalls for cattle and storage space for the harvest of arable crops and hay. The remains of an excavated farm near Gouda dating from the first half of the twelfth century give an impression of medieval fenland farmsteads. The boat-shaped building measured 11 x 18 metres and had wattle and daub walls. The roof was supported by seven pairs of vertical posts that divided the space into three aisles. The side aisles provided room for eight heads of cattle and were provided with wooden planking to prevent the animals from sinking into the soggy soil, which illustrates how wet the area still was, even though it was drained. Buildings in such environments did not survive for long, especially since the vertical posts were dug into the soil and quickly rotted. The excavated farm had already ceased being used by the second half of the twelfth century.[67]

The presence of cattle stalls and bones of cows and a horse indicate that animal husbandry was practised at the Gouda farm, but were crops cultivated as well? Despite drainage, the reclaimed fenlands remained a very wet environment. Especially during the autumn and winter months the land must have been permeated with water, which must have made the cultivation of winter crops difficult. These objections did not prevent Van der Linden from stating that until the mid-fourteenth century considerable amounts of barley and rye were cultivated in the Holland–Utrecht fenlands.[68] Most historians, such as Bas van Bavel and Jan Luiten van Zanden, have adopted Van der Linden's views.[69] Recently, however, historical geographers have raised the question of whether arable agriculture on a serious scale was ever practised in much of the fenlands. They point out that there are hardly any archaeological finds

---

[67] C. Bakels, R. Kok, L. I. Kooistra and C. Vermeeren, 'The Plant Remains from Gouda-Oostpolder, a Twelfth Century Farm in the Peatlands of Holland', *Vegetation History and Archaeobotany* 9 (2000), pp. 147–9.
[68] Van der Linden, *De cope*, p. 317.
[69] B. J. P. van Bavel and J. L. van Zanden, 'The Jump-Start of the Holland Economy during the Late-Medieval Crisis, c. 1350 – c. 1500', *Economic History Review* 57 (2004), pp. 508, 516.

that confirm Van der Linden's statement.[70] On the other hand, the count of Holland did have considerable income from corn tithes in the peat areas until well into the fourteenth century.[71] An investigation of the available evidence can bring more clarity on this point.

First, it is important, as De Bont has demonstrated, to realise that the opportunities to grow crops were not the same in all types of peat reclamations. In the eutrophic peaty plains, prospects for arable farming were bleak, because, although they were quite fertile, due to oxidation and shrinkage these low plains would soon after drainage become too wet for the cultivation of crops.[72] If any arable farming took place here, it must have been on small higher plots. Arable farming on any scale was only possible on the oligotrophic raised peat bogs that were situated several metres above sea level. Here crops could be grown, at least until the level of the land was lowered to sea level as the result of drainage. But this was still a wet environment, and it was nutrient poor. Which crops could be grown there?

Analysis of the plant remains found at the site of the twelfth-century Gouda farm confirms that arable crops were grown there. The main crops were emmer wheat and flax, with traces of barley, oats and hemp also found. Emmer was also found at other sites from this period in the Holland–Utrecht peat area, such as Vlaardingen, Dordrecht, Papendrecht and Delfgauw. Although it was not found at all medieval sites in the peatlands, emmer seems to have been a common type of cereal in the area.[73] This less demanding relative of wheat grows well on poor soils and is more resistant than wheat to fungal diseases that occur in wet areas. This, in combination with the fact that it can be sown in spring, was probably the reason why agriculturists in the fenlands preferred to sow emmer to provide themselves with bread grain. It was well suited to the humid environment. Emmer wheat was not everywhere as predominant as on the Gouda site. At the site of a farm near Delfgauw in Zuid-Holland, dated between 1150 and 1250, barley and oats were the most important grain crops, but here too emmer wheat was found, and rye was absent.[74] The scarcity of rye in the fenlands is not that surprising as rye does not do well on wet soils. It may have been cultivated during the first years following drainage, when the land was relatively dry, but

---

[70] De Bont, *Amsterdamse boeren*, pp. 41–2; G. J. Borger, 'Henk van der Linden en de historische geografie: De betekenis van het werk van prof. mr. H. van der Linden', *Historisch-Geografisch Tijdschrift* 33 (2015), p. 75.

[71] D. E. H. de Boer, *Graaf en grafiek: Sociale en economische ontwikkelingen in het middeleeuwse 'Noordholland' tussen ca. 1345 en ca. 1415* (Leiden, 1978), pp. 223–4.

[72] De Bont, *Amsterdamse boeren*, p. 42.

[73] Bakels et al., 'The Plant Remains', pp. 154–6; L. I. Kooistra, *Delfgauw vindplaats PZPD2: Een middeleeuwse boerderij met een stads sausje?*, BIAXiaal 151 (Zaandam, 2002), pp. 3–5.

[74] Kooistra, *Delfgauw*, pp. 5–6.

probably its cultivation was abandoned after a few years.[75] Barley, oats and emmer wheat, sturdy summer grains, were more suited to the local circumstances.

An excavation of a farm from the last quarter of the thirteenth century near Barendrecht in Zuid-Holland confirms this. In and around this 14 x 6 metres longhouse, emmer wheat was again found. It clearly was a mixed farm, because bones of cows, pigs, sheep and horses were discovered.[76] Although the evidential base is narrow, an impression of farms in the high-medieval fenlands can be formed. They were modest-sized family farms where mixed farming was practised.[77] Cows, pigs and sheep were kept and summer grains like barley, oats and emmer wheat were cultivated, along with fibre crops such as flax and hemp. The share of arable land in the farm operation was probably small. For the surroundings of Delft it has been estimated at a quarter of the cultivated area.[78] The remainder was mostly under grass. Due to variations in the micro relief of the reclaimed peat lands, only small and irregularly formed parcels could be used as arable. This raises the question of whether under such circumstances the fields could be ploughed.[79] It is possible that these small plots were cultivated with spades. Agricultural produce could be complemented with berries, fish and wildfowl from the still un-reclaimed parts of the peat, which provided the population also with fuel in the form of dried peat sods. Most produce was probably consumed by the farm family itself, although some butter, meat or processed flax may have been exchanged for industrial products. Opportunities for trade were still very limited, however, because the urban sector in Holland remained small until the late thirteenth century.

The archaeological data is confirmed by one of the scarce documentary sources from the thirteenth century: the accounts of an estate in Aalburg in the Land van Heusden – an area on the Holland–Brabant border with both peat and clay soils – owned by the abbey of Sint-Truiden. In

---

[75] Bakels et al., 'The Plant Remains', p. 156; J. Z. Rodengate Marissen, *Bijzondere plantenteelt*, vol. 1 (Groningen, 1907), p. 79.
[76] J. M. Moree, 'Barendrecht bouwt Carnisselande, een dijk van een wijk: Archeologisch onderzoek in een nieuwbouwwijk naar het middeleeuwse dijkdorp Carnisse in de verdwenen Riederwaard', in B. Wouda, ed., *Ingelanden als uitbaters: Sociaal-economische studies naar Oud- en Nieuw-reijerwaard, een polder op een Zuid-Hollands eiland* (Hilversum, 2004), p. 59.
[77] W. Ettema, 'Boeren op het veen (1000–1500): Een ecologisch-historische benadering', *Holland* 37 (2005), p. 255.
[78] G. H. P. Dirkx and J. A. J. Vervloet, *'Oude Leede', een historisch-geografische beschrijving, inventarisatie en waardering van het cultuurlandschap* (Wageningen, 1989), p. 45.
[79] J. Bieleman, 'De ossen sijn hier seer schoon en groot: De landbouw in Holland tijdens de Republiek', in T. de Nijs and E. Beukers, eds., *Geschiedenis van Holland 1572–1795* (Hilversum, 2002), p. 83.

1260 oats, barley and *amera* – very probably emmer wheat – were sown on the abbey's land. Oats were by far the most important crop, being 69–87 per cent of the sown area. Only small amounts of rye were sown, probably on the high parts of the stream ridges, but in the lower peaty areas only spring crops were cultivated. In 1263 the abbey held six milk cows, 14 other heads of cattle and 40 pigs and piglets on its estate. It also employed 14 draught horses.[80] Apart from the finds of horse bones on the Barendrecht and Gouda sites mentioned above, this is one of the very few indications of the kind of draught animal that was used. The only thing that can be said at this stage is that in Aalburg, Gouda and Barendrecht, in the second half of the thirteenth century, horses were employed. We do not know how widely horse traction was spread nor when it was introduced to the peat areas. Horse traction may have started here early because farms were small and research in England has shown that smaller farmers were often the first to employ horses. Unlike oxen, horses were multi-purpose animals that could not only be used for ploughing and harrowing, but also for hauling, riding and pack-animal work. Therefore, on a smallholding one horse could provide many services.[81] However, in the wet environment of the fenlands almost all transport was by boat, so the only remaining function for horses was the drawing of ploughs and harrows, which could also be done by oxen or cows. As noted above, however, it is possible that ploughs were not used at all on the small and irregularly shaped arable plots of the fenlands.

There is just as little evidence that oxen were employed as draught animals. Only in the river clay area of Utrecht, to the east of the fens, do forms of plots indicate that before 1100 large draught teams of oxen with mouldboard ploughs were used.[82] In light of the demands of the wet environment it seems at first glance reasonable to assume that mouldboard plough and horses were introduced in the fenlands at an early stage. However, considering the probably very small size of the arable operations on the fenland farms, it may also have been that farmers worked their small arable plots with a spade. We do not know enough to draw firm conclusions. To conclude, as Bas van Bavel recently did, that this period 'saw the massive introduction of heavy ploughs and horse traction' and that 'productivity was hugely increased' is speculative.[83]

---

[80] P. C. M. Hoppenbrouwers, *Een middeleeuwse samenleving: Het Land van Heusden, ca. 1300 – ca. 1515* (Groningen, 1992), pp. 255–7.

[81] J. Langdon, *Horses, Oxen and Technologal Innovation: The Use of Draught Animals in English Farming from 1066 to 1500* (Cambridge, 1986), pp. 252–3.

[82] G. Pleijter and J. A. J. Vervloet, *Kromakkers en bol liggende percelen in de ruilverkaveling Schalkwijk; in het bijzonder bij Tull en 't Waal en bij Honswijk*, Rapport Stichting voor Bodemkartering no. 1703 (Wageningen, 1983), p. 32.

[83] B. J. P. van Bavel, *The Invisible Hand? How Market Economies Have Emerged and Declined since AD 500* (Oxford, 2016), p. 152.

The data available for the fenland areas and for the Wadden Sea area at this moment do not support this claim. Total *production* was indeed hugely increased through the transformation of vast stretches of fens and marshes into farmland, but there is no evidence that the *productivity* of that farmland was also increased.

As in the Holland–Utrecht plain, farming in the Lincolnshire fenlands was predominantly pastoral, although as the population increased over the thirteenth century cereal farming became more important, especially in the higher and drier parts of the region. Crops cultivated included legumes, wheat, barley, oats and flax, and dairy products and wool were produced.[84] The main difference with the Holland–Utrecht fens was the way in which space was organised. Whereas in Holland all land whatever its soil type was divided into mixed farming units, in Lincolnshire the large fens remained in communal use. Arable farming was limited to the higher siltlands and embanked parts of the marshes. This way of farming was more in accordance with the environment. The fens were not drained so peat shrinkage was prevented and arable farming confined to the most fertile and best drained part of the village area. The floods during the winter season even improved the quality of the fens by depositing fertile sediment. Even though some land was drained and made into arable during the twelfth century, most of the vast fens remained commons used for fishing, fowling, cutting reed and sedge, gathering of fuel and grazing of cattle and sheep. The fenland communities valued these resources too highly to jeopardise them through large-scale reclamations.[85]

In Holland and Utrecht the count and the bishop controlled the fens and promoted their large-scale reclamation, but in Lincolnshire local communities controlled these wetlands and decided to preserve most of them in their mostly un-drained state. This had long-term consequences for agriculture in these areas. In Holland and Utrecht, the drained fens shrank and oxidised, leading to increasing problems in evacuating excess water. In the English fenlands farming was much more attuned to the wetland environment. Here, the resources of the fens were kept intact and even improved through flooding, and soil subsidence was prevented. Comparing the Holland–Utrecht plain with the Lincolnshire fens shows that institutional arrangements during reclamation influenced the course taken in reclamation and agrarian development.

As far as holding size is concerned, the fenlands of Lincolnshire show similarities with those of Holland. The sokemen and freemen had small farms, many of them smaller than five acres (two hectares) and only a few larger than 30 acres (12 hectares). Larger holdings were only predominant among the numerically smaller group of servile farmers with heavy labour duties. The smaller size of the holdings of the freemen

---

[84] Hallam, *Settlement and Society*, pp. 178–9, 181–2, 185.
[85] Hallam, *Settlement and Society*, p. 171.

and sokemen was caused by the custom of dividing the inheritance equally between male and female heirs, which was also the custom in the Holland–Utrecht fens.[86]

Conclusion

The construction and maintenance of the physical infrastructure of a water management system, such as dikes, sluices and drains, required a collective effort. Simple drainage ditches could be dug and maintained by an individual farmer and his family, but all other larger and more costly elements required the labour of at least the families of a village or hamlet. Although documentary evidence is scarce, all available evidence indicates that in the Low Countries, north-west Germany and England alike, the village or hamlet – and in the Frisian lands the *Landgemeinde* – was the unit that initially coordinated and monitored maintenance work. From the beginning, the farmers in the reclaimed wetlands had to cooperate to keep dry feet. Village administrators started to design rules that determined who was liable for maintenance of what stretch of dike or drain and how negligent landowners should be punished. Usually this meant that each landowner was obliged to maintain a stretch of dike or drain proportional to the land area he owned and could be fined for any neglect of his duty. From the thirteenth century such rules were codified and explicitly or tacitly recognised by the overlord.

The need to maintain the water management system collectively was the cause of strong communal elements in lowland societies, but other communal elements were conspicuously absent. Contrary to the uplands, there were no open fields that were collectively exploited by communities, and within the lowland societies on the continent common grazing land was absent or disappeared at an early date. In the English lowlands, however, commons remained in existence until well into the early modern period. The explanation for this divergence is that in the principalities of the Low Countries all waste land was claimed by the prince, in whose interest it was to have all this land drained, whereas in England local lords and communities owned the fens and appreciated the many resources the undrained fenlands had to offer. Therefore, in England the commons remained intact, and less land was transformed. This illustrates how the institutional arrangements that were in place when the Great Reclamation started strongly influenced the long-term development of settlement and agriculture in the wetlands.

As settlement in the wetlands increased, an increasing number of villages had to cooperate to ensure the maintenance of major works such as seawalls, canals and sluices. At the start, this cooperation was

---

[86] Hallam, *Settlement and Society*, pp. 213–17.

organised on a voluntary basis, but often free-riding behaviour emerged after some generations. In the main principalities of the Low Countries this problem was solved by granting the regional associations charters giving them the power to coerce the participating villages into fulfilling their duties. Thus, the voluntary associations became water authorities, specialised units integrated into the administrative organisation of the principality, but they did not completely lose their communal characteristics. Well into the nineteenth century they also remained associations of landowners. The rise of regional and local water authorities was limited to Flanders, Zeeland, Holland, Utrecht, Guelders and Overijssel. Along the Wadden Sea, regional water authorities were not chartered, because in the rural republics predominant there, there was no central authority with sufficient powers to do this. In England, the most centralised state in the area, regional associations were not granted charters either. Instead, from the thirteenth century, the Crown installed ad hoc Commissions of Sewers, royal commissioners appointed to solve conflicts arising over the maintenance of important works, with the authority to make the interested parties comply with their decisions.

Around 1300 a relatively free society had developed in the lowlands of the Low Countries. Commons and the manorial system had never existed or had disappeared. Farmers had strong property rights and no manorial obligations, and they could individually cultivate their land. Institutions designed to benefit agricultural development prevailed. There are few indications, however, that farming in the lowlands was conspicuously precocious or productive. Even in the most developed area, the Flemish coastal plain, where horse traction was introduced early and commercial crops such as madder were cultivated, cereal yields were low compared with inland areas such as southern Flanders. In the Holland fenlands, mixed subsistence farming was predominant. Beneficial institutions in themselves were not sufficient to cause development of productive agriculture. As Bas van Bavel correctly stated about Holland, 'at the beginning of the fifteenth century, there was little to foreshadow such a development'.[87] It took a series of environmental crises followed by adaptation of the institutions to accomplish that.

---

[87] B. J. P. van Bavel, 'Rural Development and Landownership in Holland, c. 1400–1650', in O. Gelderblom, ed., *The Political Economy of the Dutch Republic* (Farnham, 2009), p. 167.

3

Sinking Land and Disastrous Floods: Water Management, c. 1300–1550

Introduction

Around 1300 the expansion period of the high Middle Ages came to an end, and shortly afterwards climate change and disease caused a deep demographic and economic crisis in Europe. In addition to this, the North Sea lowlands were also confronted with environmental challenges partly caused by the large-scale reclamations of the previous period. Changing climate, storm floods and lowering of the surface level of the land led to catastrophic losses of land and lives. The problems were partly solved by less intensive land use and partly by both technological and institutional innovations. The changes of the late Middle Ages would have a far-reaching influence on later developments.

This chapter begins with an introductory section about the main demographic, economic, environmental and political developments in the North Sea area. This is followed by sections regarding the environmental problems in the lowlands, and the technological and institutional response to those problems in the diverse regions of the North Sea lowlands. The consequences these technological and institutional changes had for farming in the area will be dealt with in the next chapter.

The North Sea Area during the Late Middle Ages

Cold and wet years in the early fourteenth century marked the beginning of the Little Ice Age, a period with lower average temperatures that would last until the middle of the nineteenth century, although during these five and a half centuries colder and warmer periods did alternate. In the period under study in this chapter, the years from c. 1300 to 1380 were especially cold, followed by a warmer period that ended with a series of severe winters in the 1430s.[1] For the North Sea lowlands, situated in a moderate

---

[1] E. Le Roy Ladurie, *Histoire humaine et comparée du climat: canicules et glaciers XIIIe–XVIIIe siècles* (Paris, 2004), p. 31; C. Camenisch, 'Endless Cold: Seasonal Reconstruction of Temperature and Precipitation in the Burgundian Low

climate zone, increased cold was not the main problem. At this latitude most damage was caused by increasing precipitation. Grain can tolerate cold winters, especially if the land is covered by snow, but very wet seasons can destroy the crops, as occurred in 1315/16. For obvious reasons, such problems tend to be worse in lowlands than in uplands. The famine of the 1310s was not enough to cause a collapse of the population. That took until the 1340s, when what Bruce Campbell called 'a perfect storm' occurred, wreaking havoc in Europe. Population pressure, economic depression, warfare – the Hundred Years' War began in 1337 – and bad weather already made life very difficult for the majority of the population, but the situation became apocalyptic when the plague reached Europe from Central Asia in 1347. From the Mediterranean the disease quickly spread across the continent, already reaching England in 1348. In that country some 45–50 per cent of the population is estimated to have died in the two years from 1348 to 1350.[2] This disease, which kept recurring, in combination with the cooling climate, economic problems and warfare, would keep the European population at a low level for over a century.

How did the North Sea Lowlands fare under these tribulations? Information is scarce, especially concerning the German coastal areas, but for the Low Countries a somewhat impressionistic image can be painted. Of course, climate change affected this region too. The lowland areas along the coast were especially vulnerable due to increased storminess, which will be discussed in more detail in the next section. It is clear the Low Countries suffered from both famine and rinderpest in the 1310s. In the Flemish cities of Ypres and Bruges thousands died due to famine-related causes. In the same years, a chronicler from Utrecht wrote about the poor who ate grass and gnawed like dogs at the bones of dead cattle.[3] Very little is known about the effects of the Black Death in this area, but the Low Countries were not spared this ordeal, and the number of victims seems to have been substantial, though quantification is not possible. In Holland the later epidemics, or epizootics, also caused considerable numbers of casualties.[4]

For the Low Countries in general, population development in the late Middle Ages has been described as one of stagnation or slight decline, not as a collapse. The cities in particular managed to maintain, or even

---

Countries Based on Documentary Evidence', *Climate of the Past* 11 (2015), p. 1057; Pfister and Wanner, *Klima und Gesellschaft*, pp. 194–5, 208, 288.

[2] Campbell, *The Great Transition*, pp. 316, 328–9; Dyer, *Making a Living*, p. 272.

[3] H. Van Werveke, 'La famine de l'an 1316 en Flandre et dans les régions voisines', *Revue du Nord* 41 (1959), pp. 6–7; T. Newfield, 'A Cattle Panzootic in Early Fourteenth Century Europe', *Agricultural History Review* 57 (2009), pp. 61–2.

[4] W. P. Blockmans, 'The Social and Economic Effects of Plague in the Low Countries, 1349–1550', *Belgisch Tijdschrift voor Filologie en Geschiedenis* 58 (1980), pp. 836–9, 844–5, 861; De Boer, *Graaf en grafiek*, pp. 34, 165; Van Bavel and Van Zanden, 'The Jump-Start', p. 505.

increase, their population numbers.[5] Holland and Flanders are estimated to have lost only some ten per cent of their population and remained highly urbanised. Recently, however, this relatively optimistic view of the demographic development of the Low Countries following the Black Death has been contested by Roosen and Curtis, who claim that heavy mortality in the cities was compensated for by rural–urban migration, thus obscuring the impact of the Black Death.[6] Towns in England seem to have lost a large part of their population in the fourteenth and fifteenth centuries. The major cities in the north-western coastal area of Germany – Hamburg and Bremen – also incurred heavy population losses.[7]

From a European perspective the late-medieval Low Countries are exceptional because the region remained highly urbanised. By far the most exceptional were the two counties in the heart of the lowlands, Holland and Zeeland. In Holland, the share of the population living in towns increased from c. 14 per cent in 1300 to 33 per cent in 1400 and even 45 per cent in 1514. Important export industries – cloth, beer, brick – had developed by 1400 and Hollanders were active in fishing and shipping. In the smaller county of Zeeland, industry, fisheries and shipping grew too, and by 1550 at least one third of its inhabitants lived in cities.[8] The cities on the island of Walcheren also profited from the rise of the port of Antwerp. The population of this Brabant trade centre increased from less than 6000 in 1374 to almost 30,000 in 1496.[9]

The eastern littoral of the southern North Sea can be divided into two areas. One in which during the late Middle Ages urbanisation, trade and industry either were stable at a high level (Flanders) or increased strongly (Holland, Zeeland and Antwerp). The other, from present-day Friesland to Schleswig-Holstein, was much less urbanised and its economy remained agricultural. The 11 cities of Friesland were mostly small and agrarian, and the only major cities in the Wadden Sea area were Groningen, Bremen and Hamburg. On the western shore, the only major city, London, had incurred huge population loss due to the Black Death. Therefore, only the lowlands of Flanders, Zeeland, Holland, north-west

---

[5] Prak and Van Zanden, *Pioneers*, p. 63.
[6] J. Roosen and D. R. Curtis, 'The "Light Touch" of the Black Death in the Southern Netherlands: An Urban Trick?', *Economic History Review* 72 (2019), p. 50.
[7] Dyer, *Making a Living*, pp. 190, 298–302, 304; F. Konersmann and W. Troßbach, 'Die Bevölkerungsverluste des Spätmittelalters', in Konersmann and Troßbach, eds., *Grundzüge der Agrargschichte*, vol. 1, p. 26.
[8] Van Bavel and Van Zanden, 'The Jump-Start', pp. 505–6, 523; A. van Steensel, 'Bevolking en sociale structuren', in Brusse and Henderikx, eds., *Geschiedenis van Zeeland*, p. 215.
[9] W. P. Blockmans, G. Pieters, W. Prevenier and R. W. M. van Schaïk, 'Tussen crisis en welvaart: Sociale veranderingen 1300–1500', in *Algemene Geschiedenis der Nederlanden*, vol. 4 (Haarlem, 1980), p. 51.

Brabant, western Utrecht and western Guelders were situated close to a thriving urban economy that could stimulate their development. In these areas cities also had political influence, especially in Flanders, Brabant, Zeeland and Holland, where they were prominent members of the emerging Provincial States, which negotiated with the prince on taxation.

Characteristic for the urban network of the coastal areas of the Low Countries in the late Middle Ages and early modern period – apart from the northern provinces – was the high density of smaller and middling towns.[10] These towns exchanged goods and services with the surrounding countryside, over which they tried to exert political and economic influence. The towns had a beneficial influence on the farming sector as markets for food and raw materials produced by the rural population and because urbanites provided credit to farmers and invested in land reclamation and water infrastructure. Such investments may not just have been driven by the search for profits. The inhabitants of the towns that were situated within the reclaimed wetlands knew that both they and the country dwellers might ultimately share the same fate if water control in the countryside failed. That knowledge did not necessarily prevent them from exhibiting behaviour that was detrimental to the wetland landscape, as is demonstrated by the involvement of urbanites in destructive peat digging activities.[11]

State formation continued in the North Sea area during the late Middle Ages. The territorial princes in the Low Countries strengthened their grip on their territories and from the late fourteenth century a process of unification began. In 1384 the duke of Burgundy inherited the county of Flanders. In the 1420s Holland, Zeeland and Brabant were added to the Burgundian possessions. At the end of the fifteenth century the Burgundian lands were inherited by the house of Habsburg, which continued to add to its existing possessions. By 1543, when Guelders was added to the Habsburg lands, all territories with substantial lowland areas in the Low Countries were under the control of the Habsburgs. They strengthened the bureaucracies of the lands under their rule and introduced central administrative and judicial bodies.[12] This centralisation had limited consequences for water management. In each territory the organisations and institutions that had emerged in the Middle Ages remained in place and were developed further, so regional variations were confirmed and even deepened. The provincial authorities, rather than the central authorities, remained responsible for the supervision of

---

[10] Blockmans et al., 'Tussen crisis en welvaart', pp. 47–8.
[11] B. Ibelings, 'Turfwinning en waterstaat in het Groene Hart van Holland vóór 1530', *Tijdschrift voor Waterstaatsgeschiedenis* 5 (1996), pp. 78–80.
[12] H. De Schepper, 'De burgerlijke overheden en hun permanente kaders, 1480–1579', in *Algemene Geschiedenis der Nederlanden*, vol. 5 (Haarlem, 1980), pp. 328–36.

water authorities. In the diverse territories the prince was assisted by an increasingly professional Audit Office (*Rekenkamer*). These offices were important advisors of the prince in the field of water management.[13]

The late Middle Ages saw the end of the *Landgemeinden*, the small autonomous republics on the Wadden coast. This outcome had both internal and external causes. Feuding had always been endemic in these territories, but at the end of the Middle Ages it became more intense. In Ostfriesland the political strife resulted in the founding of a new principality uniting several *Landgemeinden* under the lead of the nobleman Ulrich Cirksena in 1464. The external cause was that the lords of the adjoining principalities tried to enlarge their territories by seizing power in these republics. The Habsburgs seized power in Friesland in 1524 and in Groningen in 1536. The count of Oldenburg conquered Butjadingen and Stadland in 1514. Finally, in 1559 the last *Landgemeinde*, Dithmarschen, had to surrender to the Danish Crown, which had controlled the duchy of Schleswig and the county of Holstein since 1460. In most cases, this did not mean that peasant freedoms completely disappeared. The population of Dithmarschen, for instance, retained its strong property rights and low taxes after the Danish conquest.[14] Butjadingen was an exception; here the rural population was subjected to an oppressive and rapacious regime that introduced *corvée* labour and other onerous duties.[15]

Around 1550 political power on the eastern shore of the North Sea rested mostly with territorial principalities. Flanders, Brabant, Zeeland, Holland, Utrecht (including Overijssel), Guelders, Friesland and Groningen were controlled by the Habsburgs. On the German Wadden coast Ostfriesland, Oldenburg, Jever and the archbishopric of Bremen were autonomous units, whereas Schleswig and Holstein were controlled by the Danish Crown. On the western shore England remained the only centralised kingdom in the area.

New Challenges

In the reclaimed fenlands the consequences of shrinkage and oxidation, caused by the drainage of the soil, began to be seriously felt in the fourteenth century, and in some of the earliest reclaimed areas even as early as the thirteenth. At the beginning of settlement in the peat areas drainage was simple: ditches were dug, and gravity made the water flow to a river. As the surface level became lower the ditches had to be deepened and the lowest parts of the area could no longer be used as

---

[13] P. J. van Cruyningen, 'Dealing with Drainage: State Regulation of Drainage Projects in the Dutch Republic, France, and England during the Sixteenth and Seventeenth Centuries', *Economic History Review* 68 (2015), pp. 423–4.
[14] Abel, *Geschichte*, p. 220; Knottnerus, 'De Waddenzeeregio', p. 57.
[15] Norden, *Eine Bevölkerung*, pp. 226–31.

arable. The process reached its most dangerous phase when the level of the peat lands became lower than that of open water. Gravity could then no longer ensure that excess water was evacuated and of course the risk of flooding was increased. A report from 1570 on a part of the Rijnland district in central Holland gives an impression of the condition of insufficiently drained fenland. On 6 June of that year much of the land was still under water and thus would yield no hay or grass. In the better parts cows were grazing, but they were splashing through the water and could only find wet and yellow grass to eat. Farmers tried to supplement the feed of the cows with water plants, but that was unhealthy for the animals.[16] Under such circumstances even stock farming was doomed.

The moment at which drainage problems became acute varied.[17] This variation was determined by several factors, such as the time at which the land had been reclaimed and the way the land was used – the more arable farming, the quicker the land will sink. Too little is known about land use, but the available data confirms that crisis phenomena can indeed first be observed in the areas that were reclaimed earliest. For Friesland, where reclamation had started first, archaeological data indicates that drainage had already become difficult in the thirteenth century and some fenland settlements were already abandoned in the fourteenth.[18] For Holland, the lowering of the surface level of the soil and the increasing trouble in evacuating excess water are mirrored by the income of the count of Holland from corn tithes. From the second half of the fourteenth century the yields of the tithes, which had still been considerable at the beginning of the century, began to decrease. This indicates that even growing summer crops was becoming impossible. In West-Friesland, in the north of the county, tithe yields decreased by 73 per cent between 1344 and 1400.[19] In the Rijnland district in the south, where reclamation had started later, the first fall in tithe income took place in the 1370s; elsewhere, such as in the Holland–Utrecht border area, it occurred after 1400.[20]

In some areas the problems were exacerbated by other human interventions in the landscape. In the late thirteenth century, for example, a dam had been constructed in the Hollandse IJssel river in the Holland–Utrecht border area, cutting off the connection between this river and the Lower Rhine. Thus it became a tidal river that began to silt up. The rising

[16] A. Bicker Caarten, *Middeleeuwse watermolens in Hollands polderland, 1407/'08 – rondom 1500* (Wormerveer, 1990), pp. 24–5.
[17] Ettema, 'Boeren op het veen', p. 255.
[18] J. A. Mol, *De Friese huizen van de Duitse Orde: Nes, Steenkerk en Schoten en hun plaats in het middeleeuwse Friese kloosterlandschap* (Ljouwert, 1991), p. 167; De Langen, *Middeleeuws Friesland*, p. 48.
[19] C. M. Lesger, *Hoorn als stedelijk knooppunt: Stedensystemen tijdens de late Middeleeuwen en vroegmoderne tijd* (Hilversum, 1990), pp. 68–9.
[20] De Boer, *Graaf en grafiek*, pp. 223–4; Van Bavel and Van Zanden, 'The Jump-Start', pp. 516–17.

Water Management, c. 1300–1550

riverbed made it even more difficult for the adjoining fenlands to drain their excess water.[21]

By the early fifteenth century, arable agriculture had collapsed in much of the fenlands, and without new technology that could pump water from the fens to open water, in the course of time animal husbandry would also become impossible and the fenlands would have to be abandoned. Indeed, butter production contracted in the Holland peat areas after 1380, indicating that dairy farmers were also in trouble. As already noted, villages in Friesland were deserted and in the southern part of Holland the population of the peat villages declined in the second half of the fourteenth century. Of course, population decreased everywhere in that period, but in the peat areas of Holland the downward turn was much stronger than in the areas with sandy or clay soils.[22]

The problems in the fenlands were aggravated due to the fact that the peat layers were not only cultivated, but also exploited for the production of fuel and salt. In Holland, Utrecht and Friesland the scale of peat cutting for fuel remained limited in the pre-1300 period. Farmers mostly cut peat sods for the needs of their own family. Further to the south, peat extraction had a more commercial and industrial character. In Flanders, the increasing demand for fuel in the emerging cities prompted the start of peat extraction on a large scale from the twelfth century onwards. The count granted vast tracts of peat lands in the coastal plain and the north of the county to abbeys, noblemen or bourgeois investors, who systematically cut away the peat layers to satisfy the insatiable demand for fuel from urban consumers and industries. These were not small operations by individual farmers, but large-scale projects. When the layers in Flanders itself threatened to become depleted, the peat moors of adjoining Brabant were attacked from 1245 onwards.[23] The peat layers in Flanders and Brabant were completely cut away in the Middle Ages and have disappeared almost without a trace.

Peat was also still exploited for salt production. This occurred especially in the south-western delta of the Meuse, Rhine and Scheldt. Peat was cut, dried and burned both within and without the areas protected by dikes. The scale of these activities increased as a salt processing industry emerged in the cities of Zeeland and northern Flanders.[24] This *moernering*, as it was called, was extremely destructive. The foreshore of the dikes, which protected them from the worst onslaught of the waves, was removed, making the cultivated land much more vulnerable in case

---

21 G. J. Borger, F. H. Horsten and J. F. Roest, *De dam bij Hoppenesse: Gevolgen voor de afwatering van het gebied tussen Oude Rijn en Hollandsche IJssel, 1250–1600* (Hilversum, 2016), p. 164.
22 De Boer, *Graaf en grafiek*, pp. 57, 60–1, 261–4.
23 Gottschalk, *De Vier Ambachten*, p. 38; Leenders, *Verdwenen venen*, pp. 142–5.
24 Van de Ven, *Man-Made Lowlands*, p. 83.

of storm surges. Although conducted on a smaller scale than in the south, peat cutting in the northern provinces also had detrimental consequences through the emergence of peat lakes that gradually increased in size due to erosion. By 1300 many parts of Holland, the area north of the IJ in particular, and the south-west of Friesland, were dotted with peat lakes.[25] In the fourteenth and fifteenth centuries peat cutting grew in Holland due to both the increasing demand from the growing cities and the decreasing profitability of farming resulting from soil subsidence. A vicious cycle was at work here: farmers tried to compensate for income loss due to the lowering of the surface level with peat digging, which in its turn led to further destruction of the landscape. As a result of this process more peat lakes were formed, such as the huge lake Haarlemmermeer to the south-west of Amsterdam.[26]

Around 1400, the prospects for the peat areas seemed bleak, but in the marshes of the Flemish coastal plain and the Wadden Sea area the situation was not much better. From the late thirteenth to late sixteenth centuries these regions were hit by a series of catastrophic floods. Major floods occurred in the Flemish coastal plain and the south-western delta in 1288, 1375, 1376, 1404, 1421 and 1424. The sixteenth century was disastrous in this area, with devastating floods in 1509, 1511, 1530, 1532, 1552 and 1570.[27] In addition, the Wadden Sea coast was hit by a series of catastrophic storm surges. Particularly infamous was the *Grote Mandränke* (Great Man Drowning) of 1362 along the German coast, which is supposed to have caused the destruction of the trade centre of Rungholt and large areas in Schleswig-Holstein.[28] Further to the west, sea inlets such as the Dollard, Leybucht, Harlebucht and Jadebusen were formed.[29] The All Saints' Flood of 1570 wreaked havoc along all of the southern North Sea coast. Many human and animal lives must have been lost due to these floods, but there is no reliable data on this. Chroniclers tend to mention grossly overestimated numbers of victims. For the worst flood of the sixteenth century, the aforementioned All Saints' Flood of 1570, a number of 100,000 victims was mentioned by chroniclers. All available data indicates that the real number was much lower, although for Friesland some 3000 human victims were counted by the local

---

[25] G. Bakker, 'Het ontstaan van het Sneekermeer in relatie tot de ontginning van een laagveengebied 950–1300', *Tijdschrift voor Waterstaatsgeschiedenis* 10 (2001), pp. 54–66.

[26] P. J. E. M. van Dam, *Vissen in veenmeren: De sluisvisserij op aal tussen Haarlem en Amsterdam en de ecologische transformatie in Rijnland 1440–1530* (Hilversum, 1998), pp. 58–81.

[27] A. M. J. de Kraker, 'Flood Events in the Southwestern Netherlands and Coastal Belgium, 1400–1953', *Hydrological Sciences Journal* 21 (2006), pp. 918–19.

[28] Meier, 'From Nature to Culture', p. 100.

[29] Behre, *Landschaftsgeschichte*, pp. 97–8.

Water Management, c. 1300–1550

authorities.[30] What we do know for certain is that tremendous amounts of land were lost. New sea inlets were formed, such as the Braakman in the Flemish coastal plain.[31] From the last quarter of the thirteenth century the eastern coast of England was also hit by floods, often the same floods that devastated the opposite shore of the North Sea, such as those of 1288, 1334 and 1375.[32]

The fact that this increased storminess began in the late thirteenth century indicates that it may have been related to the changing climate. The cooling of the Arctic Ocean resulted in an increasing temperature differential with the Atlantic Gulf Stream, which in its turn caused stronger thermal gradients in the northern oceans and increased storminess in the North Sea area.[33] Storm surges had natural causes, but that does not mean that humans were not to blame for the damage these storms caused. As in the peat lands, the environmental challenges in the marshes were to a large degree the unintended consequences of the human efforts to reclaim the land. In the south-western delta, for example, embanking of large saltmarsh areas along the sea inlets had strongly reduced the storage capacity of those inlets and thus the water level increased during storm surges. Combined with the lowering of the surface level of the land due to peat cutting, this made the region extremely vulnerable to flooding. Population loss, economic depression and warfare may have led to insufficient maintenance and thus to even more risk of flooding.[34]

The destruction of the Grote Waard area on the Holland–Brabant border can serve as an illustration of the combination of natural and human factors that contributed to land loss in the late Middle Ages. This huge polder of 54,000 hectares had become vulnerable to flooding because the soil level had been lowered due to cultivation of the peat soil and peat digging for salt production. When hit by the St. Elisabeth Flood of 1421, the dikes were breached at several places. This would not have been catastrophic, however, if the breaches had been repaired quickly.

---

[30] J. Buisman, *Duizend jaar weer, wind en water in de Lage Landen* (7 vols., Franeker 1995–2019), vol. 3, pp. 653–4.

[31] M. K. E. Gottschalk, *Historische geografie van westelijk Zeeuws-Vlaanderen* (2 vols., 2nd edn., Dieren, 1984), vol. 1, p. 162.

[32] M. Bailey, 'Per Impetum Maris: Natural Disaster and Economic Decline in Eastern England, 1275–1350', in B. M. S. Campbell, ed., *Before the Black Death: Studies on the 'Crisis' of the Early Fourteenth Century* (Manchester and New York, 1991), pp. 189–90; J. A. Galloway, 'Storms, Economics and Environmental Change in an English Coastal Wetland: the Thames Estuary, c. 1250–1550', in Thoen et al., eds., *Landscapes or Seascapes?*, pp. 385, 387.

[33] Bailey, 'Per Impetum Maris', pp. 187–8.

[34] Vos and Zeiler, 'Overstromingsgeschiedenis', p. 91; De Kraker, 'Flood Events', pp. 926–7; T. Soens, 'The Origins of the Western Scheldt: Environmental Transformation, Storm Surges and Human Agency in the Flemish Coastal Plain (1250–1600)', in Thoen et al., eds., *Landscapes or Seascapes?*, p. 304.

That did not happen because powerful lords in the area could not reach an agreement about who was to finance repair. While the dikes were still in disrepair, another flood hit the area in 1424. This time the damage to the embankments was so extensive that repair had become impossible and almost 40,000 hectares had to be abandoned for generations.[35] Dordrecht, the main city of Holland at that time, situated on the edge of the Grote Waard, survived on a small island. Although Dordrecht later managed to recover, it lost its leading position in Holland for good. A town in Zeeland, Reimerswaal, fared even worse after the floods of 1530 and 1532. Within a century it had completely disappeared.[36] The examples of Dordrecht and Reimerswaal show how the fates of town and country in the lowlands were connected.

Another human factor that contributed to land loss in the late Middle Ages was that some landlords reduced investment in maintenance of the flood defences, making them more vulnerable to floods. In the most threatened areas, they even preferred abandoning their land over continuing to invest in dike maintenance, leaving the burden of maintenance to poorer – often peasant – landowners. They invested more in safer upland areas where no high costs for water management had to be paid. From an economic viewpoint, and in a period of decreasing lease prices, that was perfectly rational behaviour, but the consequences for some lowland areas, in northern Flanders in particular, were catastrophic. Many of these landowners were abbeys or hospitals, situated in cities like Bruges or Ghent, on the edge of the lowlands but not in them, so safe from flooding.[37]

The other shore of the southern North Sea had its own share of disasters. Parts of the English coast suffered from erosion, and due to this, thousands of acres of land from Yorkshire to Sussex flooded. Storm surges exacted their toll here too, as in 1374, when sea walls in parts of Kent breached, or in 1377, when marshes in the Thames estuary in Essex were flooded. In England too, these disasters were at least partially

---

[35] T. Stol, 'Opkomst en ondergang van de Grote Waard', *Holland* 13 (1981), pp. 129–45; K. A. H. W. Leenders, 'Zoe lanck ende breedt alst oijt hadde geweest bij staende lande: Het landschap van de Grote Waard vóór 1421', in *Nijet dan water ende wolcken: De onderzoekscommissie naar de aanwassen van de Verdronken Waard (1521–1523)* (Tilburg, 2009), pp. 33, 39, 43.

[36] C. Dekker and R. Baetens, *Geld in het water: Antwerps en Mechels kapitaal in Zuid-Beveland na de stromvloeden van de 16e eeuw* (Hilversum, 2011), pp. 249–59.

[37] T. Soens, 'Floods and Money: Funding Drainage and Flood Control in Coastal Flanders from the Thirteenth to the Sixteenth Centuries', *Continuity and Change* 26 (2011), pp. 337–49; T. Soens, 'Threatened by the Sea, Condemned by Man? Flood Risk and Environmental Inequalities along the North Sea Coast, 1200–1800', in G. Massard-Guilbaud and R. Rodger, eds., *Environmental and Social Inequalities in the City: Historical Perspectives* (Cambridge, 2011), pp. 103–4.

man-made, because in the depressed circumstances many landowners or occupiers were not able or prepared to spend enough on maintenance.[38]

After painting this picture of doom, it is good to point out that there were wetlands that fared better during the late Middle Ages. The English peat fens suffered much less from soil subsidence because – as noted in the previous chapter – the fenland communities wanted to preserve them for common grazing. They were modified to improve drainage, but they were not transformed into fully drained agricultural land. However, even here drainage problems occurred. In parts of the Cambridgeshire fens both the incidence and the length of the period of flooding increased in the fourteenth century.[39] It seems that even the relatively modest modifications of the wetland landscape caused some shrinkage and oxidation. There was no need, however, to take recourse to mill drainage, so the problems seem to have remained limited. In the late Middle Ages the English fenlanders reaped the fruits of their prudent attitude. There was no serious environmental crisis and the prosperity of the fen areas even increased somewhat.[40] The fenmen could afford not to drain the fens because their parishes contained different kinds of soils. The villages and arable fields were situated on high silt belts, in front of which and towards the sea saltmarshes were situated, and behind which the fens stretched deep inland.[41] Transforming those fens was not required to facilitate making a living. In contrast, in the huge fen areas of Holland and Utrecht it was necessary to drain these fens if people wanted to use them for agricultural purposes because there was hardly any other type of soil available to cultivate apart from narrow bands of clay along the rivers.[42]

Another reason to nuance this pessimistic story is that flooding does not necessarily make land useless for mankind. Of course, the land most exposed to the sea that had been swept away completely was not fit for agricultural use anymore, but elsewhere in more sheltered places the land often reverted to more extensive uses, as in the centuries before the Great Reclamation. About a century after the 1421 disaster, the flooded part of the Grote Waard was not an inland sea, but an area thriving with activity. A commission that visited the area in 1521–3 reported that the flooded land was not only being used for fishing and fowling, but also for reed cutting, wood production (willows were planted) and grazing, and some

---

[38] E. Miller, ed., *The Agrarian History of England and Wales. III 1348–1500* (Cambridge, 1991), pp. 44, 54, 119, 125.
[39] J. R. Ravensdale, *Liable to Floods: Village Landscap on the Edge of the Fens AD 450–1850* (Cambridge, 1974), pp. 6–7.
[40] H. C. Darby, R. E. Glasscock, J. Sheail and G. R. Versey, 'The Changing Geographical Distribution of Wealth in England: 1086–1334–1525', *Journal of Historical Geography* 5 (1979), p. 261.
[41] Thirsk, *English Peasant Farming*, p. 15.
[42] Renes, 'The Fenlands', p. 105.

higher parts were even provided with low dikes enabling summer grains such as oats and barley to be grown.[43] In Nordfriesland, particularly after the 1362 flood, sedimentation by the tides resulted in the emergence of new islands, called *Halligen*. People settled on these islands but erected no dikes. The inhabitants lived on *Warften*, artificial dwelling mounds, from where they exploited the lower parts of the islands, mostly for pasture. Life on these islands was hard and dangerous, but high expenditure on seawalls could be avoided.[44] The inhabitants of the small islands of Marken and Schokland in the Zuider Zee followed the same path: from the late fourteenth century onwards they neglected dike maintenance, which had become too expensive, and settled on artificial mounds. Until 1514 half of the island of Marken was lost, but the islanders survived on their mounds and made fishing their main source of income.[45]

Although part of the flooded land could be made productive again, due to the more extensive uses of the land, it could only feed fewer people. Little is known about the fate of those for whom there was no place anymore. There are indications, however, that they were not to be envied. Following the St. Elizabeth Flood of 1424, a quarter of the population of the flooded village of Kieldrecht in Flanders were reported to have left the village as beggars.[46] After the St. Felix flood of 1530, four prosperous farmers from the Zeeland village of Mare borrowed money from an Antwerp merchant to repair the seawalls of their village. In 1532, however, the village flooded again. Two of the four farmers drowned and the other two lost everything they owned and could not repay the loan. By 1535 one of them was living in deep poverty in an upland village in Brabant.[47] The fate of smallholders or labourers was probably worse. They often lived in the lowest parts of the polders and the number of casualties among them was highest.[48]

New Technology

What options were open to the landowners and inhabitants of the reclaimed wetlands that were threatened by soil subsidence and floods? They could choose to stop investing in drainage and flood protection and

[43] I. Zonneveld, 'Schoen riet, rijs ende grienten: Begroeiing en benutting van de Verdronken Waard', in *Nijet dan water ende wolcken*, pp. 113–16.
[44] Meier, Kühn and Borger, *Der Küstenatlas*, p. 163; Meier, *Die Hallligen*, p. 15.
[45] G. J. Schutte and J. B. Weitkamp, *Marken: De geschiedenis van een eiland* (Amsterdam, 1998), p. 17; Boschma-Aarnoudse, *Boer en poorter*, p. 81; Van Popta, *When the Shore*, p. 26.
[46] Augustyn, *Zeespiegelrijzing*, p. 170.
[47] Dekker and Baetens, *Geld in het water*, p. 87.
[48] T. Soens, 'Flood Disasters and Agrarian Capitalism in the North Sea Area: Five Centuries of Interwoven History (1250–1800)', in *Gestione dell'acqua in Europa (XII–XVIII secc.)/Water Management in Europe (12th–18th Centuries)* (Florence, 2018), p. 387; Meier, Kühn and Borger, *Der Küstenatlas*, p. 110.

leave the land to its fate. This happened to much of the marshes in the Thames estuary in the late Middle Ages. Like the peat fens of Lincolnshire and Cambridgeshire these marshes were mostly uninhabited and exploited from the nearby uplands. Landowners could afford to let the embankments decay and use the land in a more extensive way, for example for fishing.[49] This was not an option for most of the reclaimed wetlands in the Low Countries, which were relatively densely populated and almost entirely lacking higher ground where people could resettle. Most people decided to stay in the wetlands, but that required adaptation to the changed circumstances and introduction of new technology. In the fenlands this meant devices that could pump water up, and in the marshes better constructed dikes that could withstand the onslaught of the waves. Introduction and diffusion of such innovations was not self-evident, because some conditions had to be met. All new technology required considerable investments, which the rural population, impoverished due to the worsened environmental circumstances, could not provide. Moreover, some new technology required institutional change to function optimally. The diffusion of new technology will be examined here for three different regions: Zeeland/Flanders, Holland and central Friesland.

Although some peaty parts of Zeeland and the Flemish coastal plain also had to deal with soil subsidence due to drainage, this was not the main problem confronting these areas in the late Middle Ages. As noted above, the area was hit by a series of devastating floods in the fifteenth and sixteenth centuries. The frequency of storms may have been caused by climate change, but the extensive damage can also be attributed to human activity, such as embanking land in the sea inlets and destroying the foreshore by digging peat for salt production.

The effects of storms were also magnified by the erosion of seawalls. This erosion was caused by the strong currents in the tidal rivers that cut out gullies up to 40 metres deep. These gullies could destroy the foreshore and eventually undermine the dike. The subsoil of the dike then became saturated with water and changed into an unstable liquid mass, which eventually slid into the river, taking a part of the embankment with it. The weakened dike then had to be abandoned and a new one had to be constructed behind it. Often, and after a while, the new dike also had to be given up, and in this way one had to withdraw step by step. Between the fourteenth and the seventeenth centuries the island of Schouwen lost 3000 hectares of land due to dike erosion.[50] It is difficult to estimate how much land was lost in the fifteenth and sixteenth centuries, but it must have been considerable. As noted above, almost 40,000 hectares were lost in the Grote Waard in 1421–4. This was by far the worst catastrophe of this period, but if we add the consequences of other storm surges and of

[49] Galloway, 'Storms, Economics', pp. 388–90.
[50] Van de Ven, *Man-Made Lowlands*, pp. 108–9.

dike erosion, an estimate of at least 100,000 hectares of land loss during the late Middle Ages does not seem exaggerated. It seems humans were incapable of stopping the advance of the sea and rivers.

However, from the fifteenth century onwards several means were developed to improve coastal defence. In the south-western Netherlands, during the fifteenth century dikes were designed with a more gentle outward profile than the relatively steep profiles previously employed as standard practice. As intended, the waves now spent their power on the long gentle slopes instead of hitting the steep slopes with full force. Another late-medieval invention were groynes, dams perpendicular to the coast intended to force currents away from the dikes. To prevent dike erosion, osier was developed, mats woven of willow and elm branches. These were sunk at the foot of the dike and then loaded with stone to keep them in their place under water. In this way they could prevent the currents in sea or rivers from undermining the dike.[51] Steep dikes were not usual everywhere during the Middle Ages. In Nordfriesland dikes had gentle slopes, but they were too low. It took a long time, however, before the height of the dikes was increased.[52]

These new technologies had one significant drawback: they were very expensive. A lot of material was required, such as earth, stone and wood. Since the coastal areas were not rich in wood or stone, a lot of this had to be bought in faraway places, so transport costs were high. Moreover, enormous amounts of labour were required to construct this infrastructure. A dike with a gentle outward profile requires not just more earth, but also much more labour than a traditional steep dike. Therefore, in the late Middle Ages this new infrastructure was built in only a few places.[53] The rural population, impoverished by floods, did not have the capital to finance this. If new infrastructure was to be constructed, it would have to be paid for by wealthy absentee landlords.

The most prominent landlords in the medieval Flemish coastal plain and Zeeland were wealthy ecclesiastical and charitable organisations, such as monasteries, abbeys and hospitals. They were reluctant, however, to invest heavily in the coastal defences. Tim Soens has demonstrated that in the late Middle Ages these landlords tended to withdraw from the coastal areas, especially from the most imperilled parts, preferring to invest in more safe inland areas rather than the coastal plain where sea defences were expensive to maintain, even without costly technological improvements.[54] Another wealthy group, the nobility, was also reluctant

---

[51] Van de Ven, *Man-Made Lowlands*, pp. 133–5.
[52] H. J. Kühn, 'Sieben Thesen zur Frühgeschichte des Deichbaus in Nordfriesland', in Steensen, ed., *Deichbau und Sturmfluten*, p. 10.
[53] Soens, *De spade*, pp. 159–60.
[54] Soens, 'Floods and Money', pp. 345–9.

to invest in dike repair and improvement.[55] Their reluctance is understandable: the maintenance costs of the dikes were often enormous and higher than the rent of the land. From an economic perspective it made sense to withdraw investments from these imperilled areas.

Flanders and Zeeland were counties with prosperous cities. Could urbanites have supported the coastal polders in the same way city dwellers in Holland supported the people in the fenlands? Research by Tim Soens has demonstrated that cities and city dwellers were reluctant to invest much in the coastal defences. The aim of most landowners was to keep investment as low as possible.[56] Some change in this respect can be observed in the sixteenth century. The city of Antwerp together with burghers from Antwerp and Mechelen invested large sums in attempts to repair the damage done to the eastern part of the island of Zuid-Beveland after the devastating floods of 1530 and 1532. They were not driven by compassion for the stricken rural population, but instead by worries that the disappearance of land in this area might lead to a change in the course of the Scheldt river that would be detrimental to the city.[57] More altruistic may have been the desperate attempt by a burgher of Aardenburg to save the island of Wulpen in the 1540s, which ultimately caused his financial ruin.[58] Urban capital was to be very important in the recovery of Zeeland and the Flemish coastal plain, but its effects were only really felt from the end of the sixteenth century.

Technically it was not very difficult to design a device capable of lifting water and thus solve the drainage problems of the Holland fens. A scoop wheel, already used in water mills, could be used for that. This wheel could be hand-driven or horse-driven, but the most powerful device for pumping up water was the wind-driven mill. Windmill technology was already known in north-western Europe from the twelfth century onwards; therefore in the late Middle Ages it was technically possible to connect a windmill with a scoop wheel and thus to construct a wind-driven drainage mill. The first of such devices was erected shortly before 1408 near the city of Alkmaar in northern Holland. Around that time the count of Holland sent a letter to two members of the board of the Delfland water authority ordering them to go to Alkmaar to inspect a mill that had been built there and that could 'throw out' water.[59] The count had seen his tithe income dwindle due to increasing drainage problems and was interested in finding solutions.

The mill near Alkmaar must have made a good impression on the men who went to inspect it, because in 1408 one of them, the nobleman Floris

---

[55] Dekker and Baetens, *Geld in het water*, pp. 56, 87.
[56] Soens, *De spade*, pp. 267–8.
[57] Dekker and Baetens, *Geld in het water*, chs. 4 and 10.
[58] Gottschalk, *Historische geografie*, vol. 2, pp. 182–3.
[59] Bicker Caarten, *Middeleeuwse watermolens*, p. 45.

van Alkemade, had one constructed close to his castle near Leiden. The next wind-driven drainage mill was built shortly before 1413 near Delft.[60] For a long time, however, these three drainage mills remained the only ones of their kind in the whole county of Holland, although the situation of agriculture in the fenlands was becoming desperate during the first three decades of the fifteenth century.[61] Only from the 1430s onwards were more mills built, and even then only in a few specific areas. Why did it take so long for the drainage mill to be adopted as the solution to the increasing problems of fenland agriculture?

A general explanation of the late adoption of drainage mills was already referred to above: wind-driven drainage mills were expensive, costing hundreds of guilders. For farmers who were already impoverished due to the fact that their land was turning into mud, such investments were not feasible. They may have tried to manage with less expensive hand-driven or horse-driven mills, but the capacity of those mills was much smaller.[62] For introduction of wind-driven drainage mills on a larger scale, capital from outside the peasant population, from the nobility or wealthy urbanites, was required. These sources were prepared to provide such capital if there was the prospect of good returns on their investment, which depended on the quality of the soil. As Petra van Dam has stated, the regional variation in the introduction of the drainage mill can be explained by two factors: soil quality and availability of urban capital.[63]

The fenlands of Holland were not one large undifferentiated peaty plain; there were considerable variations in soil types. Apart from the clay ridges along the rivers the main soil types were eutrophic peat, oligotrophic peat and clay-on-peat. *Eutrophic* peat contains minerals deposited by rivers. This type of peat is relatively fertile and suited for agriculture, but less suitable as fuel due to the mineral content. *Oligotrophic* peat on the other hand contains few minerals and is unfertile but provides very good fuel. In the *clay-on-peat* areas, rivers or the sea have deposited a fertile layer of clay on the peat. From the perspective of townsmen considering investing in drainage, clay-on-peat areas seem to be the most attractive because they promise good harvests. If these areas are situated close to the city or have good connections over water with it, that will increase the profitability of investing in their drainage. However, areas which provided good quality fuel could be interesting too. To be able to cut the peat deeper below the surface level, the land had to be drained, meaning mills were very useful in the peat extraction industry. Even when later

---

[60] Ibid., pp. 50, 54.
[61] Van Bavel and Van Zanden, 'The Jump-Start', p. 518.
[62] Bicker Caarten, *Middeleeuwse watermolens*, pp. 25–30, 159–61.
[63] P. J. E. M. van Dam, 'Schijven en beuken balken: Een sociaal-ecologische transformatie in de Riederwaard', in Wouda, ed., *Ingelanden als uitbaters*, pp. 22–4.

the technique to dredge peat from below the water table was introduced, some drainage was still required. The increasing urban demand for fuel from households and industries could make investment in peat extraction and drainage mills profitable.[64]

A closer look at the diffusion of drainage mills in fifteenth-century Holland proves Van Dam's assumptions to be correct. In the territory of the water authority of Delfland, the first drainage mills were constructed in the 1430s in the parishes to the east of the city of Delft, which was part of the huge, raised peat bog of central Holland. These five mills were not paid for by the farmers, but by the landowners in the area. In 1440 the board of Delfland ruled that the costs of the construction of the mills would have to be repaid by the users of the land in ten years' time. The users were also liable for the maintenance of the mills.[65] Although the identity of the landowners is not mentioned, it is very probable that they were burghers of Delft. Delft had an important brewing industry that required large amounts of fuel, so it is likely that the investors wanted to extract peat, and that afterwards the farmers could again reclaim the land. To the west of Delft, the landscape was different; peat soils were interspersed with sandy, better drained ridges. The closer one came to the coast, the more likely it was that light clays and sand became predominant. Drainage mills were constructed here from 1445 onwards.[66] Here good harvests could be expected from better quality land. The investments in drainage mills paid off, as is witnessed by the tithe revenue from the village of Maasland, which increased strongly in the late 1440s and the 1450s.[67] The same is valid for the Reijerwaard polders on the island of IJsselmonde to the south-east of Rotterdam. This area had flooded in 1375 and was re-reclaimed in two phases, in 1405 and 1443. While the area was submerged, a thick layer of clay had been deposited on the peat soils. This young clay was very fertile and promised bumper harvests. This must have been a stimulus to reclaim the area, employing drainage mills to keep it dry.[68]

For a long time Delfland and the Reijerwaard polders were the only areas where drainage mills were employed on a larger scale. Although the first drainage mill in the territory of the Rijnland water authority had already been erected in 1408, it took until the 1480s before they were employed

---

[64] Ibelings, 'Turfwinning en waterstaat', p. 79; Ibelings, 'Aspects of an Uneasy Relationship: Gouda and its Countryside (15th–16th Centuries)', in Hoppenbrouwers and Van Zanden, eds., *Peasants into Farmers?*, p. 263.
[65] Bicker Caarten, *Middeleeuwse watermolens*, p. 105.
[66] *Ibid.*, pp. 109, 115–16.
[67] Van Bavel and Van Zanden, 'The Jump-Start', p. 519.
[68] L. J. Pons, 'Passen en meten: De landinrichting bij de herdijking van de polders Oud- en Nieuw-Reijerwaard in respectievelijk 1404/05 en 1442/3', in Wouda, ed., *Ingelanden als uitbaters*, p. 89; Bicker Caarten, *Middeleeuwse watermolens*, p. 157.

here on a significant scale. In 1491, 13 drainage mills were counted along the river Oude Rijn to the east of Leiden. Nine of these mills had been built during a short period from 1480 to 1486.[69] Time and place are significant. The mills were concentrated on the riverbanks and pumped dry the light clays of the ridges along the river and the clay-on-peat areas behind them. This was land that was suitable for arable agriculture – tithes were levied here on oats, beans, hemp and flax – and prices were high in the last two decades of the fifteenth century. Moreover, the land was close to the major city of Leiden and the Oude Rijn river provided a cheap transport route to the city.[70] Soil quality, location and prices made investment in windmills in this area attractive to wealthy urbanites. In the rest of the large territory of Rijnland, wind-driven drainage mills remained exceptional until well into the sixteenth century.

The diffusion of the wind-driven drainage mill in Rijnland is well documented because the permits for erecting drainage mills from the sixteenth century have been preserved, and Siger Zeischka has published the number of permits per quarter century. In the first half of the sixteenth century few drainage mills were built in Rijnland and most mills that were erected were horse driven. Then, in the third quarter of the century, suddenly many more mills were constructed, and most of them, 58 from a total of 92, were wind driven. The turning point came in the 1560s. From then onwards almost all new drainage mills were wind driven. The causes of this sudden change remain unclear. Zeischka has suggested that flood problems in 1566 may have made landowners aware that only wind-driven mills possessed enough capacity to drain the land sufficiently.[71] Until then, farmers seem to have managed with small horse-driven mills.

In the northern part of Holland, the wind-driven drainage mill also spread slowly. As noted above, the first one was built here before 1408. The second one, however, was only erected in the 1460s. In the whole district of West-Friesland there were only five drainage mills in 1514, which had increased to 14 by 1544. In the whole of Holland north of the IJ only 20 drainage mills were counted in that last year, although the number may have been somewhat underestimated.[72]

From its invention in the first decade of the fifteenth century, the wind-driven drainage mill slowly spread through Holland during the fifteenth and sixteenth centuries. The first areas where the drainage mill

---

[69] Bicker Caarten, *Middeleeuwse watermolens*, pp. 74, 78–94.
[70] Van Dam, *Vissen in veenmeren*, pp. 90–2; L. Noordegraaf, *Hollands welvaren? Levensstandaard in Holland 1450–1650* (Bergen, 1985), p. 21; Van Tielhof and Van Dam, *Waterstaat*, pp. 78–9.
[71] S. Zeischka, *Minerva in de polder: Waterstaat en techniek in het Hoogheemraadschap van Rijnland 1500–1865* (Hilversum, 2007), pp. 81–3.
[72] Beenakker, *Van Rentersluze*, pp. 101–2.

was adopted successfully were those where urban investors saw opportunities to make a good profit. This could either be areas with peat of high fuel quality that could be sold in the burgeoning cities, or areas with better soils where commercial agriculture was feasible. If these areas had good connections with cities, usually by water, the chances that urbanites would invest in mills were even bigger, as the Oude Rijn area near Leiden shows. In the less endowed areas, farmers made do with small hand- or horse-driven mills.

Although the drainage mill spread only slowly, by the second half of the sixteenth century several hundreds of such mills sprinkled the fenlands of Holland. This was very different from the low centre of Friesland, where the first wind-driven drainage mill was only mentioned in 1551.[73] The failure to introduce mill drainage here in the late Middle Ages resulted in the desertion of villages and abandoning of cultivated land in the peat areas. Part of the population probably moved to the uplands eastward of the fens. By 1500, much of the Frisian fenland was deserted.[74] In neighbouring Groningen peat areas also were abandoned and people resettled on higher sandy ridges.[75]

Older historiography has attributed the failure to respond to the new environmental challenges in the northern fenlands to the political chaos resulting from increased feuding in the late Middle Ages.[76] It is indeed true that political unrest leads to decreasing investment in expensive water infrastructure. During war or civil war, less was invested in drainage mills in Holland too, and it is striking that in Friesland the construction of drainage mills did not begin until after the Habsburgs had restored order in 1524.[77] However, while both Holland and Friesland experienced political unrest in the fifteenth century, in Holland this did not prevent the diffusion of the drainage mill. Two other factors may have been more important.

Firstly, the Frisian fenlands were less extensive than those of Holland. People who had to leave the Frisian fens could settle on higher sandy land nearby and maybe even continue to use parts of the fens as meadows. Yet this was not a viable solution for most inhabitants of the vast fenland plain of Holland as their higher areas were too far away. Secondly, in both regions, investment in drainage mills was too expensive for the

---

73   M. Knibbe, *Lokkich Fryslân: Landpacht, arbeidslonen en landbouwproductiviteit in het Friese kleigebied 1505–1830* (Groningen and Wageningen, 2006), p. 232.
74   B. K. van den Berg, *Het laagveengebied van Friesland* (Utrecht, 1933), p. 15.
75   J. Zomer, *Middeleeuwse veenontginningen in het getijdenbekken van de Hunze: Een interdisciplinair landschapshistorisch onderzoek naar de paleogeografie, ontginning en waterhuishouding (ca. 800 – ca. 1500)* (Eelde, 2016), pp. 226, 274.
76   Winsemius, *De historische ontwikkeling*, pp. 130–2.
77   M. Knibbe, 'De kerk, de staat en het vredesdividend: De pachtopbrengsten van kerkelijke goederen in Friesland', *De Vrije Fries* 94 (2014), pp. 269, 272.

Figure 3. Peat landscape with drainage mill near Vlist, Zuid-Holland. Photo G. J. Dukker. Collection Rijksdienst voor het Cultureel Erfgoed. This type of landscape with pastureland divided into strips by drainage ditches and canals, with here and there drainage mills, dates from the sixteenth–seventeenth centuries.

impoverished farmers. In Holland this obstacle to the introduction of mills was removed through investments by or loans from prosperous inhabitants of the booming cities. The 11 cities of Friesland, however, were small, mostly agrarian and not really thriving. From these places, not much help was to be expected for the fens. It seems a combination of geographical circumstances, a lack of capital, and political unrest prevented the introduction of drainage mills and led to the depopulation of the central Frisian fenlands in the late Middle Ages. Only after 1550 was this area reclaimed.

As mentioned above, the erection of large numbers of drainage mills also required additional investments, such as the creation of a system to store and transport excess water, ending in the construction of sluices with the capacity to drain large amounts of water. In Holland, the capacity of discharge sluices was increased by 50 per cent between 1300 and 1600. Moreover, from the late fifteenth century onwards, more durable brick and stone was substituted for timber as building material. Whereas wood had to be replaced every 25 to 40 years, brick and stone could withstand the powers of water and weather for much longer, for a century or more.[78] By the mid-sixteenth century most major sluices in Holland, Zeeland and Flanders were constructed of brick and stone. For most parts of the sluices brick was used, which could be produced locally. Stone was not present in the Lowlands and had to be imported, which made it precious. It was therefore only used for parts that were required to be particularly strong, such as the protective corners of the sluice chambers.[79] This innovation did eventually spread to north-west Germany, with brick and stone sluices only being introduced there some two centuries later (see pp. 212–13).

Institutional Change

In much of the area under research, the late Middle Ages was a period of profound institutional change in water management. This was partly caused by political, social and economic change and partly a reaction to the environmental challenges of the fourteenth and fifteenth centuries. The development of agriculture in the early modern period was strongly influenced by these institutional changes. In this section the changes in Flanders/Zeeland, Holland, the Wadden Sea area and England will be discussed.

The most important development in late medieval Flanders was the monetisation of the maintenance duties of dikes and drains. The

[78] J. Kramer, 'Binnenentwässerung und Sielbau im Küstengebiet der Nordsee', in Kramer and Rohde, eds., *Historischer Küstenschutz*, pp. 124–6.
[79] P. J. E. M. van Dam, 'Ecological Challenges, Technological Innovations: The Modernization of Sluice Building in Holland, 1300–1600', *Technology and Culture* 43 (2002), pp. 507, 514–17.

obligation of the individual owners to maintain a parcel of a dike by performing labour and delivering materials was replaced by a rate levied on all those who owned land within the territory of the water authority. The water authority then had the maintenance work performed by specialised contractors. In Dutch this maintenance system is called *gemeenmaking*, in German *Kommuniondeichung*. Both words indicate that maintenance was no longer dealt with on an individual basis, but that it had become a responsibility of the community of landowners, organised within the water authority.[80] Originating in the late thirteenth century in the Flemish coastal plain the system spread first to Zeeland and then further to the north and east.

Monetisation had some clear advantages. More uniform and higher quality maintenance could be guaranteed. Individual owners performing maintenance were often tempted to economise on the time and materials they spent on dike maintenance, especially when incomes from farming were low. The yearly inspections were not always sufficient to detect all cases of bad maintenance. Moreover, dikes maintained by individual owners were divided into numerous small parcels, with all owners having their own way of maintaining their particular parcel, as well as their own ideas regarding the correct gradient of the dike's profile. The result was a dike that looked like 'swallow's nests unequally glued together', as a seventeenth-century dike along the Lower Elbe was contemporaneously described.[81] At the spots where the nests were supposed to be glued together the dike was most vulnerable. Another drawback of maintenance in kind was that some people were liable for dikes that were easy to maintain whereas others had to maintain stretches of dike that were very vulnerable due to their exposed position, making maintenance very costly. Since this was evidently unfair, in some regions landowners were assigned several short stretches in different places to spread the risks more evenly. This exacerbated the problem of unequal maintenance.[82]

The new system also made it possible to hire specialised workers for the maintenance of more complicated structures such as sluices. It is no coincidence that in Zeeland rates for sluice maintenance were introduced earlier than rates for dike maintenance. On the island of Zuid-Beveland, as early as the fourteenth century, a yearly rate was being levied for maintenance of sluices and canals, whereas rates for dike maintenance were only levied intermittently, usually when large investments were required after storm surges. For regular dike maintenance, performance in kind

[80] Soens, *De spade*, pp. 114–15.
[81] N. Fischer, *Wassersnot und Marschengesellschaft: Zur Geschichte der Deiche in Kehdingen* (Stade, 2007), p. 46.
[82] H. Hauschildt, *Zur Geschichte der Landwirtschaft im Alten Land: Studien zur bäuerlichen Wirtschaft in einem eigenständigen Marschgebiet des Erzstifts Bremen am Beginn der Neuzeit (1500–1618)* (Hamburg, 1988), vol. 1, pp. 616, 650.

Water Management, c. 1300–1550

was only replaced with yearly rates at the end of the sixteenth century.[83] On the nearby island of Walcheren sluice maintenance was already being financed by levying rates in the fourteenth century, whereas regular dike maintenance rates were only introduced in 1546.[84]

That performance in kind was retained longer for regular dike maintenance than for other tasks is not surprising. This was simple 'earth work', digging and transporting clay with spades and baskets – and later wheelbarrows – that could be easily performed by anyone with sufficient physical power. It did not cost the farmer any money, which was especially important for small farmers who did not have much cash income. But larger farmers could also benefit from performing maintenance in kind. Especially when prices of agricultural products were low, rates could be a burden even for larger farmers. It was then preferable for them to perform maintenance in kind, keeping expenditure on labour and materials low. This was good for the farmer's purse, but not for the quality of dike maintenance. It should be kept in mind, however, that within water authorities where maintenance was monetised, there was no automatic immunity from unwise budget cuts, as the board could exercise its power to decide to spend less money.[85]

Gemeenmaking was beneficial to large absentee landlords who could not perform maintenance themselves. Instead of leaving the work to their tenants, who required monitoring and did not always perform well, they could pay rates and leave the organisation and monitoring of maintenance to the board of the water authority. It comes as no surprise that the monetisation of water management first emerged in the Flemish coastal plain, where there were many absentee landlords, such as hospitals and abbeys and from the late Middle Ages onwards also many townsmen.[86] It is also no surprise that it only spread very slowly to the north, where peasant landowners were more numerous. Despite its obvious drawbacks, the old system of maintenance in kind maintained itself in some areas until the nineteenth and even twentieth centuries (see pp. 162–3, 198–201, 211–12 and 258).[87]

One of the consequences of the monetisation of water management was the improvement of land registration. Landowners contributed to maintenance in proportion to the area of land they owned, so for the board of the water authority it was essential to know how much land each owner possessed. Therefore, records were compiled listing all plots of land in the

---

83   Dekker, *Zuid-Beveland*, pp. 580–1.
84   Henderikx, *Land, water en bewoning*, pp. 158–9.
85   Soens, 'Floods and Money', pp. 347–9.
86   Thoen, 'Social Agrosystems', pp. 55–7.
87   N. Fischer, *Im Antlitz der Nordsee: Zur Geschichte der Deiche in Hadeln* (Stade, 2007), p. 368; M. Ehrhardt, *Ein Guldten Band des Landes: Zur Geschichte der Deiche im Alten Land* (Stade, 2003), p. 563.

territory of the water authority, indicating where they were situated, their size, the name of the owner and sometimes also the name of the tenant. The oldest preserved ledger for a part of Flanders dates from 1388.[88] From the fifteenth century onwards, hundreds of such records have been preserved. The ledgers were made by sworn surveyors, a profession that is already mentioned in the area under study from the fourteenth century. From the sixteenth century onwards, these records were increasingly often accompanied by detailed maps. Maps drawn by sworn surveyors were legally binding documents and could be used as evidence in lawsuits.[89] Ideally the measurement of the territory of the water board had to be repeated at regular intervals to register mutations. Some water boards even repeated the survey every seven years.[90] Surveying and map making were expensive, however, and sometimes postponed, which led to records becoming outdated. Despite these shortcomings, the records and maps of the water boards made the landscape much more 'legible', not only for the water board itself, but also for farmers and investors in land. So, by their careful registration of landownership the water boards contributed to the early emergence of secure property rights in the coastal provinces of the Low Countries, first in Zeeland and Flanders, and later in Holland.[91]

Another consequence of the monetisation of maintenance was that the function of the board members of the water authorities changed. They were no longer judges who pronounced a verdict of the quality of the work done by the landowners liable for maintenance and had their verdict executed. Instead, they became administrators with their own maintenance budget, dealing with contractors hired to implement the maintenance plans of the board. This meant that power was transferred from individual landowners to the board. The members of the board were elected by the landowners from their midst, but usually from the wealthiest of them: noblemen, wealthy townsmen or at least prosperous farmers.[92] That may have been another contributing factor as to why smaller farmers were not enthusiastic about monetisation.

An advantage of monetisation that people in the Middle Ages were probably not yet aware of was that it increased the creditworthiness of the water authority. Recently, Heleen Kole and Milja van Tielhof have

---

[88] Soens, *De spade*, p. 76.
[89] R. J. P. Kain and E. Baigent, *The Cadastral Map in the Service of the State: A History of Property Mapping* (Chicago, 1992), pp. 12–13, 17, 23–4.
[90] L. Hollestelle, 'Polder- en waterschapsarchieven', in A. C. Meijer, L. R. Priester and H. Uil, eds., *Gids voor historisch onderzoek in Zeeland* (Amsterdam, 1991), p. 144; J. L. van der Gouw, *De Ring van Putten: Onderzoekingen over een hoogheemraadschap in het Deltagebied* ('s-Gravenhage, 1967), p. 65.
[91] Van Bavel, *Manors and Markets*, pp. 167–9.
[92] Soens, *De spade*, pp. 200–1.

pointed out that water authorities with monetised maintenance possessed a steady stream of income. This increased the trust moneylenders had in these authorities and made it possible for water authorities to use their future income as collateral for loans. Thus, the creation of a funded or consolidated debt became possible.[93] In the early modern period many water authorities would make use of this opportunity, for example when they were in need of extra capital to repair dikes after heavy storm surges.

In Holland, far-reaching institutional changes took place in water management in the late Middle Ages. These were related to the introduction of drainage by windmills, which required coordination due to the natural tendency of people to solve their problems with the least possible effort. In the case of drainage that usually implied pumping your excess water to the neighbouring village and letting the people there fend for themselves. Moreover, since villages were often too far away from open water to directly drain their excess water themselves, windmill drainage also required a reservoir system in which the water from several villages could be stored and then transported to open water. These tasks could not usually be performed at the local level, so a regional authority had to take on the responsibility for this. Therefore, installation of drainage mills demanded both institutional and infrastructural change.[94]

In Holland, the *heemraadschappen*, the regional water authorities that had been incorporated by the counts, became responsible for the coordination of drainage. This started in Delfland, the region between Rotterdam and The Hague. In 1440 the Delfland water authority intervened in a conflict about the maintenance costs of five drainage mills near the city of Delft. It designed an agreement between the parties and decided that in the case of new conflicts the board of the authority had jurisdiction. It also prescribed that drainage mills were obliged to continuously pump when there was enough wind. This shows clearly that the Delfland authority considered drainage matters belonged under its jurisdiction. This was accepted by the landowners and mill owners who in the subsequent years always applied for permission when they wanted to erect mills or make changes in the existing organisation of water management.[95] The Rijnland water authority introduced regulation in 1460 requiring permission from its board when erecting drainage mills.[96] Often the

---

[93] H. Kole, *Polderen of niet? Participatie in het bestuur van de waterschappen Bunschoten en Mastenbroek vóór 1800* (Hilversum, 2017), pp. 123, 224–5; M. van Tielhof, 'After the Flood: Mobilising Money in order to Limit Economic Loss (the Netherlands, 12th–18th century)', in *Gestione dell'acqua in Europa*, p. 398.

[94] P. J. E. M. van Dam, 'Harnessing the Wind: The History of Windmills in Holland, 1300–1600', in P. Galetti and P. Racine, eds., *I mulini nell'Europa medievale* (Siena, 2003), pp. 46–7.

[95] Bicker Caarten, *Middeleeuwse watermolens*, pp. 104–6.

[96] *Ibid.*, p. 71.

interference of the water authorities with drainage matters was triggered by conflicts over the construction of new mills. The parties involved seem to have perceived the water authorities as neutral arbiters and accepted their verdicts. Based on this trust and on the charters the counts of Holland had granted the water authorities in the previous two centuries, these organisations became responsible for drainage affairs in most of Holland in the fifteenth century.

There was another aspect of water management in the fenlands that could not be arranged at local level. The peat lands were vast, and water often could not be pumped directly from a drained area to open water. Canals had to be dug and maintained to transport water to river or sea, and sometimes storage basins were required for occasions when sluices needed to be closed due to high tides. These canals and basins together are called *boezem* – bosom – in Dutch. Organising and maintaining such a reservoir and transport system of water requires regional coordination, and again this was delivered by the regional water authorities. They also started to raise rates for the maintenance of the canals and sluices that formed the *boezem*. The rural districts comprising the territory of the water authority had to pay their contribution towards the maintenance of the reservoir system in money.[97] As a result, in the fifteenth century the board members of the regional water authorities in Holland also changed from judges to administrators.[98]

Below the regional water authorities, a new layer of local water authorities emerged. Drainage mills were financed by groups of landowners and farmers who created a polder, an area in which the water level could be maintained artificially by the drainage mill and low embankments that separated it from the surrounding area. These polders were not administered by the village aldermen, but by a board that was elected by those who owned land in the polder. Both the geographical unit and its water authority were called polder.[99] Because the type of polder discussed here was specifically founded to maintain a drainage mill, they were often called *molenpolder* (mill polder). These new local units were supervised by the regional water authorities, who granted permits to construct drainage mills and thus to form polders. As a result of these developments, a two-tiered system of water management emerged.[100] At the local level, mill polders financed drainage mills and maintained the mill and the ditches required to transport water to the regional reservoir system. That

---

[97] Van Tielhof and Van Dam, *Waterstaat*, pp. 81, 103.
[98] Van Tielhof and Van Dam, *Waterstaat*, p. 103.
[99] H. S. Danner, B. van Rijswijk, C. Streefkerk and F. D. Zeiler, ed., *Polderlands: Glossarium van waterstaatstermen* (Wormerveer, 2009), p. 104.
[100] A. Kaijser, 'System Building from below: Institutional Change in Dutch Water Control Systems', *Technology and Culture* 43 (2002), pp. 538–9.

reservoir system was maintained by the regional water authorities, who also supervised and coordinated the activities of the mill polders.

Thus, a sophisticated system of water management emerged in the Holland fenlands. At the local level, people kept their feet and those of their cows dry by using drainage mills. At the regional level, water authorities maintained the infrastructure to move excess water safely to open water, and through their supervision provided coordination of the activities of the mill polders to prevent them from harming the interests of landowners outside of their polders. Of course, reality was more complicated, and problems and conflicts did arise, but in general this proved to be an efficient system that promoted optimal functioning of drainage mills.

During the late Middle Ages, the Frisian and non-Frisian rural republics along the Wadden coast lost their independence, a period often accompanied by war and civil war. How did the organisation of water management develop under these circumstances? Developments were quite divergent. I will concentrate here on the present-day provinces of Friesland, Groningen and Ostfriesland, and on the former *Landgemeinden* on the left bank of the lower Elbe river: Hadeln, Kehdingen and Altes Land.

As noted in the previous chapter, the *Landgemeinden* had jurisdiction over maintenance of water infrastructure (see p. 55). By the late Middle Ages, however, this did not mean that they organised maintenance themselves. That was mostly left to smaller units, such as the *grietenijen* (rural districts) in Friesland or the parishes in Ostfriesland.[101] The *Landgemeinden* only had a supervisory function and intervened in conflicts between parishes or rural districts concerning liability for maintenance. Responsibility for the upkeep of dikes and drains shifted from the regional to the local level, but remained in the hands of the general administration, there were no specific water authorities responsible for them. Major sluices were the exception. For the maintenance of this large infrastructure *zijlvesten* had been formed. In Groningen and Ostfriesland these organisations developed into water authorities; in Friesland they atrophied in the late Middle Ages.[102]

In 1464 most of Ostfriesland was united under Ulrich Cirksena, who became the first count of this territory. He and his successors codified and modified the existing regulations on water management in the principality and thus provided a legal basis for the organisations who maintained sluices, which were called *Sielachten* in Ostfriesland.[103] In the sixteenth century, the maintenance of dikes was also entrusted to water authorities, the *Deichachten*. *Sielachten* and *Deichachten* became local water

---

[101] Winsemius, *De historische ontwikkeling*, pp. 23, 33; Siebert, 'Entwicklung des Deichwesens', p. 85.
[102] Fockema Andreae, 'Dijken of wijken', p. 187.
[103] Siebert, 'Entwicklung des Deichwesens', p. 91.

authorities with a considerable degree of autonomy. Landowners elected their own officials from their midst and levied rates when required. Only when coercion was needed to force an unwilling landowner to fulfil his obligations was government asked to intervene. The counts of Ostfriesland would have preferred to have more control over the water authorities, but they lacked power and money. The Estates of Ostfriesland opposed attempts by the count to increase his power, and the landowners would only have been prepared to give up their rights in the water authorities if the count would also take over their financial obligations, but the count lacked the capital to do so.[104] In this way, autonomous water authorities emerged in Ostfriesland in the sixteenth century very much comparable with those that had developed in Flanders, Zeeland, Utrecht and Holland two to three centuries earlier. By the sixteenth century Ostfriesland was in the same stage of state formation as those other territories had been earlier in the Middle Ages: there was a prince who was strong enough to incorporate water authorities, but not strong enough to dominate them.

The case of Ostfriesland again seems to prove that the interference of some kind of central authority was crucial for the formation of water authorities. The Ommelanden, however, the small territories around the city of Groningen, proved otherwise. Here the associations for the maintenance of sluices that had emerged in the thirteenth century became real water authorities in the late Middle Ages without the support of a central authority, simply because there was no such authority. They had their own boards, their own jurisdiction and could raise their own rates. In 1531, shortly before the Habsburgs took power in the area, the three largest *zijlvesten* even reached an agreement by which they successfully excluded themselves from the jurisdiction of the courts of appeal that were then emerging in Groningen.[105] In Groningen, water authorities were completely formed from the bottom up.

The durability of the *zijlvesten* in the Ommelanden is the more surprising because the same kind of organisations that had emerged in the northern part of the adjoining province of Friesland had disappeared by the sixteenth century. In Friesland, the maintenance of sluices remained the task of loosely organised groups of towns and rural districts. The disappearance of the Frisian *zijlvesten* has often been attributed to the political chaos in Friesland during the fifteenth century.[106] However, in the Ommelanden the situation was also chaotic at that time and there the water authorities managed to survive. A plausible explanation for this may be that in the Ommelanden the *Landgemeinden* were smaller than in Friesland and thus less able to deal with drainage problems that required coordination over a larger area. The water authorities of the Ommelanden often covered

---

[104] Ibid., p. 183.
[105] Schroor, *Wotter*, pp. 40, 42.
[106] Fockema Andreae, 'Dijken of wijken', p. 195.

(parts of) the territories of more than one *Landgemeinde*. In Friesland, Oostergo and Westergo were expected to cover considerably larger areas, so drainage could continue to take place within the context and under the authority of the *Landgemeinden*. There was no need for a new type of authority there because the scale of the existing general administration was the same as the scale required for coordination of drainage.

That scale was indeed an important explanatory factor for the emergence of water authorities is confirmed when we look at the maintenance of the sea defences in Friesland. Whereas the *zijlvesten* disappeared, new authorities for dike maintenance emerged in the sixteenth century. A large part of Friesland was protected from the Zuider Zee by the 28-kilometre long *Vijf Deelen Dijk*. Around 1500 this dike was maintained entirely by the communities immediately behind the dike, whereas communities situated more inland did not have to contribute to maintenance although they also profited from the security the dike offered. Since the costs of maintenance increased in this period of coastal erosion and storm surges, the communities that were liable for maintenance tried to convince the inland villages of the need for structural support. Here, scale was a problem, the area liable for maintenance being too small. The inland communities, however, refused to cooperate. This changed owing to the devastating floods of 1530/2 and 1570, which made it amply clear that the existing maintenance system was deficient. Still, reform might not have succeeded if it had not been for two competent governors, George Schenk van Toutenburg in the 1530s and Caspar de Robles in the 1570s, who managed to persuade the parties involved to agree with compromises that led to a new organisation of dike maintenance. The inland communities now had to contribute to the maintenance of the dike, and a water authority was created that was to be responsible for ensuring completion of such maintenance.[107] This was the first water authority in Friesland.

Thus Friesland had no water authorities for drainage, but from the sixteenth century specific authorities for dike maintenance were founded and the Habsburg authorities had supported the founding of these organisations. In Groningen, on the other hand, drainage authorities had emerged and consolidated without government support. However, dike maintenance in Groningen remained in the hands of the village communities abutting the seawalls, which had neither sufficient capital nor authority to maintain the dikes effectively. This lack was probably one of the causes of the devastating floods that resulted in the formation of the Dollard sea inlet in the late Middle Ages.[108] Insufficient dike maintenance was an issue the inhabitants of Groningen struggled with until far into the early modern period, causing high numbers of casualties in the floods of 1686 and 1717 (see pp. 213–14).

[107] Van Tielhof, 'Forced Solidarity', pp. 335–7.
[108] Schroor, *Wotter*, pp. 26–7, 36–8.

Of the three rural republics along the left bank of the Lower Elbe, Hadeln had to submit to the authority of the dukes of Sachsen-Lauenburg, and Kehdingen and the Alte Land became part of the prince bishopric of Bremen. Hadeln was the only former farmer's republic where the *Landgemeinde* stayed in charge of water management. The villages in Hadeln were each liable for the maintenance of certain parts of the dikes and sluices, and they performed these tasks under the supervision of the *Stände* (Estates) of the land, which in this case were geographic units: the higher *Hochland* near the river, the lower *Sietland* behind it, and the town of Otterndorf.[109] This organisation of water management remained in place without much change until well into the nineteenth century. Hadeln was able to retain much of its medieval privileges due to it being an exclave of Sachsen-Lauenburg and the overlord having left it a high degree of autonomy. Another reason for the durability of this system may have been that it was the appropriate scale. All communities with an interest in the maintenance of the dikes and sluices shared liability for that. There was no discrepancy between the physical infrastructure and the area served by them.

In Kehdingen and the Alte Land water authorities did emerge, and the overlord played an important part in water management. In Kehdingen there were seven *Deichgerichte*, who were responsible for the maintenance of the dikes. The Alte Land was divided into three *Meilen* (miles) that maintained the physical infrastructure within that territory.[110] Most officers were elected by the landowners but in both territories the highest dignitary, the *Gräfe*, was a representative of the archbishop of Bremen, who was also responsible for the general administration in his district.[111] Moreover, there was an episcopal court of appeal for disputes on water management for both territories in Stade, the *Botting*, meaning that the bishop held the highest authority in water management issues.[112] It is clear that, although Kehdingen and the Alte Land were semi-autonomous in the late Middle Ages, the territorial lord had a strong influence on water management comparable with that of the counts of Holland and Flanders.

In England, the Commissions of Sewers remained the bodies responsible for solving conflicts regarding maintenance between landowners and for organising repairs after storm surges. Their nature and position within the governance of the English state were more clearly defined by the Sewers Act of 1532. At least from this year onwards commissioners

---

[109] Fischer, *Im Antlitz*, pp. 37, 50–3.
[110] Hofmeister, *Besiedlung und Verfassung*, p. 70; Fischer, *Wassersnot und Marschengesellschaft*, pp. 36–7, 40; Ehrhardt, *Ein Guldten Band*, p. 151.
[111] Hauschildt, *Zur Geschichte*, vol. 1, pp. 62–3.
[112] Hofmeister, *Besiedlung und Verfassung*, p. 72; Ehrhardt, *Ein Guldten Band*, p. 278.

of sewers were formally agents of the Crown.[113] From this time, the terms for which the commissioners were appointed were gradually lengthened to ten years. The laws the Commissions introduced could be made permanent after royal assent had been acquired, and although the Commissions could expire, they were often renewed.[114] Consequently, the Commissions of Sewers became more permanent. However, as John Morgan has remarked, they continued to operate a system of courts and judgements and their personnel remained unspecialised and unsalaried, whereas their counterparts in much of the Low Countries developed into powerful regional administrative organisations.[115]

The substitution of rates for performance of maintenance in kind took place in England as early as in Flanders, from the thirteenth and fourteenth centuries onwards.[116] In eastern England, however, these rates were not collected by the commissioners of sewers, but by local officials, the dike reeves. These dike reeves exercised a large degree of autonomy in the way they collected the rates and spent the proceeds.[117] As a result, the Commissions of Sewers lacked their own budget and a steady source of income. This was something they had in common with the water authorities in the Low Countries and Germany that remained dependent on maintenance in kind. An important consequence of this was that Commissions of Sewers could not borrow money.

Conclusion

During the late medieval crisis of the European economy, the North Sea Lowlands were confronted with new environmental challenges, often caused by the earlier reclamations. The inhabitants of the fenlands had to deal with peat shrinkage due to drainage, and the marshes were confronted with storm surges, the effects of which were exacerbated by reduced storage capacity caused by earlier reclamations. The challenges were partly met by a retreat, and more extensive land use which required less maintenance costs. In many regions, however, technological and institutional innovations were introduced to meet the changed circumstances.

Technological solutions included wind-driven drainage mills in the fenlands, and heavier dikes with more gentle slopes, often reinforced by osier mats or groynes, in the south-western marshes. In most cases these innovations spread only slowly, taking decades or even centuries

---

[113] Ash, *The Draining*, p. 98.
[114] Ash, *The Draining*, p. 40; Morgan, 'The Micro-Politics', p. 418.
[115] J. E. Morgan, 'Funding and Organising Flood Defence in Eastern England, c. 1570–1700', in *Gestione dell'acqua in Europa*, p. 417.
[116] Galloway, 'Storm Flooding', p. 178.
[117] Morgan, 'Funding and Organising', pp. 427–8.

until they were adopted in wider areas. The main bottleneck was a lack of capital: impoverished farmers could not afford to pay for these often very costly innovations. Urban capital could provide a solution for this. In Holland, inhabitants of the booming towns were willing to invest in the construction of drainage mills or to lend money to farmers to do so. Therefore, in Holland, the drainage mill spread in increasing numbers throughout the fenlands in the second half of the fifteenth century. In agrarian Friesland wealthy urbanites were lacking, and there the drainage mill would not be adopted until a century later. But even when there were prosperous cities, this did not always mean that town dwellers were prepared to invest in improved water maintenance. In Flanders and Zeeland, urbanites and other large landlords were reluctant to invest in expensive new dikes and groynes. Widespread adoption of these innovations was thus delayed to the second half of the sixteenth century or even later.

Institutional change was most pronounced in the peat areas of central Holland. Here the regional water authorities developed into powerful and professional administrative bodies – *hoogheemraadschappen* – that regulated water management in large areas. They granted permission – or not – to erect drainage mills and created and invested in reservoir systems to store and transport excess water to open water. At the local level, mill polders were founded that erected and maintained drainage mills. The activities of these mill polders were supervised by the regional water authorities. In this way, a two-tiered system emerged that could efficiently drain excess water from the sinking fenlands. The regional water authorities also substituted levying of rates for performance of maintenance in kind for much of the physical infrastructure for which they were responsible. Work was now performed by contractors under the supervision of the board of the water authority. The board members now became powerful and professional administrators, rather than judges. Taxation by water authorities was also introduced in Flanders from the late thirteenth century onwards and somewhat later in Zeeland and Utrecht. In England, the Commissions of Sewers could also levy rates, but they left the collection to local officials. As a result, the Commissions of Sewers lacked a continuous source of income that could be used as collateral for loans. The water authorities in the Low Countries used their income as the basis for the creation of a consolidated debt, and the ledgers they kept of the landowners and their possessions contributed to better land registration and more secure property rights.

Outside Flanders and Zeeland, monetisation of maintenance was adopted only very slowly, and in some German regions maintenance in kind survived well into the twentieth century, despite the fact that monetisation had clear advantages, such as better and more uniform maintenance and the possibility to create a funded debt. The reluctance to accept monetisation can be explained by the fact that performance in kind was preferable for smallholders who lacked capital but who were disposed of abundant labour. It was preferable to all landowners in times of economic adversity when they could economise on labour and materials, but with disastrous consequences for the quality of the sea defences. In the following chapters the reasons for resisting monetisation and the consequences of monetising or not will be studied more closely.

Institutional innovation was mostly concentrated in a small area consisting of Flanders, Zeeland, Holland and Utrecht. This was an area where strong territorial lords had granted charters to water authorities, providing them with the legal basis to extend their power. It was also the most urbanised part of the area under study. For urban landlords, paying rates was preferable to maintenance in kind. Since they did not live in the area, their tenants had to perform maintenance for them, and that caused monitoring problems. The importance of state formation and urbanisation for institutional innovation becomes clearer when Flanders, Zeeland and Holland are compared with the Wadden area, where strong overlords were absent until the end of the Middle Ages and urbanisation was very limited. The emergence of water authorities here was a tortuous process, and in some regions they never emerged. The creation of water authorities without state interference was not entirely impossible, as the emergence of the Groningen *zijlvesten* proves, but it was unusual. Moreover, in the following chapters we will see that the *zijlvesten* were much weaker organisations than the water authorities of Holland, Zeeland, and Flanders. Whereas in some parts of the Wadden area water authorities were introduced, like in Ostfriesland, monetisation of maintenance remained highly unusual here for centuries. Can this be attributed to a lack of urban influence?

4

Surviving in Times of Adversity: Agriculture in the Late Middle Ages, c. 1300–1550

Introduction

Due to disease, unfavourable climatic conditions, and political unrest, the late Middle Ages was a time of contraction for European agriculture. Reduced demand for bread grains caused farming to become less intensive and more oriented towards animal production. This is even more true for the North Sea Lowlands, which in addition to all these problems also had to deal with the consequences of peat shrinkage, coastal erosion, and storm surges. Changes in agricultural technology and productivity remained limited in this period. There was considerable change in property relations and holding size in the countryside, however, and the more abundant sources of the late Middle Ages and sixteenth century enable tracking of those changes for several regions. For water management these changes were crucial, because the landowners, and especially the larger ones, determined the way in which water management was organised as well as who was liable for maintenance. Therefore, this chapter on late medieval agriculture will focus mostly on property relations.

During the late medieval and early modern periods, farming in the North Sea Lowlands went through profound changes. Until the sixteenth century unspecialised smallholdings mostly producing for subsistence were predominant. Then, here earlier, there later, larger farms producing for the market and employing wage labourers emerged. Often these large farms were run by tenants, but sometimes also by owner-occupiers or a combination of both. The rise of this class of independent, commercial producers had a profound impact on rural society.[1] An already existing differentiation among the rural population became more pronounced, with deep rifts developing between large, specialised farmers, middling

---

[1] E. L. Jones and S. J. Woolf, 'Introduction: The Historical Role of Agrarian Change in Economic Development', in E. L. Jones and S. J. Woolf, eds., *Agrarian Change and Economic Development: The Historical Problems* (London, 1989), p. 5.

groups of artisans and shopkeepers, and mostly landless labourers. For this new rural society, the term 'agrarian capitalism' is often used, by historians of a Marxist persuasion in particular. Agrarian capitalism looms large in the work of Robert Brenner on the 'transition from feudalism to capitalism'.[2]

This book is not meant as a contribution to the seemingly everlasting 'transition debate', but it cannot be denied that profound changes in economy and society occurred during the late Middle Ages and early modern period, and historians must analyse the causes of those changes.[3] The deep changes in rural society and economy will be addressed in this and the following chapters. The forces that caused these developments will be discussed, as will the actors driving them. Were they the landlords, as Brenner and his followers would have it, or did certain elements of the rural population, such as prosperous farmers, also play a part?[4] In this chapter the developments in the 1300–1550 period will be covered for several regions: Flanders and Zeeland, Holland, the central river area, the Zuider Zee coast, the Wadden coast, and the English marshes and fens.

Flanders and Zeeland

The oldest extant records of landownership within water authorities in the Flemish coastal plain date from 1388 and 1398. In 1398, the 4500 hectares of the *Watering* Eiesluis to the north of Bruges were distributed over more than 900 owners, over 75 per cent of whom owned less than five hectares. In the Oude Yevenewatering near Oostburg in present-day Zeeland Flanders, the dominance of small landowners in 1388 was even stronger. No less than 1461 owners were recorded for almost 3600 hectares. Of those owners, 88 per cent had less than five hectares.[5] These records paint a picture of an area where agriculture was still dominated by small owner-occupiers. This changed considerably over the next centuries. For the Oude Yevenewatering the records of 1388 can be compared with those from 1550, revealing a strong decrease in the number of landowners.

---

[2] Brenner, 'Agrarian Class Structure', 'The Agrarian Roots', and 'The Low Countries'.
[3] C. Dyer, *An Age of Transition? Economy and Society in England in the later Middle Ages* (Oxford, 2005), p. 43; E. Thoen, 'Transitie en economische ontwikkeling in de Nederlanden met de nadruk op de agrarische maatschappij', *Tijdschrift voor Sociale Geschiedenis* 22 (2002), p. 149.
[4] Brenner, 'Agrarian Class Structure', p. 46.
[5] Soens, *De spade*, p. 76.

TABLE 1. Percentage distribution of landownership on the Oude Yevenewatering, 1388 and 1550. Source: Soens, *De spade*, pp. 76–7.

	1388			1550		
	Owners (N)	Owners (%)	% of land area	Owners (N)	Owners (%)	% of land area
<5 ha	1289	88	41	303	68	12
5–10 ha	118	8	24	56	13	13
10–25 ha	45	3	19	61	14	31
25> ha	9	1	16	28	6	45
Total	1461	100	100	448	100	100

As table 1 demonstrates, the reduction in the total number of landowners was almost completely due to an enormous decrease in the number of small owners with less than five hectares. This spectacular decrease was mirrored by an increase of landowners with more than 25 hectares. Their number may still not have been that large in 1550, but they now owned no less than 45 per cent of the land in the *Watering*. Similar changes, although somewhat less spectacular, can be observed in other water authorities in the Flemish coastal plain.[6]

Thanks to Tim Soens' research, the large landowners can be identified. Lords did not own much land here, because there were very few of them. The church, however, did manage to increase its possessions in the late Middle Ages. In the Oude Yevenewatering the percentage of land owned by ecclesiastical and charitable organisations increased from 20 to 29 per cent between 1388 and 1550. In the large – 17,000 ha – Blankenbergse watering to the west of Bruges, they owned 33 per cent of all land.[7] In Zeeland Flanders, around the middle of the sixteenth century, the church owned 35–40 per cent of the land. On polders to the north of the town of Hulst the clergy even owned 50 to 100 per cent of this land area.[8] Burghers of Bruges also extended their possessions in the countryside. Close to the city they could hold as much as 45 per cent of all land, further away around 20 per cent. Due to the increase of both ecclesiastical and urban landownership not much scope was left for the rural population. In 1545, in the *Watering* Romboutswerve near the town of Damme, 32.4 per cent of the land was owned by burghers of Bruges and

---

[6] *Ibid.*, pp. 80–1.
[7] *Ibid.*, pp. 92–3.
[8] Van Cruyningen, 'From Disaster', p. 248; A. M. J. de Kraker, *Landschap uit balans: De invloed van de natuur, de economie en de politiek op de ontwikkeling van het landschap in de Vier Ambachten en het Land van Saeftinghe tussen 1488 en 1609* (Utrecht, 1997), pp. 262–4.

35.5 per cent by the church.[9] This did not leave much room for farmers. However, in other parts of the coastal plain the rural population owned considerably more land. For instance, in the polders of the Castellany of Veurne in 1569, they owned 48.6 per cent of the land. Urban landowners were almost absent here.[10]

From the late thirteenth century onwards urban and ecclesiastical landlords began to lease their land to tenants on short-term leases. Owing to this development leasehold became predominant on the Flemish coastal plain.[11] It also resulted in the emergence of a class of large tenant farmers who leased increasingly larger plots. This is illustrated by the developments on the domain of St. Peter's Abbey of Ghent near the town of Oostburg. In 1281 the abbey had 40 tenants each with less than one hectare, and only one tenant with more than 15 hectares. By 1561 the number of small tenants was reduced to 20 and the number of large tenants had risen to 18. The reduction in the number of small tenants had come about during the last quarter of the fourteenth century, following the disastrous storm surges of 1375 and 1376 and the havoc wreaked by the revolt of Ghent in 1379–85. The rise of the large tenants occurred later, in the last quarter of the fifteenth century.[12] The reduction in the number of farms was also visible in the landscape. The ledgers of the Oude Yevenewatering of 1567 mention at least 40 deserted farmsteads.[13] The prevalence of leasehold does not always mean that the rural population had been 'expropriated'. In the polders of the Castellany of Veurne, 92 per cent of all land was held in leasehold in 1569, whereas the farming population owned almost half of all land there. Farmers were leasing land from each other. Plots too far away from the farm were rented to other farmers, whereas plots nearby were leased to consolidate the holding. Moreover, land was also leased from relatives who were no longer farming themselves.[14]

Apart from warfare and incorporation by large tenants, the decline of the smallholders in the coastal plain was also caused by developments in water management. The frequent storms in the late Middle Ages demanded more investment in dike maintenance and repair. In Flanders, where maintenance had already been monetised at that time, this meant that local smallholders could no longer perform their maintenance duties in kind but had to pay for maintenance from their already low incomes.

---

[9] Soens, *De spade*, pp. 97–8.
[10] P. Vandewalle, *De geschiedenis van de landbouw in de Kasselrij Veurne (1550–1645)* (Brussels, 1986), pp. 126, 134.
[11] Soens and Thoen, 'The Origins of Leasehold', pp. 35–9.
[12] Soens, *De spade*, pp. 84–5.
[13] Gottschalk, *Historische geografie*, vol. 2, pp. 253–4.
[14] Vandewalle, *De geschiedenis*, pp. 113, 135.

This contributed to their demise.[15] These changes in landownership and holding sizes not only caused a strong decrease in the number of smallholders, but also the emergence of a group of large, prosperous farmers. These farmers, described by Tim Soens, did not belong to the regional elite of the coastal plain, but they can be called village elites. They could be owner-occupiers, tenants or a combination of both. By accumulating their own properties and leases they formed farms of up to 40 or 50 hectares. Although they often leased more land than they owned, through landownership they were eligible for offices in the water authorities and could thus exert influence there.[16]

In the western part of the coastal plain the shift towards larger holdings was accompanied by a shift towards less labour-intensive forms of farming, especially animal husbandry. Whereas, around 1340, some 70 per cent of the land was used as arable, in the second half of the sixteenth century this was reduced to about 34 per cent.[17] On the polders of the Castellany of Veurne the smaller farmers held mostly dairy cows, whereas the larger farmers concentrated more on – less labour intensive – stock raising and fattening of cattle for slaughter. As a result of the rise of large farms the share of dairy cows in the total herd decreased from two thirds in the fifteenth century to about half in the early seventeenth century.[18] Therefore in this area the rise of large holdings and leasehold did indeed lead to less labour-intensive farming. Whether this was also the case in the eastern part of the Flemish coastal plain – present-day Zeeland Flanders – is unclear. Data is much more scarce for this area, but the information that can be gleaned from lease contracts and ledgers of water authorities seems to indicate that farming remained more mixed here, with considerable amounts of land under cereals and other crops, and large numbers of orchards where apples, pears and cherries were grown. Especially on the younger polders, reclaimed from the early sixteenth century onwards, grain cultivation was very important.[19]

Unlike in the Flemish coastal plain, in the small county of Zeeland, an archipelago between Flanders and Holland, a powerful group of local lords existed. However, this did not necessarily mean that they owned much land, because there was no land attached to their seigneuries. These seigneuries consisted of a collection of rights, such as the right of wind (to exploit a windmill), ferry, fishing or fowling. Usually these rights were leased by individuals and thus provided the lords with income.[20]

---

15 Soens, *De spade*, p. 260.
16 *Ibid.*, pp. 88–90, 180.
17 Vandewalle, *De geschiedenis*, pp. 166–7.
18 *Ibid.*, pp. 226–7.
19 Gottschalk, *De Vier ambachten*, pp. 255, 505–6; Gottschalk, *Historische geografie*, vol. 2, p. 253; Vervaet, 'Het goederenbeheer', p. 358.
20 A. van Steensel, *Edelen in Zeeland: Macht, rijkdom en status in een laatmiddeleeuwse samenleving* (Hilversum, 2010), pp. 127–30, 147.

Agriculture in the Late Middle Ages, c. 1300–1550

A right that did not provide regular income but could be very profitable was the right to the foreshore, which after reclamation became valuable farming land. Normally reclamation was left to entrepreneurs who paid the lord for the right to do so. For the island of Noord-Beveland some information is available on the distribution of landownership in 1333. It appears that the rural population owned some 85 per cent of all land on the island. The privileged classes – clergy, nobility and urbanites – owned the remainder.[21]

At the end of the Middle Ages, Zeeland lords did not own much land: about ten per cent of the area, roughly the same as charitable and ecclesiastical organisations. Most land owned by the nobility was leased out for terms of five to 14 years.[22] By far the most important group of landowners was formed by the farmers, who owned at least 50 per cent of the land around 1464. On some islands, however, they were confronted with serious competition from urbanites. Around 1464, for example, burghers from the major city of Middelburg owned almost 26 per cent of the land on the island of Walcheren. On the island of Schouwen, some 30 per cent of all land was owned by burghers of the city of Zierikzee in 1535.[23] Middelburg and Zierikzee were the largest and most prosperous cities in Zeeland in the late Middle Ages. Here, as in Flanders, urban landownership was on the rise, but on the island of Noord-Beveland still more than 80 per cent of the land was in the hands of owner-occupiers.[24]

Hardly anything is known about holding size or farming type in Zeeland in the late Middle Ages. It is likely that mixed farming on relatively small holdings was predominant. In the parts of the county that were not hit by the sixteenth-century floods, smallholdings remained numerous until well into the eighteenth century (see p. 169).[25] It is not clear whether a development towards larger holdings, as on the Flemish coastal plain, took place. Some social polarisation may have occurred: landless people were not unknown. In 1492, a man on the island of Noord-Beveland received a relatively mild sentence for attacking a nobleman due to mitigating circumstances: he had many small children and was a poor man who had no land, and his only income was derived from wage labour.[26] According to sixteenth-century observers, Zeeland farmers were mostly oriented towards cultivation of cereal crops. Noord-Beveland exported mostly barley at the end of the fifteenth century. The quality of Zeeland wheat was often praised and exported to cities in the

---

[21] Van Steensel, 'Noord-Beveland', p. 74.
[22] Van Steensel, *Edelen*, p. 159.
[23] *Ibid.*, p. 140.
[24] Van Steensel, 'Noord-Beveland', p. 74.
[25] Van Cruyningen, 'From Disaster', p. 253.
[26] Van Steensel, 'Noord-Beveland', p. 100.

southern part of Holland. Commercial crops such as madder were also cultivated.[27]

Holland

In sixteenth-century Holland a clear difference in property relations existed between the early-reclaimed sandy and clay soils behind the dunes and along the rivers, on the one hand, and the later-reclaimed vast peat areas on the other. In Holland overall, the nobility and the clergy did not own much land. As in Zeeland, noble lords possessed seigneurial rights but did not hold much land. The nobility probably owned less than ten per cent of all land in Holland, and the church also owned less than ten per cent.[28] However, significant regional variations existed. This can be illustrated from the ledgers created by the water authority of Rijnland in the early 1540s. For 35 *ambachten* (rural districts), with a total land area of about 55,000 hectares, records of landownership and leasehold have been preserved. In that entire area 38 per cent of the land was held in leasehold. In 13 *ambachten*, however, more than 50 per cent of the land was leased. All these districts were entirely, or for a large part, situated in areas with river clay or well-drained sandy soils. These were the oldest settled parts of Holland where the church and the nobility traditionally owned much land, which they gave out in short-term lease.[29] In the *ambachten* of Rijnsburg, Oegstgeest, Valkenburg and Katwijk, situated on river clay and sandy soils between the city of Leiden and the coast, more than 70 per cent of the land was leased. The important monastery of Rijnsburg was situated here. On the other hand, there were also eight districts where less than ten per cent of the land was held in leasehold. All of these were situated in the fenlands.[30] In Rijnland in the 1540s, therefore, farmers in most of the villages on peaty soils still owned the land the count had granted them when they reclaimed the area during the central Middle Ages. Further to the south, in the Krimpenerwaard, a region with predominantly peat soils, in the 1550s almost 90 per cent of all land was still owned by the peasantry.[31]

---

[27] Ibid., p. 53; P. R. Priester, *Geschiedenis van de Zeeuwse landbouw circa 1600–1910* (Wageningen, 1998), p. 179; M. van Tielhof, *De Hollandse graanhandel, 1470–1570: Koren op de Amsterdamse molen* (Den Haag, 1995), pp. 60–2.

[28] De Vries, *The Dutch Rural Economy*, pp. 36, 39, 42.

[29] H. F. K. van Nierop, *Van ridders tot regenten: De Hollandse adel in de zestiende en de eerste helft van de zeventiende eeuw* (Amsterdam, 1990), pp. 90–2.

[30] E. F. van Dissel, 'Grond in eigendom en in huur in de ambachten van Rijnland omstreeks 1545', *Handelingen en Mededeelingen van de Maatschappij der Nederlandsche Letterkunde te Leiden, over het jaar 1896/1897* (Leiden, 1897), pp. 152–4.

[31] Sources: Penningkohieren of Bergambacht 1556, Berkenwoude 1561, Gouderak 1553, Haastrecht 1553, Krimpen aan den IJssel 1556, Krimpen aan

There were some peat villages in Rijnland, however, where some 40 per cent of the land was leased, and all of these were situated near major cities such as Leiden, Haarlem or Amsterdam.[32] This is an indication of a process of urban encroachment in the countryside that Bas van Bavel has revealed, and which was stronger in the area to the south of Rijnland. He demonstrated, on the basis of a large collection of land tax records from around 1560, that urbanites and urban institutions owned 41 per cent of agricultural land in central Holland, the area between the IJ sea inlet and the Lek river.[33] Recent research shows that this estimate is probably somewhat too high. Around 1560, in three villages in the area between Rotterdam and The Hague, urban institutions and burghers owned about 30 per cent of the land.[34] In the Krimpenerwaard region urban landownership was negligible. All in all, it seems reasonable to estimate the share of urban landowners for central Holland at some 30–35 per cent of all land, and at 10–20 per cent in the fenlands. That is somewhat less than Van Bavel estimated but still a considerable share. To the north of the IJ, only 14 per cent of cultivated land was in urban hands, and there urban landownership had even decreased since 1514.[35]

It is unknown when burghers began to purchase land. Some information is available for the city of Leiden. The acreage owned by the patriciate of that city increased from 880 hectares at the beginning of the fourteenth century to over 1300 hectares a century later. The fourteenth-century patricians of Leiden were not interested in becoming large landlords. Few of them owned more than ten *morgen* (8.5 hectares). Most preferred land close to the city; a few invested in land reclamation projects in the southern part of Holland and in Zeeland.[36] It should be kept in mind that despite the urban 'offensive', around the mid-sixteenth century farmers still owned at least half of the agricultural land in Holland.[37] Whether burghers continued to accumulate land after the

---

de Lek 1561, Lekkerkerk 1561, Ouderkerk aan den IJssel 1561, Stolwijk 1556 and Vlist en Bonrepas 1556, transcribed by C. C. J. Lans, retrieved from www.hogenda.nl, 2 July 2019.

[32] Van Dissel, 'Grond in eigendom'.
[33] Van Bavel, 'Rural Development', p. 182.
[34] C. de Wilt, *Landlieden en hoogheemraden: De bestuurlijke ontwikkeling van het waterbeheer en de participatiecultuur in Delfland in de zestiende eeuw* (Hilversum, 2015), pp. 346, 354, 365.
[35] Van Bavel, 'Rural Development', p. 182; L. Schuijtemaker, 'Stedelijke weides: Landbouwgrondbezit van stedelingen en de conjunctuur van de Westfriese agrarische sector', unpublished BA thesis, University of Amsterdam, 2016, pp. 12–13.
[36] F. J. W. van Kan, *Sleutels tot de macht: De ontwikkeling van het Leidse patriciaat tot 1420* (Hilversum, 1988), pp. 66–8.
[37] P. C. M. Hoppenbrouwers, 'Mapping an Unexplored Field: The Brenner

1550s, and whether this had any consequences for agriculture, will be discussed in the next chapter.

The introduction of water-lifting devices during the fifteenth century had prevented the Holland peat lands from reverting to their natural state. Due to the insufficient capacity of the drainage mills, however, most land remained flooded in autumn and winter, and arable farming remained impossible in most of the fenlands. Thus, pastoral husbandry was predominant here. Only by dredging mud from ditches and spreading it on the banks of the ditches could some cereals be grown.[38] Most land was meadow or pasture. Where possible, arable crops were grown. The changes in farming that occurred during the fourteenth and fifteenth centuries resulted in a shift within the mixed farming system, not a shift from arable to animal husbandry. The share of arable crops within the farming operation was reduced, and the timing of this shift depended on the drainage situation and the availability of urban markets where dairy products could be exchanged for cereals.[39]

Arable agriculture and horticulture on a significant scale remained limited to the scarce clay or sandy soils. Although there were some larger farmers, due to the partible inheritance practised by the rural population holdings were small. There is hardly any information available on the amount of land Holland farmers held in this period, but there is data on herd sizes, and for a pastoral area that is a fine indicator of holding size. Jan de Vries concluded, mostly based on records from 1514, that the richest farmers had ten to 12 cows, average farmers four to six, and many as little as two or three. In 1477 in the area around the city of Hoorn, the average herd size was less than five.[40] More recent research on the Zeevang area to the north of Amsterdam has confirmed De Vries' conclusions. In 1506, the average herd size there was four head of cattle, which was reduced to 2.8 in 1554.[41] Obviously, a decrease in holding size was ongoing here during the sixteenth century.

The small size of the holdings meant that most farmers in Holland were not able to maintain their families from the proceeds of their land. Therefore, they had all kinds of non-agricultural by-employments to supplement their incomes: peat cutting, fishing, fowling, transport, shipping and all kinds of industrial activities.[42] This resulted in a remarkably non- agricultural Holland economy. Around 1500 only about

---

Debate and the Case of Holland', in Hoppenbrouwers and Van Zanden, eds., *Peasants into Farmers?*, p. 44.

[38] De Vries, *The Dutch Rural Economy*, p. 71.
[39] Ettema, 'Boeren op het veen', pp. 249–55.
[40] De Vries, *The Dutch Rural Economy*, pp. 70–1.
[41] C. Boschma-Aarnoudse, *Tot verbeteringe van de neeringe deser stede: Edam en de Zeevang in de Middeleeuwen* (Hilversum, 2003), p. 260.
[42] De Vries, *The Dutch Rural Economy*, pp. 68–9.

25 per cent of total labour input was in agriculture, 40 per cent in industry, 20 per cent in services, and 15 per cent in fishing.[43] In the countryside, about one third of labour input around 1550 was in industrial activities such as brick making and bleaching.[44] Early sixteenth-century Holland farming, therefore, was characterised by smallholdings practising mostly pastoral farming in combination with several non-agricultural activities. Most rural dwellers were not specialised farmers but instead made a living from several sources of income.

A consequence of this orientation towards animal husbandry was that Holland was unable to provide its population with sufficient bread grains. These were imported, first mostly from northern France, the southern Netherlands and northern Germany, and then from the mid-sixteenth century onwards mostly from the Baltic.[45] It was not only the urban population who had to be fed with imported cereals, but also many small pastoral farmers, who could not produce sufficient grain themselves. To facilitate this, more and more markets emerged in the fenlands where dairy products could be exchanged for cereals.[46] Fenland farmers were forced to exchange their butter and cheese for bread grain, therefore becoming market dependent, as Brenner correctly concludes.[47]

The Central River Area

Bas van Bavel has applied an approach based on Brenner's ideas to the Low Countries, and the Guelders river area in particular. There, on the estates of large landowners, a class of large commercial tenant farmers employing wage labour emerged in the late Middle Ages. These leaseholders and their landlords invested in capital-intensive and labour-extensive farm units producing for urban markets. Large tenants needed to decrease wage costs so they could compete on the market for lease land.[48] The large tenants could emerge because large landlords were prepared to lease their land to them and to let them accumulate leases. Where smallholders owned most of the land, agrarian capitalism could not emerge. There were no large chunks of land to be leased, so no big holdings could be formed. This is evidenced by the case of Holland, where agrarian capitalism arose late. It was only around 1600, after

---

[43] J. L. van Zanden, 'Taking the Measure of the Early Modern Economy: Historical National Accounts for Holland in 1510/14'', *European Review of Economic History* 6 (2002), p. 137.

[44] Van Bavel and Van Zanden, 'The Jump-Start', pp. 523–4.

[45] Van Tielhof, *De Hollandse graanhandel*, pp. 1–2, 121.

[46] Prak and Van Zanden, *Pioneers*, p. 67.

[47] Brenner, 'The Low Countries', p. 310.

[48] B. J. P. van Bavel, 'People and Land: Rural Population Developments and Property Structures in the Low Countries, c. 1300–c. 1600', *Continuity and Change* 17 (2002), pp. 21–5.

urbanites had purchased vast areas of land from the rural population, that large farms could be formed and rented to capitalist tenants.[49]

Van Bavel's views are convincing and attractive. He does not perceive the rural population as helpless victims of lords, geography and the tides of the economy, but instead as individuals who shaped their own futures. The tenants are agents of change through creating large holdings and turning them into profitable enterprises. However, Peter Hoppenbrouwers has correctly remarked that we should add the larger owner-occupiers to this group, primarily because there is no clear-cut distinction between tenants and owner-occupiers.[50] A son of a tenant could become an owner-occupier and vice versa. Secondly, a farmer could be both owner-occupier and tenant, and sometimes even landlord, when he leased land to another farmer.[51] Daily life tends to be messy. Moreover, owner-occupiers had to compete on the land market as well, and purchasing land is even more expensive than leasing it. Therefore, both the larger owner-occupiers and tenants who employed wage labourers had good reasons to keep wage costs low.

Now let us take a closer look at Van Bavel's account of the developments in the central river area. This area in Guelders and Utrecht was the only part of the lowlands in the Netherlands that had been thoroughly manorialised, but by the middle of the fourteenth century the manorial system had almost completely disappeared from this area.[52] The former manorial lords retained control of their land, however, and in the second quarter of the fourteenth century started leasing out the land on short-term contracts. This meant that leasehold became predominant in the river area, because the rural non-noble population owned only just over 25 per cent of the land in the Guelders part of the river area, c. 1300, and by 1550 this had been somewhat reduced to 22 per cent. Around 1550 the nobility and the church each owned about a third of the land and town dwellers a meagre 11 per cent.[53] Near Utrecht the distribution of landed property was similar in 1461, with the nobility owning 31 per cent, the church 36 per cent and urbanites and the rural population each 13 per cent.[54] Property relations in the river area therefore were very different

[49] Van Bavel, 'Rural Development', pp. 184–5; B. J. P. van Bavel, 'Structures of Landownership, Mobility of Land and Farm Sizes: Diverging Developments in the Northern Part of the Low Countries, c. 1300–c. 1650', in Van Bavel and Hoppenbrouwers, eds., *Landholding and Land Transfer*, p. 139.
[50] Hoppenbrouwers, 'Mapping', pp. 47–8.
[51] J. Cronshagen, *Einfach vornehm: Die Hausleute der nordwestdeutschen Küstenmarsch in der frühen Neuzeit* (Göttingen, 2014), pp. 222–3.
[52] Van Bavel, *Manors and Markets*, pp. 86–7.
[53] Van Bavel, *Transitie en continuïteit*, pp. 427, 479.
[54] J. H. Huiting, 'Domeinen in beweging: Samenleving, bezit en exploitatie in het West-Utrechtse landschap tot in de Nieuwe Tijd' (unpublished PhD dissertation, University of Groningen, 2020), p. 220.

from those in the coastal areas, where either burghers or farmers owned much of the land.

Large tenant farmers who marketed substantial amounts of agricultural products could profit from increasing food prices after c. 1500 and enlarge their holdings by accumulating leases.[55] The average size of tenancies of the abbey of Mariënweerd near Culemborg increased from somewhat less than seven hectares in 1442 to 13.7 hectares in 1580. The number of small tenants with less than five hectares decreased strongly in the period before 1550; middling tenants were reduced in number after 1550. Around 1580, large farmers with 40 *morgen* (c. 35 hectares) or more held a significant share of the abbey's land. Such large tenants could also be found on the estates of other landlords in the western river area. Around 1570 there were about 90 large tenant farmers with more than 40 *morgen* in Van Bavel's research area. Together with the large farms exploited by the landlords themselves, these large tenants cultivated some 40 per cent of the land. These large tenant holdings were not directly formed by the lords, but by the tenants through accumulation of leases. Indirectly the lords supported the large tenants by having good farm buildings constructed for them, investing in improvement of the soil by marling, and better water management. Because of increasing wealth, these large tenants managed to continue leasing their holdings for decades or sometimes even generations.[56] The development in the river area to the west of Utrecht was similar, although the trend towards large holdings appears to have started here earlier. In 1461, the average holding size here was just over 20 *morgen* (c. 17 hectares). Here too, large farms of 25 to 50 *morgen* were formed through accumulation of leases, sometimes combined with plots owned by the tenant himself. Landownership was fragmented, but holdings were consolidated.[57]

To beat their competitors on the lease market for land, the large tenants needed to optimise their profits. The most feasible way to do this was through lowering the cost of labour, which was a heavy burden for these farms. Wage costs were about the only costs farmers could influence. The level of other expenditure, such as taxes or the costs of water management, could not be influenced by them. Therefore, from the fifteenth century a shift took place from arable farming to less labour intensive pastoral activities such as horse breeding and fattening of oxen. The area of land under crops was reduced; in the district of Beesd and Rhenoy, for instance, this went from 48 per cent of the total acreage in

---

[55] Van Bavel, 'People and Land', p. 22.
[56] Van Bavel, *Transitie en continuïteit*, pp. 574–5, 584–7; B. J. P. van Bavel, 'Land, Lease and Agriculture: The Transition of the Rural Economy in the Dutch River Area from the Fourteenth to the Sixteenth Century', *Past and Present* 172 (2001), pp. 31–2.
[57] Huiting, 'Domeinen in beweging', pp. 548–51.

1364 to 33 per cent around 1400.[58] Moreover, within the remaining arable part of the farms, a demanding crop like wheat had to make place for less demanding oats, which could also be used as fodder for the expanding pastoral share of the farms. Crop yields decreased. During the period 1400–1570, the yields of wheat were reduced from 16 to 13 hectolitres per hectare and of oats even more so, from 39 to 22 hectolitres per hectare. According to Van Bavel, this was caused by the tenants cutting labour costs, leading to less weeding, less efficient ways of manuring and harvesting, etcetera.[59]

Even though this is a plausible and convincing story, it is not the whole story. There is something paradoxical about the behaviour of Van Bavel's landlords and especially of the large tenants. On the one hand they invested in improvement of water management and soil improvement through expensive activities such as marling – fertilising the land by mixing the topsoil with calcium-rich marl – which must have been intended to improve crop yields. Yet on the other hand tenants apparently were so loath to spend money on labour that a large part of any improved yields was lost again due to insufficient care for the crops. Van Bavel is right in assuming that it was rational for large tenant farmers to try to keep labour costs low, but the characteristics of the wetland landscape played a part here too. The Dutch river area descends gently from east to west, where the rivers reach the sea. This means that the eastern part of the area is the highest part and will suffer least from flooding. It is also the part with the best soils. High stream ridges with light well-drained clays are predominant here, whereas the west has mostly heavy impermeable clay that is almost impossible to cultivate.[60] As a result, arable farming is much more difficult in the western than in the eastern part of the region. This is confirmed by Van Bavel's own findings: in the westernmost village of Dalem, with its low waterlogged soils, there was very little arable, and only oats were cultivated, whereas in the eastern village of Hier, with its higher well-drained soils, almost half of the land was arable, and around 1540, half of that was sown with demanding, labour-intensive wheat.[61] This difference was caused by the varying constraints of the wetland landscape, not by farmers in Dalem being more cost-conscious than their counterparts in Hier.

The decline of crop yields can also be at least partly attributed to hydrological factors. During the central Middle Ages, dikes had been

---

[58] D. R. Curtis, *Coping with Crisis: The Resilience and Vulnerability of Pre-Industrial Settlements* (Farnham, 2014), pp. 162–3.
[59] Van Bavel, *Transitie en continuïteit*, pp. 589–606.
[60] H. P. de Bruin, 'Wording van het land', in H. P. de Bruin, ed., *Het Gelders rivierengebied uit zijn isolement: Een halve eeuw plattelandsontwikkeling* (Tiel, 1988), p. 14; P. Brusse, *Overleven door ondernemen: De agrarische geschiedenis van de Over-Betuwe* (Wageningen, 1999), p. 205.
[61] Van Bavel, *Transitie en continuïteit*, pp. 589, 592.

constructed along all the rivers to protect agricultural land from flooding. An unintended side-effect of this was that the riverbeds contained within the dikes began to silt up. Before the construction of embankments, river floods deposited sediment over an entire area, but afterwards this sediment was only deposited within the riverbeds and so eventually the level of those riverbeds became higher than the level of the surrounding land. This in its turn caused seepage: river water seeped to the land behind the dikes. As a result, the land closest to the dike tended to get flooded for part of the year and much land became heavily polluted with weeds. Fighting these weeds was one of the biggest problems of farming in the Guelders river area until well into the nineteenth century.[62] These problems must have started in the late Middle Ages, after the closing of the dikes. Although landlords and farmers tried to prevent damage by the construction of drainage mills from the middle of the fifteenth century onwards,[63] seepage must have caused part of the decline of crop yields. The changes in agriculture in the river area were not just caused by farmers wanting to cut labour costs, but also by farmers adapting to the constraints imposed by the landscape.

The Zuider Zee Coast

On the southern and eastern coasts of the Zuider Zee a layer of fertile marine clay had been deposited that was used for grazing and haymaking by farmers living on higher sandy ridges or on dwelling mounds – here called *pollen* or *belten*. During the fourteenth century, several of these areas, such as the polders of Arkemheen and Oosterwolde in Guelders and Mastenbroek in Overijssel, were embanked. This was partly caused by the fact that the coasts of the Zuider Zee were more threatened by flooding in this period.[64] There was also an economic motive for embanking the coastal polders, however. From the fourteenth century, breeding and fattening of oxen to provide meat to the population centres in the west and south of the Low Countries became increasingly important, and thus the pastures along the Zuider Zee increased in value, not only for the rural population, but also for urban and noble investors aiming to protect valuable grazing from flooding. Therefore, several areas on the Zuider Zee coast were embanked. Mastenbroek, a peat area, was reclaimed from 1364 onwards. This new polder was situated close to Zwolle, an important centre of the international cattle trade.[65] In 1356, dikes were constructed around the Arkemheen polder in the north-west

---

[62] Brusse, *Overleven door ondernemen*, pp. 204–5.
[63] J. D. H. Harten, 'De invloed van de mens op het landschap', in De Bruin, ed., *Het Gelders rivierengebied*, p. 23.
[64] Van de Ven, *Man-Made Lowlands*, p. 103.
[65] Kole, *Polderen of niet*, p. 158.

of Guelders. There, the pasture was used for grazing of oxen, and the duke of Guelders himself owned land.[66]

Transforming the valuable grasslands on the Zuider Zee coast through drainage and erection of dikes seems economically rational, but there were also other options. In 1364, the same year the reclamation of Mastenbroek started, the bishop of Utrecht granted ownership of the Kampereiland, the salt marshes and mudflats situated in the delta of the IJssel river in front of Mastenbroek, to the nearby city of Kampen. The magistrates of Kampen decided not to transform this landscape, but to modify it. Low dikes of about one metre high were constructed. These dikes were too low to protect the area from the winter floods, and that was intentional. These floods deposited essential fertile sediment and silted-up the area, a process accelerated by the planting of reeds and rushes. The farmsteads on Kampereiland were built on artificial mounds of 2.5 to 3 metres high to protect people and animals from the floods.[67] The city of Kampen was not protecting this area *against* nature, but working *with* nature to create valuable land, and at a considerably lower cost than by draining and embanking it. We do not know why the rulers of Kampen decided to follow this divergent course, but the case of Kampereiland shows that transformation of wetlands was not the only way to make them economically more valuable, and maybe in many cases not even the most rational.

The Wadden Coast

For the Dutch province of Friesland, the oldest data on the distribution of landownership and holding size can be derived from land tax records of 1511. Data is available for four *grietenijen* (rural districts), covering approximately ten per cent of the land area of mainland Friesland. These four districts were all situated in the marine clay area, the most prosperous part of the province. Monasteries, abbeys and other ecclesiastical organisations were the most significant landowners here in 1511. In the districts of Leeuwarderadeel, Ferwerderadeel and Hennaarderadeel these landlords owned 44, 40 and 39 per cent of the land, respectively. In Baarderadeel they owned less, 27 per cent of the land, probably because there were no major monasteries in this district.[68] The average for the four districts is 38 per cent. All other landowners were individuals, most

---

66 W. Hagoort, *Het hoofd boven water: De geschiedenis van de Gelderse zeepolder Arkemheen 1356 (806) – 1916. Gemeenten Nijkerk en Putten* (Nijkerk, 2018), pp. 47, 66.

67 A. M. J. de Kraker, 'Sustainable Coastal Management: Past, Present and Future or How to Deal with the Tides', *Water History* 3 (2011), pp. 154–5.

68 T. J. de Boer, 'De Friesche grond in 1511 (Leeuwarderadeel – Ferwerderadeel volgens het register van den aanbreng)', in *Historische Avonden* (Groningen, 1907), p. 109; O. Postma, *De Friesche kleihoeve. Bijdrage tot de geschiedenis van den cultuurgrond vooral in Friesland en Groningen* (Leeuwarden, 1934), p. 45.

of them living in the countryside. In the early sixteenth century it was not easy to distinguish between owner-occupier farmers and the Frisian nobility, the *hoofdelingen*. De Boer, who published the tax records of Leeuwarderadeel and Ferwerderadeel, simply assumed that those who owned more than 30 hectares could be considered as *hoofdelingen* because it seemed unlikely that ordinary sixteenth-century farmers would own more than 30 hectares. These wealthy individuals owned in both districts 32 per cent of the land. Some of them lived in towns, but most lived on their estates in the countryside.[69] Smaller landowners in town and country owned only 24 to 28 per cent of the land in these two districts.

The recent publication of De Langen and Mol, based on the same source of 1511 for the districts of Franekeradeel and Hennaarderadeel, reached somewhat different conclusions. They found that the nobility owned almost 28 per cent of all farms in Franekeradeel, but only six per cent in Hennaarderadeel. This difference between the two districts can be explained by the higher quality of the light, friable soils of Franekeradeel, which constituted a more attractive investment. In Hennaarderadeel more than half of all holdings were farmed by owner-occupiers.[70]

TABLE 2. Owner-occupiers and leaseholders in four districts in Friesland, 1511. Sources: De Boer, 'De Friesche grond', p. 106; Postma, *De Friesche kleihoeve*, p. 45.

	N	%
Full owner-occupier	158	11
More than half owner-occupier	61	4
Less than half owner-occupier	223	16
Full tenant	945	68
Unknown	10	1
Total	1397	100

The large landowners did not cultivate all of their land themselves, although some monasteries still exploited part of their land directly, as did several members of the nobility. The abbey of Mariengaarde, for example, in 1511 exploited 121 of its 649 hectares in Leeuwarderadeel and Ferwerderadeel directly.[71] Most of the land owned by monasteries,

[69] De Boer, 'De Friesche grond', p. 112.
[70] De Langen and Mol, *Friese edelen*, pp. 122, 126.
[71] De Boer, 'De Friesche grond', pp. 106, 108; De Langen and Mol, *Friese edelen*, p. 128; G. de Langen and J. A. Mol, 'Vroege benedictijner kloosterboerderijen in Zuidwest-Friesland', in Nieuwhof and Buursma, eds., *Van Drenthe tot aan 't Wad*, p. 168.

however, was leased out for short terms of five to nine years, at least from the fifteenth century onwards.[72] Sixteenth-century Friesland was a region characterised by leasehold, as table 2 shows. Most Frisian farmers were full tenants or individuals who owned some land but leased more. The preponderance of leaseholders was even bigger when one considers that monasteries who had land in direct exploitation were counted among the owner-occupiers.

The records from 1511 also provide information on holding size. Table 3 shows that almost two thirds of farms in the four districts were small-holdings or family farms with less than 20 hectares. Very large farms with more than 40 hectares were still very unusual, and 90 per cent of holdings were smaller than 30 hectares. A document from 1479 described owners of 100 *pondematen* (37.5 hectares) or more as *hoofdelingen* (noblemen). Therefore, the largest farms were probably mostly noble estates. This document also provides insight into herd size: the largest tenant farmers were estimated to possess at most 20 cows.[73] According to Gilles de Langen and Hans Mol, Frisian agriculture in the high Middle Ages was dominated by substantial family farms with holdings of c. 70 *pondematen* (26 hectares) on average. The smaller holdings only began to emerge when markets grew during the late Middle Ages. There are indications that the number of small farms continued to increase until c. 1550.[74] Therefore, what table 3 shows is *not* a stage in a development towards agrarian capitalism, but a farming sector in which small farms are on the rise at the cost of large and middling holdings.

TABLE 3. Distribution of farms by holding size in four Frisian districts, 1511. Sources: De Boer, 'Friesche grond', p. 103; Postma, *Friesche kleihoeve*, p. 44.

	N	%
<5 ha	145	10
5–10 ha	194	14
10–20 ha	568	41
20–30 ha	355	25
30–40 ha	96	7
40> ha	39	3
Total	1397	100

[72] Mol, *De Friese huizen*, pp. 177, 179.
[73] De Langen and Mol, 'The Distribution', p. 30.
[74] Ibid., p. 35; Ph. Breuker, *Het landschap van de Friese klei* (Leeuwarden, 2017), p. 149.

These farms on the Frisian clay were producing for the market.[75] They did not have much choice, because the heavy, badly drained clay was not very suitable for arable farming. Therefore, like the smallholders in Holland, they had to exchange meat and dairy products for cereals to feed their families. Mixed farming was only possible on the lighter clays closer to the Wadden coast. In the south-western part of the Frisian clay area the situation was different. Here very small farms prevailed; farmers seldom held more than three cows. In the same way as their counterparts in Holland, these smallholders had to combine dairy farming with non-agricultural pursuits such as fishing and shipping.[76] As in Holland, all these activities were of a commercial nature, and the Frisian smallholders were not subsistence farmers. There are also indications that they intensified farming throughout the sixteenth century. In the previous chapter it was already noted that the central fenlands of Friesland were mostly abandoned by the early sixteenth century. As far as the peat lands were still being exploited, they were used for hay production by farmers living on higher sandy ridges.

In the adjoining province of Groningen, ecclesiastical institutions also owned vast tracts of land. After the Reformation, the property of the monasteries was confiscated, and according to a late sixteenth-century inventory, their property amounted to 24,000 hectares, about one eighth of the land area of the province. The largest landowner was the abbey of Aduard with 5800 hectares.[77] More detailed information is only available for the Westerkwartier, the most western part of Groningen. In many villages there, most land was owned by ecclesiastical institutions, the abbey of Aduard in particular. Elsewhere, owner-occupiers owned a considerable share of the land, whereas the amount of land owned by the nobility was relatively modest. Still, most farmers were tenants; only some 30 per cent were owner-occupiers around 1550. In 1540, average holding size here was 15–20 hectares, with most farms being smaller than 25 hectares, although there were also a few very large farms with up to 50 hectares.[78] Animal husbandry prevailed in this area. Fattening of oxen and breeding of horses or grazing of young horses that had been bred in the nearby sandy areas were important activities here; arable farming was marginal. Horses and oxen were exported to the large population centres to the west and south: Holland, Flanders and the Rhineland.[79]

---

[75] Knibbe, *Lokkich Fryslân*, pp. 68–9.
[76] Ibid., p. 162.
[77] J. F. Benders, *Een economische geschiedenis van Groningen: Stad en lande, 1200–1575* (Assen, 2011), p. 44.
[78] H. Feenstra and H. H. Oudman, *Een vergeten plattelandselite: Eigenerfden in het Groninger Westerkwartier van de vijftiende tot de zeventiende eeuw* (Leeuwarden, 2004), pp. 32–9, 43–7.
[79] Benders, *Een economische geschiedenis*, pp. 248, 316–17; Feenstra and Oudman, *Een vergeten plattelandselite*, pp. 45–6.

Over a century ago, Friedrich Swart provided a survey of holding size and property relations in the Wadden area in the sixteenth century which still has value today and was recently updated by Otto Knottnerus. At the time Swart wrote, the German marshes were characterised by middling and large holdings of 20 to 100 hectares.[80] Sixteenth-century records for the German Wadden coast from Ostfriesland to Dithmarschen show that at that time in most regions holdings were not as large as they would be around 1900. An exception was the island of Nordstrand, where large farms with over 40 hectares already held a quarter of all cultivated land as early as 1436.[81] The most usual farm size in the German part of the Wadden Sea area was 20–30 hectares, with a few larger holdings of up to 60 hectares.[82] This is similar to the situation in Friesland and Groningen. Contemporaries noted a tendency towards increasing farm sizes. On the Ostfrisian peninsula this was driven by the farmers themselves, who exchanged plots of land through leasing and thus tried to create large contiguous holdings. Swart has demonstrated how this happened by studying a detailed land tax record from the village of Hohenkirchen in the territory of Jever from 1542. In Hohenkirchen only 22.5 per cent of the land was cultivated by owner-occupiers, 12.9 per cent was leased from the church, 19.5 per cent from the lord of Jever and no less than 41.5 per cent from other farmers.[83] What happened here was that farmers were concentrating their landholdings close to their farmsteads. Plots far away were leased to others, and in exchange plots nearby were leased from other farmers. For instance, farmer Umme Jabben owned 91 *Gras* (one *Gras* is c. 1/3 hectare), of which he himself cultivated only five *Gras*, the rest being leased by three other farmers. Nevertheless, Umme had a large holding, because he also leased 120 *Gras* from neighbouring farmers.[84] In this way, the most substantial farmers managed not only to concentrate their holdings around the farm buildings, but also to enlarge them. This was made more possible because land of smaller farmers often served as collateral for loans provided by their more prosperous colleagues, to whom the land reverted when the loan was not repaid.[85] However, it was during the seventeenth and eighteenth centuries that the strongest increases in holding size were to occur.

In Hohenkirchen farmers owned most of the cultivated land. In this respect it can be considered as representative of the German Wadden coast in the sixteenth century. In several regions farmers owned most of the

---

[80] Swart, *Zur friesischen Agrargeschichte*, pp. 226–7.
[81] Knottnerus, 'Yeomen and farmers', p. 156.
[82] Swart, *Zur friesischen Agrargeschichte*, pp. 231–9.
[83] Swart, *Zur friesischen Agrargeschichte*, pp. 240–1.
[84] Ibid., p. 241.
[85] Ibid.

land. Along the west coast of Schleswig almost all farmers were owner-occupiers.[86] In Hadeln they held almost all the land; in Kehdingen they had most of the land, although the local nobility owned about one third; in the Alte Land farmers also owned most of the land, although around the mid-sixteenth century, noblemen, the church and burghers from Stade and Buxtehude bought land from farmers who were confronted by an agrarian depression.[87] In some areas farmers' landownership was on the decline though. In chapter one it was already observed that the farmers in lowlands near Bremen lost the free ownership of their land after the defeat of the Stedinger in 1234. In this area, urban landownership increased in importance during the late Middle Ages. In the Hollerland district, burghers of Bremen owned half of the cultivated land and in other areas around the city they were also expanding their possessions during the fifteenth century.[88] Burghers of Emden also increased their possessions in the environs of the city from the late fifteenth century onwards.[89] Exceptional was the development in the districts of Butjadingen and Stadland. These former rural republics were conquered by the count of Oldenburg in 1514. Partly by expropriation of farmers, but mostly by reclaiming new land, this territorial lord managed to get hold of about half of the land in the area: some 20 per cent on the old polders and all land on the new polders.[90]

Along the German Wadden coast, short-term leasehold was less important than in the Low Countries. Those farmers who were not owner-occupiers mostly held their land on long-term leases. Particular to this area (including Groningen) was that the tenant usually owned the farm buildings, which gave him a strong bargaining position with the landlord. In the long run this strong position of the tenants led to the development of hereditary lease for fixed rents, especially in Groningen, where this type of lease was called *beklemrecht*, and in Ostfriesland, where it was called *Beherdischheit*.[91] This development began in the sixteenth century but was only completed during the seventeenth and eighteenth centuries.

Farming in the German coastal areas was predominantly pastoral, although cereals – especially barley and oats – were also cultivated. In

[86] B. Poulsen, 'Rural Credit and Land Market in the Duchy of Schleswig c. 1450–1660', in Van Bavel and Hoppenbrouwers, eds., *Landholding and Land Transfer*, p. 206.
[87] T. Schürmann, *Die Inventare des Landes Hadeln: Wirtschaft und Haushalt einer Marschengesellschaft im Spiegel überlieferter Nachlassverzeichnisse* (Stade and Otterndorf, 2005), pp. 10–11; Fischer, *Wassersnot und Marschengesellschaft*, p. 27; Hauschildt, *Zur Geschichte*, p. 154.
[88] Knottnerus, 'Yeomen and Farmers', p. 167; Hofmeister, *Seehausen und Hasenbüren*, pp. 123–4.
[89] Swart, *Zur friesischen Agrargeschichte*, pp. 205–8.
[90] Swart, *Zur friesischen Agrargeschichte*, p. 280; Knottnerus, 'Yeomen and Farmers', p. 166.
[91] Knottnerus, 'Yeoman and Farmers', p. 161, Swart, *Zur friesischen Agrargeschichte*, p. 243.

the Holsteinische Elbmarschen arable farming was even predominant.[92] In several regions, such as Ostfriesland, the cultivation of bread grains was insufficient to feed the population and rye had to be imported from Westphalia. Oats and barley, however, were exported in significant amounts from Ostfriesland, the Lower Elbe area and the west coast of Schleswig-Holstein to Holland during the fifteenth and sixteenth centuries.[93] The most important export products of the German marshes were horses and fattened oxen for the northern German and later the Dutch markets.[94] All of this indicates that farming in the Wadden area was thoroughly commercialised. This is also evidenced by the account books that farmers on the west coast of Schleswig kept in the first half of the sixteenth century, some of which have been preserved.[95]

The English Marshes and Fens

If there was any area in the North Sea lowlands in the sixteenth century that was characterised by agrarian capitalism in the Brennerian sense – the three-tiered version with commercial landlords, capitalist tenant farmers and wage labourers – it was Romney Marsh. In the first half of the fifteenth century this still was an area relatively densely populated by smallholders and family farmers who practised mixed farming. By the mid-sixteenth century it had been depopulated and most land was held by large farmers who leased their holdings from commercial landlords.[96] Ambitious men, initially from the farming population, but from c. 1520 onwards mostly from the gentry, made use of the demand for meat from London and for wool from the Wealden cloth industry to create large holdings where they held huge numbers of sheep or cattle. The plots of smallholders were amalgamated into bigger holdings, sometimes by applying ruthless methods to drive out the original tenants, sometimes resorting to driving herds of cattle over their arable fields and thus destroying the crops. The perpetrators of these clearly illegal acts seem to have wielded so much power locally that it was impossible for the victims to find legal redress

---

[92] Knottnerus, 'Yeomen and Farmers', p. 153; Lorenzen-Schmidt, 'Reiche Bauern', pp. 31–3.
[93] Van Tielhof, *De Hollandse graanhandel*, pp. 76–80.
[94] Knottnerus, 'Yeomen and Farmers', p. 153; Poulsen, 'Rural Credit', p. 206.
[95] Poulsen, 'Rural Credit', p. 205.
[96] M. Gardiner, 'Settlement Change on Denge and Walland Marshes, 1400–1550', in J. Eddison, M. Gardiner and A. Long, eds., *Romney Marsh: Environmental Change and Human Occupation in a Coastal Lowland* (Oxford, 1998), pp. 133–40; G. Draper, 'The Farmers of Canterbury Cathedral Priory and All Souls College Oxford on Romney Marsh, c. 1443–1545', in Eddison, Gardiner and Long, eds., *Romney Marsh*, pp. 118–23; Dimmock, *The Origin of Capitalism*, pp. 302–3.

Agriculture in the Late Middle Ages, c. 1300–1550

and so they were forced to sell their holdings.[97] The result was a society dominated by large graziers, usually from a gentry background.

As with all areas along the southern North Sea coast, Romney Marsh was hit by storm disasters during the late Middle Ages. Regardless of the high costs, landlords in the area were prepared to invest in the repair of the sea walls and even in the embankment and reclamation of new land.[98] This was not standard behaviour for landlords on the English coast. The marshes in the Thames estuary were also hit by floods in the late Middle Ages, the effects of which were exacerbated here by the fact that during the tenth to thirteenth centuries many marshes had been embanked, forcing the river into a narrower channel, increasing tidal heights and thus the risk of flooding.[99] Maintaining the existing coastal defences during a period of frequent breaches was expensive, and even more so when labour became incearingly scarce after the Black Death. This motivated landlords to cease investment in dike maintenance. Naturally, this increased the risks of flooding in the marshes, and certain forms of land use became impossible. Arable agriculture had to be abandoned, but other activities such as fishing and especially grazing could be intensified.[100] On Canvey Island, for example, which was partially embanked, the dikes were left to decay in the fifteenth century.[101]

For the Thames estuary James Galloway concluded that 'Flooding created new opportunities as well as dangers, and could act as a catalyst to land-use change'.[102] At first glance, it seems reasonable to expect that the ways in which the inhabitants and landowners in the coastal areas responded to these opportunities and challenges depended on the social and economic circumstances of the locality. However, although those circumstances on Romney Marsh were not that much different from those on the marshes of the Thames estuary – both were dominated by large landowners and commercial graziers – on Romney Marsh the sea defences were maintained and repaired after storm events, and in the Thames estuary they were not. It seems the geographical situation was of importance here. The marshes along the Thames were not very wide and could be exploited from the nearby uplands. Romney Marsh was a large, wide area and although its population diminished from the mid-sixteenth century onwards, there were still towns and villages situated in the marsh that were in need of flood protection. For this reason, and probably also because the costs of investment in the sea walls were still more than

---

[97] Dimmock, *The Origin of Capitalism*, pp. 292–4.
[98] Draper, 'Farmers', p. 113.
[99] Galloway, 'Storm Floods', pp. 182, 187.
[100] *Ibid.*, pp. 180–7.
[101] *Ibid.*, p. 181; B. E. Cracknell, *Canvey Island: The History of a Marshland Community* (Leicester, 1959), p. 13.
[102] Galloway, 'Storm Floods', p. 188.

compensated for by the proceeds from pastoral farming, landlords and tenants decided to continue paying for dike maintenance.

On Romney Marsh and the Thames marshes large-scale, extensive pastoral farming became predominant. This was not the case in all the English marshland areas. Joan Thirsk has demonstrated that in Lincolnshire, marshland farming, although also mainly of a pastoral nature, was more characterised by middle-sized holdings that combined intensive animal husbandry with arable farming. On the coastal manors not more than 25 to 40 per cent of the land was used as arable, but on the more inland 'middle marsh', crops were cultivated on 60 to 75 per cent of the land (excluding common pasture). According to sixteenth-century probate inventories, two thirds of farmers on the Lincolnshire marshes held at least ten acres (four hectares) and about half held more than 20 acres (eight hectares) of arable. About a third of the arable was under wheat, which was grown for the market.[103] Although arable husbandry was far from negligible, livestock was kept in considerable numbers by all marshland farmers. The saltmarshes outside the sea defences were the domain of sheep flocks. Several farmers in the coastal parishes held flocks of over 100 sheep. On the marshes within the seawalls, cattle were raised and oxen fattened. Oxen and corn were both destined for the urban markets of Durham and London. On average, farmers in the marshes held 12 head of cattle; herds of more than 30 animals were unusual.[104]

Mixed commercial farming in the Lincolnshire marshes resembled that found on the opposite shore of the North Sea. The main difference was that in the Lincolnshire marshes farmers had access to considerable commons on which to graze their animals, whereas in the Low Countries and north-west Germany common pasture had all but disappeared by the fourteenth century. This difference between England and the eastern shore of the North Sea was even more striking where the fenlands were concerned. In the parishes of southern Lincolnshire, it was not just the un-embanked saltmarshes that were used for common grazing, but also the vast fens situated inland from the villages.[105] These commons were the basis of a strongly pastoral-based form of husbandry. Arable farming was only practised on the higher ridges where the villages were situated, as in southern Lincolnshire, or on 'isles', sandy outcrops in the fens, such as the Isle of Axholme and the Isle of Ely. Two thirds of the Cambridgeshire fenland village of Willingham, for example, consisted of common fens.[106] The abundant grazing offered by the fens was the basis of the fenland

---

[103] Thirsk, *English Peasant Farming*, pp. 62, 75–6.
[104] Thirsk, *English Peasant Farming*, pp. 69–72; J. Thirsk, *The Agrarian History of England and Wales. Vol. IV, 1550–1640* (Cambridge, 1967), p. 35.
[105] Thirsk, *English Peasant Farming*, pp. 15, 19.
[106] M. Spufford, *Contrasting Communities: English Villagers in the Sixteenth and Seventeenth Centuries* (Cambridge, 1974), p. 121.

economy. Although more pastorally oriented than the mixed farming of the marshes, the cattle herds of the fenland farmers were smaller than those of the agriculturists in the marshes. Medium herd size was ten heads of cattle in the Lincolnshire fens compared with 12 in the marshes.[107] This is not surprising, since the area of land held by fenland farms was small in comparison with the holdings in the marshes.[108] In 1575 in the village of Willingham half of the tenants held c. 20 acres (eight hectares) and most of the other half held smallholdings of less than five hectares. Only two villagers had more than 20 hectares.[109]

The relatively small holding size in the fens can be explained by the efficient way in which the fenlanders exploited the variegated resources of the fens, which made smallholdings viable. With a few heads of cattle, grazing rights, cutting of reed, sedge and peat, a smallholder was able to maintain his family. Smallholding was to remain predominant here until well into the modern period. In the Isle of Axholme small holding size was also the result of partible inheritance, which created division of holdings. Smallholders could survive here through processing of hemp or flax, which was retted in stagnant pools in the fens. When holdings became too small, they were sold. This resulted in a lively land market in which small plots were traded. These plots were often purchased by yeoman farmers, on whose death the enlarged holding was again split up. Farming in the fens was relatively small-scale, but that did not mean it was subsistence farming.[110] Dairy products, cattle, hemp and flax were produced for the market, even by smallholders.

Conclusion

During the late Middle Ages, farming in the North Sea lowlands was confronted with many disasters: disease, warfare, and especially floods and peat shrinkage. Much land was lost, but in many places the land that had been won during the Great Reclamation was maintained as agricultural land. Some areas, in the Thames estuary, in the south-west of the Netherlands or in central Friesland, were deserted, but that did not mean they lost all economic functions. They were still used for grazing or haymaking by people living on the edge of the uplands.

In the areas that remained settled, there was a development towards more commercial, large-scale farming, especially on Romney Marsh, the Dutch central river area, the Flemish coastal plain, and – to a lesser degree – the Frisian clay areas. The emergence of these large commercial

---

[107] Thirsk, *English Peasant Farming*, p. 70.
[108] *Ibid.*, p. 42.
[109] Spufford, *Contrasting Communities*, pp. 134–6.
[110] P. F. Fleet, 'The Isle of Axholme, 1540–1640: Economy and Society' (unpublished PhD thesis, University of Nottingham, 2002), pp. 130, 159–60, 174–8.

holdings was caused by the growth of markets for products such as meat, wool and industrial crops, and by developments in water management. Many small farmers were unable to pay for dike repairs after the many floods of this period and had to give up farming. Others suffered from the most important institutional innovation of the period: monetisation of the maintenance of physical infrastructure. From then on, they had to pay money to organisations instead of performing the maintenance themselves, and they often lacked the means to do so. Their land was absorbed by the holdings of the emerging large commercial farmers.

Apart from Romney Marsh, where the gentry played the most important role, most new large farmers appear to have come from relatively modest rural backgrounds. Often, they were yeoman farmers owning some land themselves, to which they added tenancies, although there were also pure tenants. The shift towards large-scale commercial farming came from within the peasantry.

The fact that this type of farming emerged in some regions should not allow us to forget that in several other regions – the Lincolnshire and Cambridgeshire fens, the fens of the Holland–Utrecht plain – smallholders and family farmers remained predominant. In Holland, an area that was threatening to get flooded due to the lowering of the surface level, this was made possible through both technological – the introduction of the drainage mill – and institutional – the emergence of the *heemraadschappen* as coordinating authorities in drainage – innovations. The construction of expensive drainage mills was enabled by financial support from urban capitalists, who began to expand their landed property in the late Middle Ages. For the time being, however, the Holland countryside remained an area dominated by small farmers.

5

The Second Reclamation: Urban Capital in the Countryside, c. 1500–1700

Introduction

There is an overlap in time between this chapter and the previous two, because although peat shrinkage and storm surges continued to plague the inhabitants of the North Sea Lowlands, the first half of the sixteenth century also witnessed the beginning of a new period of expansion. This expansion was so strong that I will call the sixteenth and seventeenth centuries the era of the Second Reclamation of the North Sea Lowlands. This Second Reclamation was less impressive than the Great Reclamation of the central Middle Ages as far as land area was concerned, but it required much more capital than the earlier reclamation phase and it fundamentally changed property relations and farming systems in the lowlands. The strongest drive for this change was the growth of population and economy in the Low Countries, first concentrated around the port of Antwerp, later in the northern Low Countries, the emerging Dutch Republic. Like the preceding periods, the sixteenth and seventeenth centuries will be covered in two chapters. This chapter will demonstrate how Dutch capital, expertise and institutional innovations made the Second Reclamation possible. Chapter six will be dedicated to the changes in property relations and farming following this second wave of reclamations. Firstly, a survey of the history of the North Sea area from 1500 to 1700 provides context.

The North Sea Area in the Sixteenth and Seventeenth Centuries

At first glance the two centuries after 1500 do not seem to offer propitious circumstances for a strong expansion of agriculture. The period was characterised by warfare and also included some of the coldest phases of the Little Ice Age. This was partly caused by periods of extremely low solar activity reducing the amount of solar radiation, leading to lower temperatures. The first decades of the sixteenth century were part of such a solar minimum, as was the second half of the seventeenth century. The period from 1520 to the late 1560s was relatively warm, however, but followed by cold and wet years until 1602. Apart from the 1630–60 period, the seventeenth century was relatively cold.[1]

The sixteenth century especially was very stormy, culminating in the 1570 All Saints' Flood, that wreaked havoc along the entire southern North Sea coast.[2] The seventeenth century was also stormy. For Flanders and the south-western Netherlands, Adriaan de Kraker identified three particularly stormy periods during this century, 1588–1613, 1624–40 and 1660–72.[3] It is striking, however, that these storms did not prevent large-scale reclamation of coastal marshes, and that unlike the All Saints' Flood they had relatively few victims and caused little permanent damage. This, at least, is the pattern in most of the Low Countries. Along the Wadden Coast storm surges could still be very devastating. On 11 and 12 October 1634, for example, a heavy storm hit the coast of Schleswig-Holstein, taking thousands of lives and completely destroying the large island of Nordstrand.[4] In 1686 a flood in Groningen killed about 1600 people and 9000 head of cattle and destroyed some 650 houses and barns.[5] A similar heavy storm four years earlier that had flooded 200 polders in the south-west of the Netherlands killed only about 100 people and the damage was repaired quickly.[6] There appears to be a difference between a western area stretching from Flanders to Friesland, where after 1570 the consequences of storm surges remained limited, and an eastern region, stretching from Groningen to Nordfriesland, where floods could still be catastrophic. This chapter will provide explanations for that divergence.

The most important political development in the area under study in the first half of the sixteenth century was the unification of most of the Low Countries, including all coastal territories, under Habsburg rule. The centralisation, which the Habsburgs tried to introduce, quickly provoked resistance from the nobility and cities whose privileges were threatened by it. In combination with the aversion many inhabitants of the Low Countries felt for the ruthless persecution of Protestants by the Habsburg authorities, this caused the Dutch Revolt. This revolt started in the 1560s

[1] D. Degroot, *The Frigid Golden Age: Climate Change, the Little Ice Age, and the Dutch Republic, 1560–1720* (Cambridge, 2018), pp. 41–3; D. Degroot, 'Climate Change and Society in the 15th to 18th Centuries', *WIREs Climate Change* 9 (2018), p. 2; Pfister and Wanner, *Klima und Gesellschaft*, pp. 221–2, 292–3.
[2] De Kraker, 'Flood Events', p. 926.
[3] A. M. J. de Kraker, 'Storminess in the Low Countries, 1390–1725', *Environment and History* 19 (2013) p. 166.
[4] Meier, Kühn and Borger, *Der Küstenatlas*, pp. 104–12.
[5] M. K. E. Gottschalk, *Stormvloeden en rivieroverstromingen in Nederland/Storm Surges and River Floods in the Netherlands* (3 vols., Assen 1971–7), vol. 3, pp. 371–7; Buisman, *Duizend jaar*, vol. 5, pp. 134–8.
[6] Gottschalk, *Stormvloeden*, vol. 3, pp. 293–356; Buisman, *Duizend jaar*, vol. 5, p. 94.

and led to the emergence of a new state on the North Sea shore, the Dutch Republic, a federal state consisting of seven principalities in the northern part of the Low Countries that had managed to gain their independence from the Habsburgs.[7]

Political turbulence was not limited to the Low Countries: England experienced a civil war in the 1640s, and from 1618 to 1648 the Thirty Years' War ravaged much of the Holy Roman Empire. Oldenburg remained neutral and was spared from occupation or destruction by the warring parties. Ostfriesland, Schleswig-Holstein, and the archbishopric of Bremen, however, had to endure occupation and plunder by armies moving through their territories. The most important political consequence of the war for north-west Germany was that the archbishopric of Bremen was conquered by Sweden and at the Peace of Westphalia became Swedish territory as the duchy of Bremen. The other major Scandinavian power, Denmark, in 1667 gained control over Oldenburg at the death of its last count, Anton Günther.[8] However, Denmark had lost control over much of Schleswig-Holstein, after King Christian III and his stepbrothers divided the duchies in 1544. During the seventeenth century the duchy of Schleswig-Holstein-Gottorf became an independent state.[9]

Whereas the Scandinavian states tried to centralise and reform water management, such innovations were not attempted in the Dutch Republic. The Revolt was primarily a conservative backlash by local and regional elites against centralisation. Politically this extremely decentralised federal state with its strong attachment to local and regional 'freedoms' was a relic from the Middle Ages. The federal government in The Hague was only responsible for defence and foreign policy. All other policy, including drainage policy, was left to the provincial Estates, which had usurped the sovereignty from the prince. Economically, however, it became the powerhouse of north-western Europe in the early modern period. When the economy of Europe started on a new period of expansion at the end of the fifteenth century, initially it was the southern Netherlands which mostly profited from this. Textile industries in Flanders and Brabant boomed and Antwerp became the trading centre of northern and western Europe.[10] The Revolt brought an end to this prosperity. The countryside

---

[7] C. M. Lesger, *Handel in Amsterdam: Kooplieden, commerciële expansie en verandering in de ruimtelijke economie van de Nederlanden ca. 1550 – ca. 1630* (Hilversum, 2001), pp. 110–13.

[8] Cronshagen, *Einfach vornehm*, pp. 46, 52; M. Ehrhardt, 'Eine kleine Territorialgeschichte der Region Unterweser', in Bickelmann et al., eds., *Fluss, Land, Stadt*, pp. 170–1, 173.

[9] A. Bantelmann, A. Panten, R. Kuschert and T. Steensen, *Geschichte Nordfrieslands* (Bredstedt, 1996), pp. 109–12.

[10] J. de Vries and A. van der Woude, *The First Modern Economy: Success, Failure, and Perseverance of the Dutch Economy, 1500–1815* (Cambridge, 1997), pp. 24, 359; Lesger, *Handel in Amsterdam*, p. 27.

of Brabant and Flanders was ravaged by warfare for decades. Moreover, the integrated economic system with Flanders, Brabant, Holland and Zeeland at its core was destroyed when the Low Countries were split up from the 1580s onwards. The Dutch blockaded the Flemish coast and 'closed' the Scheldt, the connection of Antwerp with the sea, which meant that they forced ships sailing to Antwerp to break bulk in Zeeland and taxed the loads.[11]

For the southern Netherlands, the Revolt and the ensuing separation from the northern Netherlands were catastrophes from which they only slowly recovered. For the newly independent Republic it meant the beginning of a boom period that would last for almost a century. Amsterdam took over Antwerp's role as centre of European trade, the Republic had the largest merchant navy in Europe, industry boomed in cities such as Leiden, Haarlem and Delft, and a colonial empire was created in Asia, Africa and the Americas. For a long time even the contraction of the European economy that started in the first quarter of the seventeenth century did not hinder the growth of the Dutch economy. Only from 1663 onwards did the economy of the Dutch Republic finally also start on a downward phase.[12]

The growth and prosperity of the Dutch cities was the basis of the Second Reclamation. A considerable part of the capital amassed in these cities was spent on infrastructural projects such as land reclamation and drainage, both within and outside the Republic. Dutch capital was the driving force of the Second Reclamation, in the Netherlands and north Germany and partly even in England and in France.[13]

The Second Reclamation

Although eventually vast stretches of land were reclaimed, the Great Reclamation of the central Middle Ages was the result of many piecemeal drainage projects. Medieval reclamation was also labour intensive, almost all the work being done by men and women with spades and baskets. In these respects, the drainage activities of the early modern era were quite different. Although there were some small projects, most were large or even huge, covering hundreds to thousands, and in a few cases in England even tens of thousands, of hectares of land. They were still labour intensive, but now the work was no longer done by groups of farm

---

[11] Lesger, *Handel in Amsterdam*, pp. 119–23; De Vries and Van der Woude, *The First Modern Economy*, p. 371.
[12] J. de Vries, *The Economy of Europe in an Age of Crisis, 1600–1750* (Cambridge, 1976), pp. 4, 17, 21; De Vries and Van der Woude, *The First Modern Economy*, p. 673.
[13] De Vries and Van der Woude, *The First Modern Economy*, pp. 27–36, 670; R. Morera, *L'assèchement des marais en France au XVIIe siècle* (Rennes, 2011).

families, but by armies of wage labourers with spades and wheelbarrows, assisted by horses and carts. The main difference was the amount of capital required. During the Second Reclamation huge sums of money were invested because wage labour was used and because many of the projects were technically more demanding. The most daunting projects, such as the drainage of large peat lakes in Holland, could only be accomplished by spending large sums on the construction of dozens of drainage mills. This is why the economic boom of the cities in the west of the Dutch Republic was crucial for the success of the Second Reclamation: these cities provided the capital for the undertakings. Indirectly, they also stimulated the success of the reclamations, because the demand for food stuffs and raw materials in the growing cities caused a rise in the prices of agricultural products and thus made expensive reclamation projects economically worthwhile because the landowners in the new polders could now demand high rents.[14] It is not surprising, then, that the Great Reclamation petered out after the Dutch economy began to stagnate in the 1660s. As table 4 shows, after 1650 the area of land reclaimed was much reduced. Only in Zeeland was much land still being reclaimed in the second half of the seventeenth century, but almost all of that land was drained in the 1650s and 1660s; after 1670 land reclamation ground to a halt in this region as well.

TABLE 4. Land reclamation in the Dutch and German lowlands, 1500–1699 (in hectares). Sources: Knottnerus, 'Culture and society', p. 149; Van Cruyningen, 'Régulation des eaux', p. 63.

	1500–49	1550–99	1600–49	1650–99	Total
Schleswig-Holstein	13,000	28,300	20,200	9800	71,300
Lower Saxony	11,900	16,100	8700	5700	42,400
Friesland/Groningen	19,400	10,300	6300	4600	40,600
Holland	4800	18,200	40,900	7300	71,200
Zeeland	5300	5700	28,800	20,300	60,100
Northwest Brabant	4400	6500	6400	3400	20,700
Total	58,800	85,100	111,300	51,100	306,300

If we add to the data shown in table 4 an estimate of the land area

[14] De Vries and Van der Woude, *The First Modern Economy*, p. 211.

drained in present-day Belgium, for which no comprehensive data is available – although on the left bank of the Scheldt river alone some 7500 hectares of marshes were reclaimed during the seventeenth century – we can estimate the total land area reclaimed on the southern shore of the North Sea during the sixteenth and seventeenth centuries at about 320,000 hectares.[15] Compared with the Great Reclamation of the central Middle Ages, when in the Netherlands alone an estimated 700,000 to 800,000 hectares were reclaimed, that appears to be rather modest. How impressive the Second Reclamation really was becomes clear when we take a look at the sums of money that were invested. The most expensive were the lake drainage projects, mostly in northern Holland. The costs of these have been estimated at 9.5 to 13 million guilders. Per hectare the costs varied from 290 to over 500 guilders and in one – admittedly extreme – case even almost 960 guilders.[16] Most other reclamations concerned coastal marshes, which were technically simpler to drain, but if we assume that the costs of marsh drainage were on average 150–200 guilders per hectare, based on preserved accounts of several projects, that means that some 42–56 million guilders were spent on reclaiming 280,000 hectares of marshes.[17] All in all some 60–70 million guilders must have been spent on reclamation of wetlands in the Low Countries and north-west Germany during this period. That sum, which is about ten times the amount of money invested in the United East India Company in 1602,[18] includes the construction of dikes, canals, sluices and mills, but not the construction of farm buildings, labourers' cottages, hedges and ditches in the new land.

Why was reclamation in the sixteenth and seventeenth centuries so costly? In the case of lake drainage it is clear. Water needed to be pumped from the lake into a surrounding canal (*ringvaart*). For the digging of that canal land had to be expropriated, on which large sums had to be spent.[19] Drainage mills were also very expensive. To pump the water from a lake, large numbers of drainage mills were required. Since these

---

[15] T. Soens, P. De Graef, H. Masure and I. Jongepier, 'Boerenrepubliek in een heerlijk landschap? Een nieuwe kijk op de Wase polders als landschap en bestuur', in B. Ooghe, C. Goossens and Y. Segers, eds., *Van brouck tot dyckagie: Vijf eeuwen Wase polders* (Sint-Niklaas, 2012), pp. 21–2; A. De Hoon, *Mémoire sur les polders du rive gauche de l'Escaut et du littoral Belge* (Brussels, 1853), pp. 9–10.

[16] H. van Zwet, *Lofwaerdighe dijckagies en miserabele polders: Een financiële analyse van landaanwinningsprojecten in Hollands Noorderkwartier, 1597–1643* (Hilversum, 2009), pp. 316, 474.

[17] P. J. van Cruyningen, 'Profits and Risks in Drainage Projects in Staats-Vlaanderen, c. 1590–1665', *Jaarboek voor Ecologische Geschiedenis* (2005/2006), p. 139.

[18] De Vries and Van der Woude, *The First Modern Economy*, p. 385.

[19] Van Zwet, *Lofwaerdighe dijckagies*, p. 86.

mills could only lift water for about 1.5 metres, rows (*gangen*) of two to four mills were required to pump the water away. Each mill in the row lifted the water for a short distance until finally the last one could lift it to the ring canal some four to five metres above the bottom of the lake. For the drainage of the 4800-hectare lake Schermer, for instance, 52 mills were required, and each mill cost about 6000 guilders.[20] Marsh drainage had also become more expensive because, especially in the south-west of the Netherlands, dikes with gentle outward slopes were constructed. This technique was already known in the Middle Ages, but became standard practice in the early modern period. The waves spent their power on the gentle upward slope of the dikes, thus reducing the risk of breaches of the dikes. They required much more earth and work to construct, so although they did provide more safety they were also much more expensive.[21] The same was true for the dikes that were protected by wooden palisades that spread along the Wadden coast in the sixteenth century.[22]

    Almost all of this money was invested by town dwellers. In the first half of the sixteenth century burghers of cities in Brabant and Flanders were the most prominent investors in land reclamation. After the floods of 1530 and 1532, for instance, burghers of Antwerp and Mechelen invested heavily in the construction of new flood defences.[23] When the economic centre of the Low Countries shifted to the north during the second half of the sixteenth century, investors from cities in Holland and Zeeland became more important. The list of investors in lake drainage in northern Holland reads like an address book of merchants, urban *regenten*, and high officials of the provincial and federal governments in The Hague. Marsh drainage in the south-west was also financed by urbanites, here mostly from Middelburg and Dordrecht, whereas in northern Holland burghers of Amsterdam were preponderant. Urbanites from Holland and Zeeland were also active outside of their own provinces. The drainage of the Bildt, a vast marsh area in the north of Friesland which was reclaimed in 1505, was mostly financed by capitalists from Holland.[24] Urbanites from Holland and Zeeland were even prepared to invest in 'enemy

---

[20] Ibid., pp. 173, 179.
[21] Van de Ven, *Man-Made Lowlands*, pp. 135–6; P. J. van Cruyningen, 'Land en water', in P. Brusse and W. Mijnhardt, eds., *Geschiedenis van Zeeland II 1550–1700* (Zwolle, 2012), p. 20.
[22] O. S. Knottnerus, 'Culture and Society in the Frisian and German North Sea Coastal Marshes (1500–1800)', in S. Ciriacono, ed., *Eau et développememt dans l'Europe moderne* (Paris, 2004), p. 146.
[23] Dekker and Baetens, *Geld in het water*, pp. 260–306.
[24] Van Zwet, *Lofwaerdighe dijckagies*, pp. 489–509; Van Cruyningen, 'Profits and Risks'; Van Cruyningen, 'From Disaster', pp. 252, 130–7; H. Schoorl, *Isaäc Le Maire: Koopman en bedijker* (Haarlem, 1969); K. Kuiken, 'Van "copers" tot compagnons: De aannemers en aandeelhouders van het Bildt 1505–1555', *Jaarboek van het Centraal Bureau voor Genealogie* 57 (2003), pp. 90–2.

territory'. During the Twelve Years' Truce (1609–21) the reclamation of several polders in the district of Hulst, which was under the control of the archdukes Albert and Isabella at that time, was financed by inhabitants of cities in Zeeland and Holland, among them even Stadholder Prince Maurice of Orange-Nassau.[25]

Dutch capital not only paid for land reclamation in the Low Countries, but also in north-west Germany. The princes of the small states in this area wanted to reclaim fertile coastal marshes, but they and their subjects lacked sufficient capital, so Dutch investors were called in.[26] Count Anton Günther of Oldenburg, for instance, had planned to finance reclamations from the receipts of a toll on the river Weser, but these proved to be insufficient. Instead, the reclamations were sometimes paid for by local dignitaries and in other cases by Dutchmen. In 1649, for example, the count reached an agreement with the Holland *regent* Anthony Studler van Zurck who was to drain the large polder of Schweiburg.[27] The re-reclamation of a part of the submerged island of Nordstrand was granted in 1652 to a group of Dutch investors by the duke of Schleswig-Holstein-Gottorf.[28] The German princes were aided in this by the stream of members of religious minorities who fled the war-stricken Low Countries and who brought with them not only capital but also technical expertise.[29] It is impossible to measure how large the share of the Dutch in the investments in land reclamation in north-west Germany was, but it appears to have been considerable.

Merchants and other urbanites from Holland and Zeeland were looking for investment opportunities for the profits they made from trade and shipping. Land was considered a relatively safe investment and thus was a welcome addition to the investment portfolio. Moreover, handsome profits could be made in land reclamation. For those who wanted to make a quick profit in marsh reclamation in Zeeland, it was possible to double the invested sum by selling the new land a year after drainage, as examples from 1644 and 1661 show. Most considered participation in a land reclamation project as a long-term investment and became

---

[25] Van Cruyningen, 'Land en water', p. 40; Soens et al., 'Boerenrepubliek', pp. 24, 26.

[26] Knottnerus, 'Culture and Society', pp. 148–9; Knottnerus, 'De Waddenzeeregio', pp. 64–5.

[27] P. J. van Cruyningen, 'State, Property Rights and Sustainability of Drained Areas along the North Sea Coast, Sixteenth–Eighteenth Centuries', in B. J. P. van Bavel and E. Thoen, eds., *Rural Societies and Environments at Risk: Ecology, Property Rights and Social Organisation in Fragile Areas (Middle Ages–Twentieth Century)* (Turnhout, 2013), pp. 190–1.

[28] Bantelmann et al, *Geschichte Nordfrieslands*, p. 138.

[29] O. S. Knottnerus, 'Deicharbeit und Unternehmertätigkeit in den Nordseemarschen um 1600', in Steensen, ed., *Deichbau und Sturmfluten*, pp. 64–7; Müller and Fischer, *Das Wasserwesen*, vol. 2/3, p. 27.

Figure 4. Borsselepolder, Zuid-Beveland, Zeeland, 1973. Photo L. M. Tangel. Collection Rijksdienst voor het Cultureel Erfgoed. This polder was reclaimed in 1616, financed with urban capital. Large arable fields and farms with huge wooden barns for storage of cereals characterise the landscape in this area.

landlords. They could also make handsome profits. The land in a polder in Zeeland Flanders drained in 1639 at a cost of 150 guilders per hectare, was valuated in 1665 at 360 to 450 guilders per hectare, and of course the owners also received rental income from that land.[30] Despite the high costs, lake drainage could still be profitable. In the Beemsterpolder, drained in 1607–12 at 290 guilders per hectare, after a hesitant start, gross rents reached 26 to 29 guilders per hectare and around 1640 even about 50 guilders.[31]

Investing in land reclamation was not without risk. Although investors tried to reliably estimate the costs beforehand, expenditure could run out of hand. An infamous example is the Bottschlotter Werk, a series of attempts to close off the Dagebüller Bucht in Nordfriesland from the sea. Between 1583 and 1647 several consortia of Dutch engineers and investors

[30] Van Cruyningen, 'Profits and Risks', p. 133.
[31] Van Zwet, *Lofwaerdighe dijckagies*, pp. 357–8.

spent huge sums of money on this probably far too ambitious project. Eventually they had to give up without having reclaimed any land.[32] But also less ambitious projects could face unforeseen financial problems. For example, newly constructed dikes could suddenly breach due to storms or could be destroyed by enemy troops. In addition, the soil quality of the new land could be found to be disappointing. Coastal marshes always had fertile marine clay soils, but the quality of the soil of a lake bottom was difficult to gauge, although investors did try to take samples before starting any drainage. In most cases, however, the quality of the soil of the lake bottoms turned out to be disappointing. The projectors had hoped to create polders that were fit for arable farming and to profit from the high cereal prices in the first half of the seventeenth century. In most drained lakes, however, the clay soils were too heavy and impermeable, and often also too polluted with weeds, to be suitable as arable. Farmers in the drained lakes eventually turned to dairy farming and fattening of oxen, which, owing to the demand from the cities in Holland, was also profitable. This turned the draining of lake Beemster into a financial success. As drainage costs increased and prices of agricultural products stagnated, draining lakes became less and less lucrative, and following the very expensive draining of lake Starnmeer in 1643 interest in drainage waned.[33] Interest in less costly marsh drainage continued for somewhat longer, but when cereal prices began to decrease in the 1660s, and the best and most easily drainable marshes had been reclaimed, investors lost interest in marsh reclamation as well.[34]

In England, piecemeal reclamation of coastal marshes took place during the sixteenth and seventeenth centuries. In 1613 it was estimated that during the preceding decades some 1900 hectares of marsh had been reclaimed along the Lindsey coast in Lincolnshire.[35] Here and there, some more large-scale marsh drainage occurred as well. In 1622 landowners contracted with a Dutch merchant living in London to embank Canvey Island. The 'undertaker' was to have a third of the acreage of these landlords if he succeeded in providing the island with seawalls that would permanently protect Canvey Island from flooding. With financial backing from investors in the Dutch Republic, the island was embanked in 1623.[36] As mentioned in the previous chapter, during the late Middle Ages this island was not protected by dikes and was mainly used for extensive sheep farming. The motivation for the construction of seawalls

---

[32] Müller and Fischer, *Das Wasserwesen*, vol. 3/2, pp. 132–56; Bantelmann et al., *Geschichte Nordfrieslands*, pp. 140–1.
[33] Van Zwet, *Lofwaerdighe dijckagies*, pp. 373, 377, 384–5, 478.
[34] Van Cruyningen, 'Profits and Risks', p. 128.
[35] Thirsk, *English Peasant Farming*, p. 63.
[36] Cracknell, *Canvey Island*, pp. 20–1.

must have been the wish to use the land on the island more intensively, probably for arable husbandry.

Historians have understandably paid most attention to the huge fen drainage projects from the second half of the seventeenth century. The most famous drained fen areas were Hatfield Chase, situated on the border of Lincolnshire and Yorkshire, measuring about 70,000 acres (c. 28,000 hectares), and the Great or Bedford Level of the Fens to the north of Cambridge, with a land area of 312,000 acres (over 126,000 hectares).[37] Together with some smaller projects in southern and northern Lincolnshire, some 210,000 hectares may have been drained.[38] The outcome of these fen drainage schemes, however, was ambiguous. On the continent, the result of such projects was straightforward: marshes and lake bottoms were permanently transformed into arable and pasture. Some of the fen drainage schemes in England were outright failures, while others achieved only partial and temporary success. This had to do with the exceptional nature of these projects, due to their huge scale and the many and extremely opposed interests involved. In the Low Countries and north-west Germany, Dutch engineers could apply knowledge that was based on centuries of experience with drainage. Nobody had any experience, however, with the draining of such vast tracts of land as the English Fens. Admittedly, the fens of the Holland–Utrecht plain were even more vast, but they had been drained in a piecemeal fashion. Cornelius Vermuyden, the Zeeland engineer who played an important part in both the Hatfield Level and the Bedford Level drainage, was moving into *terra incognita*. Although his plans were basically sound, it was inevitable that he made mistakes.[39] Moreover, since most of these fens had peat soils, shrinkage and oxidation soon became major problems in the drained fens. The Crown, the major promoter of these projects, was theoretically right in assuming that it was better to implement a regional drainage scheme, because smaller-scale local solutions often caused problems in other parts of the fen areas.[40] Knowledge and technology, however, had not yet reached a level that made such large projects really feasible. It is hardly surprising, therefore, that the most successful project was the smallest one, the Ancholme Level (18,000 acres or 7200 hectares).[41]

That something needed to be done about the drainage systems of the fens was clear to most, because from the second half of the sixteenth

---

[37] Ash, *The Draining*, pp. 147, 190.
[38] Darby, *The Draining*, pp. 147, 151; M. Albright, 'The Entrepreneurs of Fen Drainage in England under James I and Charles I: An Illustration of the Uses of Influence', *Explorations in Entrepreneurial History* 8 (1955), p. 62.
[39] L. E. Harris, *Vermuyden and the Fens: A Study of Sir Cornelius Vermuyden and the Great Level* (London, 1953), pp. 46–7, 136–8; Darby, *The Draining*, p. 258.
[40] Ash, *The Draining*, pp. 54–5.
[41] Albright, 'Entrepreneurs', pp. 59, 62.

century flooding had become more persistent and severe.[42] Floods were no longer just beneficial, depositing nutrient-rich silt, but had also become destructive and reduced the growing season because the water often stayed too long on the land.[43] Everybody agreed that drainage had to be improved, but there the consensus between the interested parties ended. The fenlanders wanted to control the floods to maintain their pastoral farming system. They also realised that implementation of a drainage scheme was not a one-off investment; it would require continuous investment in maintenance.[44] The Crown and the projectors wanted to get rid of the floods once and for all and change the fens into an arable region. What made the plans of the projectors even less palatable to the fenmen was that the investors in drainage were to be rewarded with a share of the drained land. This meant that the fenmen had to give up part of their commons, which provided most of the grazing for their cattle, at a time when population growth meant there was already a shortage of pasture.[45]

Drainage projectors often portrayed the fens as useless wasteland. However, the fens provided essential resources to the local population, which was not willing to give those up. Since they had rights of common on the fens, their interests could not be ignored. The Crown tried riding roughshod over the objections of the fenmen, but that eventually failed. The commoners responded with litigation and in many cases literally put up a fight. Labourers working for the drainers were attacked, newly cut ditches were filled in again, dams were destroyed, and so on.[46] The outbreak of the Civil War in 1642 provided an opportunity for the fenland commoners to regain their commons and undo much of the drainage schemes. In this way, fens in southern Lincolnshire such as East and West Fen and Wildmore Fen reverted to their former state as common grazing land.[47] After the Civil War, drainage work was resumed in the Great Level in 1649 and completed four years later. The Great Level, the Hatfield Level, and the smaller Ancholme Level in northern Lincolnshire appear to be the only projects that enjoyed a degree of success, although in the Hatfield Level drainers and commoners only made peace in 1719 and the Ancholme Level had to be drained a second time in the 1660s.[48] Across

[42] Darby, *The Draining*, pp. 1–8; Ash, *The Draining*, pp. 49–53.
[43] J. Bowring, 'Between the Corporation and Captain Flood: The Fens and Drainage after 1663', in R. W. Hoyle, ed., *Custom, Improvement and the Landscape in Early Modern Britain* (Farnham, 2011), pp. 240–1.
[44] E. D. Robson, 'Improvement and Environmental Conflict in the Northern Fens, 1560–1665' (unpublished PhD dissertation, University of Cambridge, 2018), pp. 64–5.
[45] Ash, *The Draining*, p. 48; Thirsk, *English Peasant Farming*, p. 112.
[46] For the resistance of the fenmen see Ash, *The Draining*, ch. 7; K. Lindley, *Fenland Riots and the English Revolution* (London, 1982), passim.
[47] Darby, *The Draining*, p. 151.
[48] J. Korthals Altes, *Sir Cornelius Vermuyden: The Lifework of a Great Anglo-Dutchman*

all areas the results of drainage were mixed. Some places had indeed become drier, but others flooded even more. Moreover, the well-drained parts now no longer profited from the fertile silt that had been deposited by the floods in the past. An additional problem was that the navigability of some rivers was reduced.[49]

Initially, as in north-west Germany, Dutch investors were prepared to spend large sums on fen drainage in England. The c. £70,000 required for the drainage of Hatfield Chase in the years 1627–34 was almost entirely provided by investors from cities in Holland and Zeeland. Due to mistakes and mismanagement by Vermuyden, and the sabotage and litigation by the commoners of the surrounding villages, this sum may have increased to £200,000. The Dutch investors lost faith in Vermuyden and took over the responsibility for the completion of the drainage scheme in 1630.[50] Vermuyden's reputation in Holland and Zeeland suffered from these setbacks and the people planning the draining of the Great Level at this time realised they would not be able to attract Dutch capital for their enterprise if they continued to work with Vermuyden. Therefore, he was excluded from the first stage of the drainage of the Great Level in the 1630s. It seems, however, that the Dutch had lost faith completely in the opportunities to make a profit from fen drainage in England, because eventually only one Dutchman was prepared to risk money on the Great Level project.[51] Therefore the £400,000 to £500,000 that was spent on drainage of this area between 1631 and 1656 was almost entirely provided by English investors.[52]

Both in England and in the Low Countries and north-west Germany huge sums were spent on the reclamation of wetlands in the early modern period and vast areas were drained. In England, however, the results of this Second Reclamation were more limited than on the continent. Several of the fen drainage projects failed, and the outcome of those that did not fail, such as Hatfield Chase or the Great Level, were often ephemeral. Much drained land reverted to its former state within a few decades. This was due not only to technical mistakes made by the drainers, but also to institutional shortcomings. The large-scale drainage projects of the early modern period, and the gargantuan English fen drainage schemes in particular, made higher demands on the organisation of these schemes,

---

   in Land-Reclamation and Drainage (London and 's-Gravenhage, 1925), p. 128; Albright, 'Entrepreneurs', pp. 59–60.
49 D. Summers, *The Great Level: A History of Drainage and Land Reclamation in the Fens* (Newton Abbott, 1976), p. 94; Robson, 'Improvement', pp. 179–80, 189.
50 P. J. van Cruyningen, 'Dutch investors and the Drainage of Hatfield Chase, 1626 to 1656', *Agricultural History Review* 23 (2017), pp. 27–32.
51 M. Albright Knittl, 'The Design for the Initial Drainage of the Great Level of the Fens: An Historical Whodunnit in Three Parts', *Agricultural History Review* 55 (2007), pp. 29–31.
52 Summers, *The Great Level*, p. 85; Ash, *The Draining*, pp. 207, 264.

and affected the way in which the consequences for the interests of third parties in and around the drained areas were handled. Especially in England it proved to be difficult to reconcile a host of clashing interests.

State Regulation of Drainage

When large areas of land were drained many interests were involved, not just those of the drainers. In the case of lake drainage in Holland, for instance, the interests of fishermen, skippers and reed cutters were threatened, and indirectly but no less importantly, those of cities that might lose trade routes and markets due to drainage. In the fenlands, commoners lost grazing, could no longer cut reed, sedge or peat, and lost opportunities for fishing and fowling. Old land near newly reclaimed land could lose the means to drain excess water directly to open water. All of this provided a potential source of endless conflict, especially when the rights of third parties were protected by law. Litigation, riots and sabotage could seriously threaten the viability of drainage schemes. The Second Reclamation was possible not only because sufficient capital was available, but also because institutions were designed – with greater or lesser success – that aimed to find a balance between all interests involved. The emerging states were responsible for designing these institutions. That situation was not without risks, because the interests of the state – which desired more taxable land and subjects – tended to run parallel with those of the drainers and therefore the state might have tended to ignore any damage to the interests of third parties. This section will show how states and territories in the Low Countries, England and north-west Germany attempted to perform this delicate balancing act.[53]

In Holland, Zeeland, Flanders and Brabant the territorial princes had possessed the rights to the wastelands and the foreshore since the central Middle Ages (see p. 45); in Friesland they usurped this right at the beginning of the sixteenth century. This had two important consequences for the implementation of drainage schemes. In the first place, there were no longer rights of common to these lands. Those who used them for fishing or grazing could only do so by leasing them from the prince. When the prince wanted to have the area reclaimed, he could simply renounce the contract. This means commoners here had no recourse to courts through which to defend their rights. Therefore, promoters of drainage schemes in the Low Countries did not have to fear litigation by disgruntled commoners.

In the second place, the fact that the prince – and later the provincial Estates – possessed the rights to foreshore and wastelands provided the legal basis for the regulation of drainage projects. Those who wanted to

---

[53] Unless indicated otherwise this section is based on Van Cruyningen, 'Dealing with Drainage'.

drain such areas needed permission, and the prince could make his permission dependent upon certain conditions. Permission, conditions and usually a grant of tax exemptions to promote the reclamation of land were recorded in a document called *octrooi* (patent). From the fifteenth century onwards, procedures were developed in which the provincial Audit Offices advised the prince on the permission, and the conditions under which it was to be granted. They researched archives, inspected the area and interviewed people with local knowledge. This enabled them to add conditions to the patent that protected interests of third parties. For instance, when a marsh was drained, the drainers had to provide means to evacuate excess water from the old land adjoining the marsh, usually by digging a drain and constructing a sluice. For marsh drainage such conditions were sufficient to prevent conflicts and litigation.

Lake drainage was more complicated because more conflicting interests and powerful actors were involved. It proved difficult to reconcile all interests, and already by the 1590s complaints were voiced by drainers about the high costs of protracted litigation that threatened the viability of their projects. A brilliant solution – almost certainly devised by Grand Pensionary Johan van Oldenbarnevelt – was found in 1607, when a patent was granted for drainage of Lake Beemster in northern Holland. Here the interests of very powerful parties, such as the regional water authority and the cities in the area, were involved. A large lake like this – over 7000 hectares – played an important role in navigation, and cities defended their navigational interests fiercely.[54] Sometimes even the militia was ordered to demolish works the city magistrates did not approve of. The 1607 patent prevented such actions by including a clause that obliged the drainers to provide *contentement* (compensation) to all third parties whose interests were somehow harmed by implementation of the project. If the drainers and other stakeholders could not reach an agreement about compensation, arbitration by members of the Supreme Court of Holland and Zeeland was to take place. No appeal was possible from the decision of the arbiters, plus it was forbidden to start procedures in other courts.[55] In this way, third parties were compensated for losses while at the same time violence and time-consuming, drawn-out litigation were avoided. After 1607 the *contentement* clause became a standard feature of lake drainage patents in Holland, and in 1633 the Estates of Friesland introduced similar legislation in their province. Usually, the drainers were prepared to compensate losses generously, and both they and other stakeholders were satisfied. It could take lengthy negotiations to satisfy all parties, but once an agreement had been reached, drainage could be

---

[54] D. Aten, 'Als het gewelt comt …': Politiek en economie in Holland benoorden het IJ 1500–1800 (Hilversum, 1995), pp. 66–90.

[55] A. J. Thurkow, 'De overheid en het landschap in de droogmakerijen van de 16ᵉ tot en met de 19ᵉ eeuw', *Historisch-Geografisch Tijdschrift* 9 (1991), p. 49.

implemented quickly. Most stakeholders accepted the outcome of the negotiations, and it was impossible to sabotage drainage schemes by drawing them into a quagmire of endless litigation.

In medieval England, the Crown had not claimed the rights to the foreshores and wastelands. In the 1570s Elizabeth I attempted to claim the foreshores, but that does not seem to have met with success.[56] As a result, the rights to foreshore and wasteland belonged to the lords of the adjoining manors, *and* to the tenants of those manors, who had rights of common there. Since these commons played a crucial part in the local pastoral economy and in the survival strategies of the poor, both poor and more well off fenlanders had ample motivation to contest drainage schemes. Moreover, owing to their rights of common, their legal position was much stronger than that of their counterparts in the Low Countries. The commoners, and other stakeholders, could go to court to defend their interests.

On the coast of north-west Norfolk, the Le Strange family implemented several small-scale drainage schemes during the first half of the seventeenth century, assisted by a Dutch expert, Jan van Ha(e)sdonck. Although it only concerns somewhat less than 100 hectares, the Le Strange case is instructive, because it demonstrates the difficulties investors were up against when they wished to reclaim wetlands. As Elizabeth Griffiths has very rightly remarked: 'Sorting out the property rights was a prerequisite to any expensive drainage scheme on the marshes.' In the Low Countries this 'sorting out' would have been part of the procedure to grant a patent undertaken by the provincial Audit Office or the Council of State. In England, however, the drainers and their neighbours had to do the sorting out themselves, which could lead to costly lawsuits. With one neighbour, the Le Stranges indeed got involved in litigation which lasted for over a decade and was very expensive.[57] If the small-scale drainage schemes of the Le Stranges could lead to conflicts, it is to be expected that the enormous fen drainage schemes of the second quarter of the seventeenth century caused their promoters even more headaches.

Promoters of fen drainage in England had to deal with many, often conflicting, interests of manorial lords and tenants. This made drainage schemes legally extremely complicated, especially in the case of very large-scale fenland drainage projects. Contemporaries were aware of the potential for conflict, which is why the Dutch-English engineer Humphrey Bradley in his 1593 proposal to drain the Great Level pleaded for an Act of Parliament that would sort out the conflicting interests and

---

[56] J. Thirsk, 'The Crown as Projector on its own Estates, from Elizabeth I to Charles I', in R. W. Hoyle, ed., *The Estates of the English Crown 1558–1640* (Cambridge, 1992), pp. 313–14.

[57] E. Griffiths, *Managing for Posterity: The Norfolk Gentry and their Estates c. 1450–1700* (Hatfield, 2022), ch. 8, quotation from p. 158.

divide the drained land between landlords, commoners and drainers. This good advice was not followed. Elizabeth I, James I and their ministers left decisions on drainage proposals to the Commissions of Sewers in the fens, which tried to arbitrate in case of conflicts and to reach consensus. This proved to be all but impossible due to conflicts of interest in the fen regions and within the Commissions of Sewers.[58] Any large-scale drainage scheme of the fens seemed to be doomed due to conflicting interests.

This may have been the reason why Charles I opted for a different policy to his predecessors. He attempted to coerce recalcitrant landlords and commoners into cooperating with drainage projects. The Crown tried to prevent opponents from commencing litigation and some of them were even prosecuted. In 1637 Attorney-General Sir John Banks proposed the creation of a system of county courts with powers to act against opponents of agricultural improvement, and the most important form of improvement at that time was wetland reclamation. As Hoyle remarks, this meant subordinating 'private property rights to the god of improvement'.[59] Banks' proposal was not implemented, but the Crown continued to follow a top-down policy characterised by a striking lack of consideration for the interests of the commoners. These often had to relinquish two thirds of their commons: one third to the investors in drainage, one third to the manorial lord (which was often the Crown itself). It was unlikely that increased quality of the drained land would compensate for the loss of two thirds of the commons, especially since drainage did not always have the expected effects. Sometimes the land became too dry, or flooded more often than before drainage. The resulting discontent could be controlled so long as Charles I's personal rule prevailed, and commoners and opposing landlords could be forced into submission. As the country slid into Civil War in the 1640s, however, this proved to be no longer possible. As described in the previous section, in most cases the commoners regained possession of their commons and destroyed much of the infrastructure that had been constructed. In only three cases – the Great Level, the Hatfield Level and the Ancholme Level – were the projects successfully revived in the 1650s and 1660s.

That the ruthless way in which the Crown dealt with the interests of the commoners was an important cause of the ultimate failure of many drainage schemes is proved by the exceptional cases in which the commoners' complaints were taken seriously into account. In Hatfield Chase, unrest and litigation were mostly limited to the eastern part of the Level, where the Crown had ridden roughshod over the rights of the villagers. In the western part, the Council of the North had ruled

---

[58] Ash, *The Draining*, pp. 111–12.
[59] R. W. Hoyle, 'Introduction: Custom, Improvement and Anti-Improvement', in Hoyle, ed., *Custom, Improvement and the Landscape*, p. 23.

positively for the commoners and forced Vermuyden and his backers to improve the design of the works in order to protect the villages situated to the west of the Level. This may have cost the investors a large sum of money, but it also prevented costly litigation and riots. Whereas in the eastern part of the Level litigation lasted for decades and infrastructure was destroyed in 1642, in the western part the investors could peacefully enjoy the possession of their lands following the verdict of the Council in 1630.

The German Wadden coast shared certain characteristics with the Low Countries. By the end of the Middle Ages commons had all but disappeared from this area. The rights to the foreshore, however, had not been usurped by territorial lords in north-west Germany as early as in the Low Countries. In some territories, such as Kehdingen and Dithmarschen, this remained unchanged. Elsewhere, however, princes started claiming the foreshore in the early modern period. The dukes of Schleswig-Holstein-Gottorf, for instance, only claimed this right in 1612; until then the villages adjoining the marshes had possessed the right to the foreshore. From 1612 onwards, they were required to petition for a patent from the duke. The 1612 resolution did not mean that the villagers lost their right to reclaim marshes, but they could no longer independently decide when they would do so. If they decided to wait until an improvement in the economic situation meant they would have sufficient capital, they ran the risk that the duke would meanwhile grant a patent to foreign – often Dutch – investors and they would lose their marshes.[60]

German princes, such as the counts of Ostfriesland and Oldenburg and the duke of Schleswig-Holstein-Gottorf, had probably borrowed the concept of an *Oktroy* (patent) for drainage from the Dutch.[61] What they do not seem to have borrowed was the Dutch habit of negotiating before the grant of a patent to provide compensation for other stakeholders. Patents were sometimes granted without much consideration for the rights of third parties. This was especially clear when in 1652 a patent was granted to a group of Dutch investors who wanted to re-reclaim a part of the island of Nordstrand which had been flooded 18 years earlier. The rights of the farmers who had owned land on the island were declared void and they did not receive any compensation. Their land was granted to the Dutch investors. The duke promised to protect the investors from protests and molestation by the islanders. Unsurprisingly, on Nordstrand and elsewhere on the Schleswig coast relations between the Dutch and

---

[60] M. L. Allemeyer, *Kein Land ohne Deich: Lebenswelten einer Küstengesellschaft in der frühen Neuzeit* (Göttingen, 2006), pp. 138–9.
[61] O. S. Knottnerus, 'Die Verbreitung neuer Deich- und Sielbautechniken entland der südlichen Nordseeküste im 16. und 17. Jahrhundert', in C. Endlich, ed., *Kulturlandschaft Marsch: Natur Geschichte Gegenwart* (Oldenburg, 2005), pp. 165–6.

the local population were strained.[62] Unlike their English counterparts, the disgruntled inhabitants of the German Wadden coast were not able to regain their land because no civil war occurred; there was no accompanying collapse of state power which might have provided them with the opportunity to do so.

Drainage Companies

Although a few large drainage schemes in Flanders in the early sixteenth century were financed by one wealthy individual,[63] most of these projects were too large to be paid for by just one person. Moreover, since drainage projects carried large risks – one heavy storm could destroy all investments – most investors preferred to spread their money over several projects. Already in the late Middle Ages people were searching for ways to form consortia of investors for large reclamation schemes. In the 1330s, for example, the drainage of the Zwijndrechtse Waard in southern Holland was granted to a group of 16 investors who were not only to be rewarded with landownership, but who were also each granted a seigneurie in a sixteenth part of the new land.[64] From the sixteenth century onwards investors organised themselves in 'companies', first in the Low Countries and later extending this type of organisation to other parts of the North Sea shores.

Drainage companies were forerunners of the modern limited companies. They may have derived from the *partenrederijen* (partnerships) that were practised in shipping. Drainage companies issued shares that were expressed in a certain area of land, usually 50 or 100 *morgen* or *gemeten*. Those who bought shares were granted possession of that area of land when the project was completed. The shares were transferable and could be passed on to heirs. The investors did not pay a fixed sum for the shares, but were instead bound to pay all expenses for reclamation of the amount of land for which they held shares. Therefore, they ran a price risk. If for some reason the costs of reclamation were much higher than expected, they would have to pay proportionally more. This price risk was not negligible: new dikes were very vulnerable to storms and were sometimes completely wiped away, forcing the company to construct new dikes. The shares were paid for in instalments; if the project took more money and time to complete than expected, more and higher instalments had to be paid.[65]

The investors appointed a board responsible for the implementation of the plans. That board was chaired by a *dijkgraaf*, who was usually

---

[62] Müller and Fischer, *Das Wasserwesen*, vol. 2/3, p. 27; Allemeyer, *Kein Land*, pp. 152–3.
[63] Soens, *De spade*, pp. 226–33.
[64] Kuiken, 'Copers', p. 80.
[65] Van Cruyningen, 'Profits and Risks', p. 132.

a technical expert from outside the group of investors. He was often rewarded with free shares in the land that was to be reclaimed. All other officers on the board, especially the treasurer and bookkeeper, were elected from the participants, to ensure that they had control over the finances of the undertaking.[66] Once the project was completed, the drainage patent usually stipulated that the drainage company be transformed into a water authority with the right to levy rates from the landowners in the new polders. The members of the board were to be elected by and from the landowners in the polder, although often the smaller landowners were excluded from the franchise. The minimum requirement for the right to vote could be ownership of up to 25 hectares, thereby excluding all but the largest landlords.[67]

From the second half of the sixteenth century almost all larger drainage schemes were organised as companies. This way of organising drainage also spread to northern Germany and England. Many reclamation schemes on the German Wadden coast were financed by groups of – often Dutch – investors, organised into companies. In Oldenburg the count tried to pay for drainage projects himself, using the receipts from a toll on the Weser river, complemented with borrowed money. In this way, 15 large-scale drainage projects were implemented between 1574 and 1596. Later, in the seventeenth century, however, Oldenburg's drainage was left to companies of local and Dutch investors. Elsewhere on the German Wadden coast, drainage companies were active too; until the middle of the seventeenth century most of them were Dutch.[68]

In England as well, companies were formed to finance the large fen drainage schemes, although Vermuyden at first tried another model in Hatfield Chase. He promised to pay for drainage himself and finance this from sales of land that was to be drained. The costs of drainage exploded, however, due to technical shortcomings and conflicts with the commoners of the adjoining manors, and Vermuyden was constantly short of money. In 1630, after four years of problems and conflicts, the investors who had bought land in the Hatfield Level decided to form a drainage company to take over the financial responsibility for the projects.[69] After this debacle it comes as no surprise that the 14 men who agreed to drain the Great Level in 1631 formed a drainage company based on principles that had proved themselves in the past. The company had 20 shares; six

---

[66] Ibid.; Van Zwet, *Lofwaerdighe dijckagies*, pp. 53–8.
[67] Beekman, *Het dijk- en waterschapsrecht*, p. 1213; Van Zwet, *Lofwaerdighe dijckagies*, p. 56.
[68] W. Knollmann and H. Bauer, *Die Oldenburger Seekante im 17. Jahrhundert: Zur Geschichte des II. Oldenburgischen Deichbandes* (Oldenburg, 1995), p. 47; Van Cruyningen, 'State, Property Rights', p. 191; Knottnerus, 'Culture and Society', pp. 146–9.
[69] Van Cruyningen, 'Dutch Investors', pp. 28–9.

participants took two shares, the remaining eight each one share. The participants promised to 'bear and sustain their proportionable parts of the charge of the said work ... and shall pay and disburse such monies as shall be necessary and required'. They were to be recompensed for these expenses with 'the proportionable part and portion' of the 95,000 acres (c. 38,500 hectares) of drained land that were to be allotted to the earl of Bedford, the 'undertaker' of the project.[70] This is a lucid formulation of the principles that formed the basis of all drainage companies. The main difference between this company and others was the huge size of the share. This also demonstrates the risks run by the participants in drainage companies. In this case, each share may have required the payment of the enormous sum of £20–25,000.

Technological and Institutional Innovation: The Diffusion of Drainage Mills

As far as water management technology was concerned, the early modern period was not characterised by many new inventions. Groynes, osier mats, brick sluices and wind-driven drainage mills already dated back to the late Middle Ages, but only became wide-spread during the sixteenth and seventeenth centuries. These inventions took a lot of capital. That capital could be provided by inhabitants of the booming cities, but often it also took the design of new institutions to make investments safe enough for urbanites to consider them. I will demonstrate here how the availability or lack of urban capital, and the way in which water management was organised on the local and regional scale, influenced the diffusion of the drainage mill in two peat areas, those of central Holland and central Friesland.[71]

The Second Reclamation did not remain limited to those marshes, lakes, and fens that were drained by companies of investors. Less spectacular at first glance, but not much less important, was the improvement through mill drainage of fenlands that had been reclaimed during the central Middle Ages, but had since become waterlogged due to peat shrinkage. Here too, impressive sums were invested to improve the quality of agricultural land and make it at least suitable for grazing. As demonstrated in chapter three, drainage mills were introduced in Holland in the early fifteenth century, but it took until the sixteenth century before they were diffused on a broader scale. There were two obstacles to rapid

---

[70] S. Wells, *The History of the Draining of the Great Level of the Fens called the Bedford Level; with the Constitution of the Bedford Level Corporation* (2 vols., London, 1830), pp. 114–15.

[71] Unless indicated otherwise, this section is based on P. J. van Cruyningen, 'Water Management and Agricultural Development of the Coastal Areas of the Low Countries, c. 1200 – c. 1800', in *Gestione dell'acqua in Europa*, pp. 63–80.

diffusion: the absence of institutions that could guarantee that the new infrastructure would work efficiently and that money lent for the erection of drainage mills would be repaid; and lack of capital. The solutions to these problems were, as described in chapter three, coordination of water management by the regional water authorities, and the formation of mill polders (*molenpolders*) at the local level, in which landlords and farmers cooperated to construct and maintain drainage mills.

From the perspective of the availability of capital, the most important feature of the mill polders was that they could raise rates from the landowners within the polder, therefore ensuring a regular yearly income that could serve as collateral for loans (see pp. 102–3).[72] This made it attractive for wealthy urbanites to lend money to mill polders for the erection of drainage mills. Within the territory of the Rijnland water authority, drainage mills were built in three waves, from 1480 to 1495, 1550 to 1570 and 1630 to 1635.[73] During the first two periods it was mostly small, often horse-driven mills, that were erected. In the last period, small mill polders were merged, and the resulting larger polders invested in the construction of powerful, wind-driven mills that could pump large areas dry. This occurred not only in Rijnland, but also in Amstelland, to the south of Amsterdam. There, for instance, in 1637 the c. 1300 ha polder Ronde Hoep was created, and 36 small water mills were replaced by three large ones.[74] The latter were octagonal mills with a cap that could be turned to the wind from outside. These octagonal mills had much larger sails, making them more powerful and therefore able to pump up much more water, as the example of Ronde Hoep demonstrates. Mills of this type were developed in the late sixteenth century.[75] The money required for building such mills was often borrowed from prosperous town dwellers. Investments could be considerable. The drainage of a 160-hectare mill polder near Leiden in the 1650s cost 28,000 guilders. The polder borrowed 70 per cent of the money needed to construct the mill and the banks, drains and ditches. A polder near Amsterdam needed over 9000 guilders to build a new drainage mill in 1641. Most of this money was borrowed from the *Burgerweeshuis* (Civil Orphanage) in Amsterdam. In 1674, again this polder borrowed money in Amsterdam to construct a new mill.[76]

Polders in central Holland could make use of the enormous amounts of capital available in the cities to invest in improvement of the land. They could do so because maintenance duties had been monetised, meaning that they had a regular income, which provided lenders with security.

---

[72] Van Tielhof, 'After the Flood', p. 398.
[73] Van Tielhof and Van Dam, *Waterstaat*, p. 182.
[74] T. Stol, 'Schaalvergroting in de polders in Amstelland in de 17^e en 18^e eeuw', *Tijdschrift voor Waterstaatsgeschiedenis* 3 (1994), p. 17.
[75] Van de Ven, *Man-Made Lowlands*, p. 133; Zeischka, *Minerva in de polder*, p. 62.
[76] Zeischka, *Minerva in de polder*, pp. 200–1, 204–6.

In the seventeenth century an organisation like the *Burgerweeshuis*, for example, preferred safe investments that guaranteed a steady income, preferably real estate but also loans. The presence of a regular source of income made it attractive for such institutions and for wealthy individuals to lend money to mill polders.[77]

Although there are indications that drainage mills may already have been present in Friesland in the late fifteenth century, it was only after 1524 – when the Habsburgs finally gained control of Friesland and restored peace and order – that landlords and farmers seriously began to invest in drainage, both on the clay lands and in the fens.[78] Around 1550, drainage mills were for the first time mentioned in probate records.[79] The fact that they were mentioned in such records shows these were privately owned mills, not mills owned and managed by mill polders. They were of necessity also small mills, because the large mills that were usual in Holland were too expensive for individual landowners. This did not change until well into the nineteenth century; around 1810 Friesland was dotted with over 2000 small drainage mills, especially in the fen part of the province, although there were many in the clay areas as well. Frisian mills did not drain large polders like in Holland, but relatively small plots surrounding farm buildings. In the central peat area of Friesland, farms were built on higher strips of clay-on-peat or sandy ridges. A section of the adjoining fens was then provided with banks and pumped dry by a small drainage mill. The rest of this area consisted of lakes, canals and undrained land.

The sophisticated Holland water management system with its powerful mills clearly provided opportunities for considerably more productive agriculture than the Frisian system. So why didn't the Frisians follow the example of their neighbours? This may be partly explained by the geography of Friesland. The centre of this province was characterised by a large number of lakes, canals and rivers that together formed a natural reservoir system, the 'Frisian bosom'. There was no need to create and maintain an artificial reservoir system as in Holland. Each farmer or landowner could dispose of his excess water by pumping it into one of the numerous lakes or rivers. In the clay areas in particular, this led to a considerable improvement in the productivity of the land, but in the peat areas much of the land remained extremely wet. Still, this does not completely explain why the Frisians were satisfied with a less efficient and productive system than the Hollanders. Lack of both capital and regional water management organisations can likely explain the difference.

---

[77] A. E. C. McCants, *Civic Charity in a Golden Age: Orphan Care in Early Modern Amsterdam* (Urbana and Chicago, 1997), p. 173.

[78] Breuker, *Het landschap*, pp. 90, 93; Knibbe, 'De kerk', pp. 268–9, 272.

[79] Knibbe, *Lokkich Fryslân*, p. 232.

There were 11 cities in Friesland. However, compared with Holland the Frisian cities were mostly small and agrarian.[80] Friesland lacked the booming ports and industrial cities that characterised Holland and Zeeland. Therefore, there was a lack of abundance of urban capital in Friesland, making it difficult to find urban credit to found mill polders. Founding such polders was made even more difficult as there were no regional water authorities responsible for the coordination of the functioning of the drainage mills. The associations for the maintenance of sluices and canals that existed here in the Middle Ages had never developed into water authorities as in Holland (see pp. 106–7). Under these circumstances, Frisian landowners opted for second best: they erected small mills, pumped their excess water into the reservoir of central Friesland and hoped for the best. As will be shown in the next chapter, for a while this solution functioned quite well.

A drawback of the Frisian water management system was that there was no single authority liable for the maintenance of the reservoir system. There were only loose groups of rural districts and towns that were together responsible for certain elements such as sluices or stretches of canals. These groups had no budget and no authority to raise rates. They were entirely dependent upon the enthusiasm of the members for their maintenance duties, and that was often less than satisfactory. Canals and sluices were silted up or even deliberately blocked, and when the provincial authorities tried to organise repair, the towns and districts responsible tried to make themselves invisible.[81] From the early seventeenth century the province itself took over the maintenance of some canals and sluices, but only on an ad hoc basis. Until well into the nineteenth century, the province of Friesland did not have the financial means to take over maintenance.

Oligarchisation and Institutional Stagnation

Not all developments in the sixteenth and seventeenth centuries were positive; the Second Reclamation had its darker sides. Better drainage of the fenlands caused more peat shrinkage, and reclamation of marshes diminished the storage capacity of sea inlets and tidal rivers, thus increasing the risk of flooding. Here I want to concentrate on an additional negative development in these two centuries that was at least partly caused by the Second Reclamation: the oligarchisation of water management. Water authorities had never been inclusive. People with little or no land seldom or never had the right to vote, and although women theoretically could vote in the general meeting, they were seldom present and almost never

---

[80] J. A. Faber, *Drie eeuwen Friesland: Economische en sociale ontwikkelingen van 1500 tot 1800* (Wageningen, 1972), p. 415.

[81] Winsemius, *De historische ontwikkeling*, pp. 198, 200, 208–24.

held office.[82] From the sixteenth century onwards the water authorities became even less inclusive. The companies that drained marshes and fens were usually composed of tight-knit groups of urban patricians.[83] They sometimes sold part of the land after completion of the reclamation scheme, but most of them continued to own the land which had been allotted to them. In many new polders they owned some 80 per cent of the land (see p. 168), meaning they dominated the board of any new water authority. Moreover, as observed above, smaller landowners were not permitted to vote on policy or for the election of officers, and as a result the administration of the new land was usually in the hands of a small oligarchy of wealthy landlords.

Oligarchisation was not necessarily always a negative development. Wealthy urbanites owned most of the land, and proportionally paid for most of the maintenance costs, and it is not unreasonable that they had a proportionally large say in the administration. After all, it was in their interest that the rates they paid were spent efficiently. There was a tendency, however, on the older lands, for offices to be held by members of elites who were not actively involved in water management but just acquired these offices for the revenue they provided. This especially occurred in some regions on the Wadden coast.[84] One of the worst examples of this phenomenon was in the province of Groningen. Here the right to hold office in water management was traditionally attached to certain farmsteads, the owners of which were also liable for dike maintenance. From the sixteenth century onwards, however, these offices were sold separately from the farms. The purchasers of the offices perceived them purely as a source of income and were not really interested in efficient water management. Abuse of office, for example by unjustly exacting fines for supposed flaws in dike maintenance, was far from unusual. Appeal from such decisions with the provincial court was forbidden.[85] Effectively, in Groningen, administration of water infrastructure was in the hands of a closed elite that was not susceptible to control either from below – by the farmers liable for maintenance – or above – by the provincial authorities. This was not conducive to good maintenance, and neither was the habit in some northern German territories such as Butjadingen of granting wealthy landlords exemption from dike maintenance, and letting the farming population carry most of that burden. In Butjadingen some farmers had to maintain up to ten stretches of dike. Perhaps to compensate for this, the fines for bad maintenance were reduced in the seventeenth century. The result of all this weak

[82] Van Tielhof, *Consensus en conflict*, pp. 118–23.
[83] Van Zwet, *Lofwaerdighe dijckagies*, ch. 7; Van Cruyningen, 'Profits and Risks', pp. 130–2.
[84] Soens, 'Resilient Societies', p. 171.
[85] Schroor, *Wotter*, pp. 40–4.

management was that Butjadingen was hit by devastating storms in 1602, 1610, 1615, 1626, 1628, 1634, 1638, 1643, 1651, 1663, 1671, 1685 and 1686.[86]

Under such circumstances, it is hardly surprising that monetisation did not make much headway after c. 1600. Monetisation meant paying rates to the often rapacious elites who manned the boards of the water authorities and leaving the organisation of water management to those elites. When farmers continued to perform dike maintenance in kind they could at least control maintenance themselves rather than leave it to wealthy individuals they did not trust. During this period, monetisation remained limited to the areas where it had already been introduced in the late Middle Ages, to the new polders which had been reclaimed by drainage companies, and to a few areas where exceptional circumstances forced farmers and landowners to accept the new system. In the polder Arkemheen in Guelders, for instance, monetisation was introduced in 1611 after a series of devastating floods convinced the provincial Estates of the need to force the polder to adopt new regulations to confront the increased challenges and substitute rates for performance of maintenance in kind.[87] The records of the polder show how problematic dike maintenance in kind could be. In 1494, a 300-metre stretch of dike needed to be maintained by 32 owners. The plots they had to maintain varied in length from 152 metres to 55 centimetres.[88] The many tiny plots made efficient dike maintenance next to impossible.

As the cases of Arkemheen and others show, often government intervention was required to introduce monetisation. At the beginning of the eighteenth century, the paying of rates and contracting out of work was the standard method of maintenance in most water authorities along the southern North Sea littoral from Flanders to Friesland. In England as well, water management was monetised from at least the sixteenth century.[89] To the east of the Dutch province of Friesland, from Groningen to Nordfriesland, maintenance in kind remained predominant. Here, monetisation was only on the new polders reclaimed by drainage companies. It is assumed that farming in the eastern area was less commercialised than in the western part and that this resulted in a lack of cash, which caused farmers to prefer maintenance in kind.[90] The problem is that very little is known about the history of agriculture in Groningen and the north-west German polder areas prior to 1700, so this is difficult to verify. A more plausible explanation is that the farmers in

---

86 Norden, *Eine Bevölkerung*, pp. 212, 222.
87 Van Tielhof, 'Forced Solidarity', p. 335; Hagoort, *Het hoofd*, pp. 165–6.
88 Hagoort, *Hoofd boven water*, pp. 90–1.
89 Morgan, 'Funding and Organising', pp. 424–7; S. Hipkin, 'Tenant Farming and Short-Term Leasing on Romney Marsh, 1587–1705', *Economic History Review* 53 (2000), p. 650.
90 Van Tielhof, *Consensus en conflict*, p. 98.

these areas lacked trust in greedy elites who pocketed nice incomes from their offices and arranged tax exemptions for themselves. Therefore, farmers preferred to keep maintenance in their own hands and opposed monetisation. Moreover, when economic circumstances were adverse, they could economise on maintenance costs. The elites in their turn also had good reasons to oppose change, because they could profit from any fines they imposed on those liable for maintenance. Does this mean that there was more trust in the western area? That notion is plausible. Monetisation had been introduced there in the late Middle Ages, and by the seventeenth century it had proved that it functioned reasonably well. Moreover, the landowning elites in Holland and Zeeland may have been less rapacious than their counterparts in Groningen and north-west Germany, since there are no known cases of large-scale exemptions of rates from this region. Also, the sale of offices was unusual here. Office-holders in water authorities were in most cases elected from the landowners within the authority's territory. In summary, farmers in the western area may indeed have had more reasons to trust the elites that controlled the water authorities and leave the collection of rates and contracting out of maintenance work to those elites.

The introduction to this chapter mentions the variegated effects of floods in the western and eastern parts of the southern North Sea littoral in this period, with floods being much more destructive in the eastern part. It is tempting to see this as a result of the way in which dikes were maintained: in the west by professional contractors supervised by benevolent elites, and in the east by parsimonious farmers supervised by rapacious elites. This is of course a very exaggerated picture of the differences, and there are many more factors that determine the magnitude of the damage caused by storm surges, such as the power and direction of the wind, temperature, the exact time when the dikes breach (when people are surprised in their beds at night there are always more victims), etcetera. Still, it is striking that in the seventeenth century really devastating floods with many victims only occurred in the eastern area from Groningen to Nordfriesland. Insufficient maintenance was very probably one of the causes of these localised disasters.

Conclusion

The sixteenth and seventeenth centuries witnessed a wave of land reclamation and improvement in the North Sea lowlands. Although when measured in hectares this Second Reclamation was more modest than the Great Reclamation of the Middle Ages, it required much more capital because projects were larger and technically more demanding. The economic engine driving most of the Second Reclamation was the booming urban economy of the northern Netherlands, Holland and

Zeeland in particular. Capital from the cities in these provinces financed not only drainage projects within the Dutch Republic but also in Germany and England. To raise the capital for these costly projects, consortia were formed in which groups of investors cooperated. These 'companies' originated in the Netherlands, perhaps inspired by the partnerships practised in shipping. Investors held transferable shares in such companies and paid a proportionate share of the costs of drainage.

The degree of success of drainage schemes was not only determined by agricultural prices or the quality of the land that was drained: institutions also mattered. In the Low Countries and north-west Germany there were no commoners with rights of usage who could oppose projects. Moreover, in the Low Countries procedures were developed that made it possible to provide compensation for third parties whose interests were harmed by drainage, without running the risk of becoming engaged in long drawn-out litigation. In England, local communities had rights of common in the fen areas, which they fiercely defended in court and in the field. What made things worse was that Charles I unsuccessfully tried to coerce landlords and commoners into accepting the drainage schemes without designing equitable rules for compensation of losses incurred by drainage. When his personal rule collapsed in the 1640s, the villagers repossessed their commons, and most fen drainage projects failed. Dutch investors had pronounced judgement about Carolean fen drainage at an earlier date: they no longer engaged in English projects after 1630.

The combined effects of urban investment and diverging institutions can also be observed in the way in which the shrinking peat soils were improved. In Holland, urban capital provided the means to build large drainage mills, and the regional water authorities provided the regulation required to make these large mills work optimally. In Friesland urban investment was lacking, and farmers and landowners erected large numbers of small mills that only drained small plots of land. Another reason that large mills were not built here was that water authorities that might have been able to coordinate their functioning were absent. Here the medieval legacy becomes obvious: in Holland the count had created powerful regional water authorities; in Friesland regional cooperation atrophied in the late Middle Ages. This divergence had far-reaching consequences until well into the nineteenth century.

A negative development during the sixteenth and seventeenth centuries was the increasing oligarchisation of the boards of the water authorities. Where offices were held by large landlords who had an interest in good water management, such as in much of Holland, Zeeland and Flanders, this did not necessarily have serious consequences. Where offices were sold and powerful elites were granted exemptions from maintenance duties, such as in parts of the Wadden coast in Groningen and north-west Germany, the consequences were more grave. These elites cared less about good maintenance and tended to abuse their offices. It is therefore not surprising that farmers in this area refused to monetise dike maintenance by paying rates to rapacious elites that might squander the money. The consequence of this was sub-optimal maintenance that may have contributed to the high numbers of casualties claimed by storm surges in this area, much higher numbers than in the more western regions where dike maintenance was monetised.

6

Risen from the Waves: Agriculture after the Second Reclamation, c. 1550–1700

Introduction

The Second Reclamation strengthened and accelerated a process that was already discernible in several regions in the late Middle Ages: the development towards large-scale, commercial, 'capitalist' farming. Of course, this tendency was strongest in the areas that had been reclaimed after c. 1550 and where new property relations emerged. Therefore, they will receive most attention, but the important question of whether this development also reached earlier reclaimed wetland regions such as the Holland–Utrecht fenlands will not be neglected. For this period, the development can be traced in more detail than for the Middle Ages, because, for the Dutch/Flemish regions in particular, data on holding size and distribution of landownership becomes much more abundant. This provides an opportunity to examine the progress of large-scale commercial farming in the wetlands and its causes. In particular, the case of Holland is interesting in this respect. According to Bas van Bavel property relations in the long run tended to be stable, although Holland in his view, was the exception to this rule, due to an impressive shift which took place there towards urban landownership. He assumes that those urban landowners introduced capitalist farming.[1] Or is Jan de Vries right and did the 'rural population [play] an active and major role' in the transformation of the countryside?[2] Five regions will be discussed: Flanders, Zeeland and the South Holland Islands; Holland; Guelders and Overijssel; the Wadden coast; and the English marshes and fens. The chapter concludes with a discussion on the question of whether the developments in these lowland areas can be characterised as the emergence of agrarian capitalism.

Agrarian capitalism is a difficult term. If one asks ten historians for a definition, one runs the risk of getting at least 11 different answers. Robert Brenner provides a useful definition of agrarian capitalism: 'classically',

---

[1] Van Bavel, 'Rural Development', pp. 170–80, 184.
[2] J. de Vries, 'The Transformation to Capitalism in a Land without Feudalism', in Hoppenbrouwers and Van Zanden, *Peasants into Farmers?*, p. 82.

agrarian capitalism is 'large-scale tenant farming on the basis of capital, improvements and wage labour'.[3] According to Jane Whittle, the orientation towards the market and the form of the labour employed are the most important elements of this definition, not the nature of tenure.[4] As we saw in chapter four, there were no clear dividing lines between owner-occupiers, tenants, and all kinds of combinations of these categories in the Lowlands. Therefore, it is wise to follow Whittle's advice and not assign too much weight to tenure. Market orientation, wage labour, and capital and improvements will be considered here as the most important characteristics of agrarian capitalism.

Flanders, Zeeland and the South Holland Islands

In this section, the islands in the southern part of Holland have been included with Flanders and Zeeland because agriculturally they have much more in common with those territories than with mainland Holland. In the seventeenth century most of the coastal plain and islands stretching from Calais to Rotterdam consisted of marine clay soils. The peat soils that had existed here in the Middle Ages had been almost entirely swept away by the floods of the fourteenth to sixteenth centuries. Those floods had most heavily hit the northern, now mostly Dutch, part of the area. Many wetlands had to be re-reclaimed in this area, not least because much land had been flooded for military reasons during the late sixteenth century, when the Dutch rebels tried to stop the advance of the Habsburg armies by inundating the land. In this south-western part of the Netherlands at least 103,000 hectares of land were reclaimed between 1550 and 1700, between 40 and 50 per cent of the total land area around 1700.[5] The Flemish coastal plain now situated in France and Belgium was mostly spared these inundations. However, it also suffered from the atrocities during the Eighty Years' War in the late sixteenth century, and yet it recovered quickly during the first half of the seventeenth century.[6]

It is to be expected that the areas needing to be completely reconstructed were also the ones that witnessed the biggest changes. As noted in chapter five, most investors in land reclamation in this period were urbanites, so it may be expected that in the newly (re-)reclaimed areas traditional landowners such as the nobility, church and peasantry had to make room for town dwellers. That this was indeed the case can be illustrated with a comparison (in table 5) between property relations in two parts of the Flemish coastal plain in 1665/1670: the area around

---

[3] Brenner, 'Agrarian Class Structure', p. 25.
[4] J. Whittle, *The Development of Agrarian Capitalism: Land and Labour in Norfolk 1440–1580* (Oxford, 2000), p. 9.
[5] Van Cruyningen, 'From Disaster', p. 250.
[6] Vandewalle, *De geschiedenis*, pp. 42–8, Vandewalle, *Quatre siècles d'agriculture dans la région de Dunkerque 1590–1990: une étude statistique* (Gent, 1994), p. 18.

Dunkirk, which had not been flooded, and Zeeland Flanders, which had been completely destroyed by floods and was reconstructed after 1600.

TABLE 5. Percentage distribution of landownership in two areas on the Flemish coastal plain, 1665/1670. Sources: Nationaal Archief, Den Haag, Raad van State nr. 2145 I; Vandewalle, *Quatre siècles*, pp. 137–43.

	Zeeland Flanders (1665)	Polders around Dunkirk (1670)
Hectares	27,945	27,391
Town dwellers (%)	65	20
Farmers (%)	18	38
Church (%)	7	20
Nobility (%)	6	22
State (%)	4	0

Whereas on the older polders of Dunkirk the peasantry was still the most important group of owners, in the newly reclaimed area of Zeeland Flanders, urban landownership dwarfed that of other groups. This trend is confirmed by data from other new polder areas in Zeeland and the South Holland Islands, where urban landownership was also clearly predominant.[7] On some of the new polders, urbanites owned 80 to 90 per cent of the land.[8] Clearly, on these new polders, property relations emerged that were very different from those of the late Middle Ages. Therefore, here we have a second area, apart from mainland Holland – assuming Van Bavel is indeed right about Holland – where a transformation of property relations occurred during the sixteenth and seventeenth centuries. One phenomenon cannot be entirely explained by the sixteenth- and seventeenth-century reclamations: the reduction in the share of the land owned by the church in Zeeland Flanders. This was mostly due to the confiscation of church lands by the new, Protestant, Dutch Republic.[9] Most territories along the North Sea coast adopted some form of Protestantism over the course of the sixteenth century and almost everywhere a 'Dissolution of the Monasteries' took place. The confiscated land fell into the hands of the state and in most cases ultimately into the hands of private owners. Flanders, which remained part of the Catholic Habsburg empire, was the only major exception to this rule.

Such a transformation in property relations may have been the basis upon which a shift towards more large-scale, commercial agriculture occurred. The available data indeed indicates that this happened. It was

[7] Van Cruyningen, 'From Disaster', p. 251.
[8] Van Bavel, 'Rural Development', p. 84.
[9] Van Cruyningen, 'From Disaster', p. 251.

not only limited to the new polders. In the Flemish coastal plain, a shift towards the creation of larger holdings was already discernible in the late Middle Ages, as noted in chapter four. On the older polders of the Dunkirk area, for instance, by the late seventeenth century large holdings – defined as units over 25 hectares – held about 60 per cent of all agricultural land.[10] In several early-reclaimed areas in Zeeland, however, between half and two thirds of all land was used by smallholders and family farms with less than 20 hectares. Really large farms with more than 40 hectares were highly exceptional there.[11] Large farms were strongly predominant on the newly reclaimed polders. Smallholders and family farmers seldom held more than 20 per cent of the land there and in some cases even less than ten per cent. Very large farms with more than 40 hectares held between 40 and 75 percent of the land on the new polders.[12]

It is tempting to attribute the rise of these large farms to urban capitalists leasing their land in large units to tenants. That did indeed occur but was not always the case. Few urban landlords owned large contiguous estates. Many of them simply did not own that much land at all, and those who did often owned land on many different polders.[13] Reclaiming polders was a risky enterprise, and it was wise not to put all one's eggs in one basket by investing in a single project. Even those who did own a large tract of land on a polder did not always lease it to one tenant. A burgher of the city of Dordrecht, for example, who owned 20 hectares of land on the nearby polder of Zuid-Beijerland in 1659, had split it into five plots, all leased to different tenants. Most large holdings on this polder had been formed by farmers, often combining owner-occupied land with several tenancies or simply accumulating tenancies. Of the 12 farms of over 40 hectares on this polder, only two had one single owner. The largest holding, with 95 hectares, was composed of seven plots rented from seven different urban landlords.[14] The same can be observed on other newly reclaimed polders. Therefore, the development towards the formation of large, 'capitalist', farms was not just driven by urban landlords, but also by tenant farmers.

By itself, urban landownership did not cause the rise of large, commercial farms on the new polders; it only did so in combination with the lively land and lease markets that emerged here.[15] Those large farms employed sizeable numbers of wage labourers. The disasters of

---

[10] Vandewalle, *Quatre siècles*, p. 121.
[11] Priester, *Geschiedenis*, pp. 700–3.
[12] Van Cruyningen, 'From Disaster', p. 254.
[13] *Ibid.*, pp. 252–3.
[14] C. Baars, *De geschiedenis van de landbouw in de Beijerlanden* (Wageningen, 1973), appendix 18.
[15] B. J. P. van Bavel, 'The Organization and Rise of Land and Lease Markets in Northwestern Europe and Italy, c. 1000 – 1800', *Continuity and Change* 23 (2008), pp. 14–15, 39–43.

the late Middle Ages had destroyed the original peat and clay-on-peat soils, and during the years when the land had been submerged the floods had deposited deep layers of fertile marine clay, which was very suitable for arable farming. As arable farming tends to require more labour than animal husbandry, the large arable farms employed greater numbers of labourers, particularly since they concentrated on relatively labour-intensive crops, such as wheat, rapeseed, madder and flax. Moreover, the land was cultivated intensively, with much ploughing and weeding. The high quality of the soil compensated for the high labour costs. Most of the labour force was local, although during the harvest it was supplemented with seasonal labourers from inland Flanders. Society on the new polders of the south-west was therefore relatively polarised, with a small group of large farmers at the top and a large group of labourers at the bottom. Many of those labourers were not proletarians, however, because they often had their own smallholding and owned one or two cows.[16]

On the older polders, especially in the part of the Flemish coastal plain that remained under Habsburg rule, the situation was somewhat different. As noted above, farms were somewhat smaller on the old polders than on the new, although large farms were not absent. They also tended to be less focused on arable farming than the farms on the new polders. Whereas on the new polders often 75 per cent or more of the land was used as arable, on for example, the old polders of the Castellany of Veurne only about 35 per cent of the land was under the plough – the rest was pasture. Farms concentrated more on cattle holdings here, with dairying being predominant in the area around Veurne.[17] The main cause of this difference in the orientation of farming was the variation in soil type. The heavy old clay soils of the Castellany of Veurne and other old polders were impermeable and difficult to cultivate and therefore not very suitable for arable farming. The average yields of 11 hectolitres per hectare of wheat in the Veurne area in the first half of the seventeenth century would have been considered dismal by the farmers of the new polders, who were used to 20–25 hl/ha.[18] Therefore, the farmers on the old polders wisely stuck to dairy farming. Substantial farms did exist on the old polders and the productivity of the dairy cows was high, so they produced considerable surpluses for the urban markets, but it was on the new polders, where the floods had previously wreaked havoc, that large-scale production for the market really took root.

Holland

According to Van Bavel, Holland was the only territory in the Low

---

[16] Van Cruyningen, 'From Disaster', pp. 255–6.
[17] Vandewalle, *De geschiedenis*, pp. 164, 166, 220–6.
[18] *Ibid.*, p. 193; Van Cruyningen, 'From Disaster', p. 256.

Agriculture after the Second Reclamation, c. 1550–1700

FIGURE 5. Harrowing and sowing of rapeseed on Isabellapolder near Aardenburg, Zeeland. Collection ZB Beeldbank Zeeland. Even as late as 1949 many labourers and horses were employed on the arable farms on polders reclaimed around 1650.

Countries where a transformation in property relations occurred during the sixteenth and seventeenth centuries. On the newly reclaimed land on the South Holland Islands, and the drained lakes in northern Holland, this was indeed the case. For the South Holland Islands the previous section has indeed demonstrated that large-scale commercial farms employing large numbers of wage labourers emerged there following the Second Reclamation. More daring is Van Bavel's assumption that this type of farming also emerged in the fenlands of mainland Holland. The basis for this idea is that a transformation of property relations occurred in central Holland during the sixteenth and early seventeenth centuries. As seen in chapter four, around the mid-sixteenth century, urbanites and urban institutions owned approximately one third of the agricultural land in central Holland. Van Bavel assumes that town dwellers continued buying up land from the farming population and that by 1620 they owned 50–55 per cent of that land.[19]

The development Bas van Bavel sketches for central Holland is plausible. The second half of the sixteenth century was a difficult period for farmers

---

[19] Van Bavel, 'Rural Development', pp. 182–4; B. J. P. van Bavel, P. J. van Cruyningen and E. Thoen, 'The Low Countries 1000–1750', in Van Bavel and Hoyle, eds., *Social Relations*, p. 175.

in this region. Costs of water management were high because the lowering surface level of the land due to peat shrinkage demanded costly investments in drainage mills and canals. During the first phase of the Revolt against the Habsburgs in the 1570s, enormous damage was inflicted on the countryside of central Holland. In 1574 all inhabitants of the districts of Rijnland, Delfland, Schieland and Alblasserwaard – together being most of central Holland – were ordered to leave the countryside and to take their cattle and other goods with them. Later that year, the first three districts were deliberately flooded for months to prevent the Habsburg army from conquering the city of Leiden. Two years later much of the area was still under water. It took the water authorities at least six years and huge amounts of money to repair the damage.[20] Much of this had to be paid for by the peasantry. Farmers had access to well-functioning capital markets, but according to Van Bavel they lacked sufficient collateral to take out loans and therefore could not compete on the land markets with the prosperous burghers of the booming cities.[21] However plausible Van Bavel's reasoning may be, it is very problematic because it is not based on any documentary evidence as far as the 1560–1620 period is concerned. He assumes that the pre-1560 development simply continued after that year. The validity of his assumptions has already been questioned by Petra van Dam.[22] There is much less evidence for the first half of the seventeenth than for the mid-sixteenth century, but the evidence that is available now does not support Van Bavel's hypothesis. In the peat areas in particular it is very difficult to discern urban engrossment at the cost of the rural population.

Milja van Tielhof studied five peat villages in Rijnland where the destructive *slagturven* – peat mining under the water table – was practised. What she found here was not the emergence of large-scale urban landownership, but the *morcellement* of already small-scale peasant landownership through partible inheritance. Extreme cases were Alkemade, where the number of landowners increased from 316 in 1543 to 807 in 1680, and Ter Aar, where there were 124 landowners in 1543 and no less than 399 in 1600. In these villages in particular, the owners of very small holdings of less than one *morgen* (0.85 hectares) increased strongly in number, whereas the number of larger landowners with ten *morgen* or more decreased. As stated, this was mostly caused by the splitting up of peasant holdings, but

---

[20] M. 't Hart, *The Dutch Wars of Independence: Warfare and Commerce in the Netherlands, 1570–1680* (London and New York, 2014), pp. 103–6; W. J. Diepeveen, *De vervening in Delfland en Schieland tot het einde der zestiende eeuw* (Leiden, 1950), pp. 156–7.

[21] Van Bavel, 'Rural Development', pp. 188–9.

[22] P. J. E. M. van Dam, 'Fuzzy Boundaries and Three-Legged Tables: A Comment on Ecological and Spatial Dynamics in Bas van Bavel's *Manors and Markets*', *TSEG/The Low Countries Journal of Social and Economic History* 8/2 (2011), pp. 103–13.

also by large landowners selling land. Monasteries especially sold land in these villages, supposedly because they did not want to destroy land by peat mining.[23] These five villages are certainly not representative of all of central Holland. They represent the part of the peat area where peat mining was at its most destructive, leading to the formation of lakes and in some cases to the disappearance of entire villages. It is not surprising that urban investors were not interested in buying land in such places. Only in the eighteenth century, when the lakes were drained again, did urban investors show interest in acquiring land in these villages.[24]

Of course, these tragic peat villages with their peasantry industriously sawing off the branch they were sitting on are not the kind of places where large-scale commercial farming can be expected to emerge. More promising is Delfland, with its mix of sandy, peat and clay-on-peat soils. Here, Carla de Wilt compared property relations in three large villages in the early 1560s with several years in the seventeenth century. The results are shown in table 6.

TABLE 6. Percentage of land owned by town dwellers and urban institutions in three villages in Delfland, 1561/2–1635, 1654 and 1663. Source: De Wilt, *Landlieden en hoogheemraden*, pp. 346, 354, 365.

Village	Soil type	% of land urban owned			
		1561/2	1635	1654	1663
Berkel	Peat	16	.	.	17
Maasland	Clay on peat	53	.	34	.
Monster	Sand	20	32	.	.

This data shows that in the Delfland area there was no clear development towards increasing urban landownership. That the percentage of land owned by urbanites hardly increased in the peat village of Berkel is not surprising, because urban investors were not very interested in low quality peat land. However, it is very surprising that urban landownership decreased strongly in Maasland with its good clay-on-peat soils. The only village where town dwellers increased their share of the land was Monster, but even there the increase was not spectacular. Of course, these are only three villages, but they are close to major cities and two of them have relatively good soils. This is the kind of area where one would expect urban landownership to increase.

---

[23] M. van Tielhof, 'Turfwinning en proletarisering in Rijnland 1530–1670', *Tijdschrift voor Sociale en Economische Geschiedenis* 2/4 (2005), pp. 113–16.
[24] Van Tielhof and Van Dam, *Waterstaat*, pp. 242–9.

Just over the border with Utrecht, the polder Bunschoten, a medieval peat reclamation, is situated, where oxen were fattened in the pastures. The share of urban landownership increased here from c. 30 per cent in 1501 to c. 40 per cent in 1548 and then remained stable until 1626.[25] For the Rijnland villages of Oegstgeest and Rijnsburg, situated on a mix of marine clay, sand and peat on the outskirts of the city of Leiden, detailed data is available for 1588, several years after the traumatic events of the 1570s.[26] In the 1540s most land here was held in leasehold from convents, chapters or charitable institutions, or from noblemen or noblewomen. This had not changed by 1588. Although convents and many other ecclesiastical institutions were abolished after the Reformation, their possessions were not sold but destined for new charitable uses. The possessions of the abbeys of Rijnsburg and Leeuwenhorst that owned much land in these two villages were now managed by the *Ridderschap*, the corporation of the Holland nobility, to provide support for younger daughters of noblemen. The income from estates of other convents was used to finance the new university at Leiden.[27] In Oegstgeest, the inhabitants of the two villages together cultivated 1367 *morgen* (1162 hectares) in 1588.[28] Of this land they owned 22.2 per cent; the rest was leased, most of it from former convents, chapters, churches and charitable institutions. Seven per cent was leased from institutions in Leiden and half of that from burghers of this city. One nobleman, the lord of Wassenaar, rented 79 *morgen* to villagers of Oegstgeest and Rijnsburg, more than all burghers of Leiden together. Here too, we do not observe a shift towards more urban landownership. This is confirmed by data from probate inventories of Leiden patricians. Of the estates of 49 patricians who died during the seventeenth century, on average less than one sixth was invested in land. They invested much more in urban real estate and bonds.[29] It appears the urban elite was not that interested in purchasing land. This is confirmed by recent research on landownership in three villages in the vicinity of Leiden, where the share of urban-owned land did not increase from the mid-sixteenth to the late-seventeenth centuries.[30] All in all, the available data does not provide much evidence for a strong increase in urban landownership between

---

25 Kole, *Polderen*, p. 79.
26 'Morgenboek van Rijnsburg 1588', transcribed by E. M. Olden Pieters; 'Register van de verboeking van landen onder Oegstgeest 1588', transcribed by A. van der Tuijn, retrieved from www.hogenda.nl, 12 July 2019.
27 Van Nierop, *Van ridders tot regenten*, pp. 111–13; Fockema Andreae, *Warmond, Valkenburg en Oegstgeest*, p. 38.
28 The record of Rijnsburg is not always clear about ownership.
29 D. J. Noordam, *Geringde buffels en heren van stand: Het patriciaat van Leiden, 1574–1700* (Hilversum, 1994), p. 91.
30 B. Hilkens, 'Land Inequality, the Land Market, and Urban Capital in Early Modern Holland: A Case Study, Hazerswoude 1555–1684', paper presented at the Feeding the Citizens Conference, Ghent, 12 April 2024.

1560 and 1620. In the village of Sloten, under the smoke of Amsterdam, the civil orphanage of that city bought much land, but this rather seems to be the exception to the rule.[31]

For once, the bourgeoisie is not seen to be rising. That does not mean, however, that Holland agriculture was not transformed. That a transformation had indeed taken place is demonstrated by a unique source, a cattle census for Delfland from 1672, which provides information on the size and composition of the herds of 754 cattle owners in this district. With its mix of peat, clay-on-peat and sandy soils, this area can be considered representative for central Holland. Table 7 presents some data from this source. Cattle farming had changed profoundly between the early sixteenth century and 1672. Around 1514, the largest farmers had 10–12 head of cattle; in 1672 they owned more than 40 head. In that last year, almost 400 farmers held herds of 21 to 76 head of cattle, numbers that would have been inconceivable 150 years earlier. In one of the Delfland villages, Wateringen, the average number of cattle was 3.3 in 1514; in 1672 that had increased to 19.8.[32] For Delfland overall the average was 21 in 1672.

TABLE 7. Herd size of cattle owners in Delfland, 1672. Source: Oud Archief Hoogheemraadschap Delfland nr. 1544, transcribed by A. van der Tuijn, retrieved from www.hogenda.nl, 9 August 2020.

Herd size	Number of owners	%
1–5 head	129	17
6–10 head	103	14
11–20 head	139	18
21–40 head	316	42
41–76 head	67	9
Total	754	100

In the second half of the seventeenth century, farming in central Holland was no longer characterised by smallholders who needed extra-agricultural sources of income to survive, but instead by substantial, specialised farmers. The composition of their herds shows that they were dairy farmers, because they held only 47 heads of young cattle per 100 cows, which was hardly sufficient to reproduce their own herds. This is confirmed by the fact that farmers from this region yearly had to purchase cows from Friesland and Groningen to replenish their herds.[33]

---

31 Zeischka, *Minerva in de polder*, pp. 136–7.
32 De Vries, *The Dutch Rural Economy*, p. 70.
33 B. Andreae, *Betriebsformen in der Landwirtschaft: Entstehung und Wandlung von Bodennutzungs-, Viehhaltungs- und Betriebssystemen sowie neue Methoden ihrer*

Delfland in 1672 offers strong evidence in favour of Jan de Vries' specialisation model. How did this transformation come about, and if the role of urban landlords was limited, who were the principal actors? The answer to this question was already provided by Jan de Vries: the rural population played an active and major part. Not all of the rural population, however. The records of Oegstgeest and Rijnsburg from 1588 show that from a total of 114 holdings, 11 were larger than 30 *morgen* (25.5 hectares). One of these substantial farmers leased his 36-*morgen* farm from the secularised abbey of Rijnsburg. The holdings of the others were composed of plots owned by themselves and land leased from between three and 13 burghers, noblemen, churches or charitable institutions. One of the largest farms in Oegstgeest was run by Lijsbeth Cortswagers, a woman who owned about 11.5 *morgen* herself and leased another 33 *morgen* from 13 different owners.[34] It appears a relatively wealthy layer of the rural population was able to accumulate owner-occupied and leased land to create large holdings. Therefore, this kind of accumulation of owned and leased land to form large holdings was not just limited to the new polders in the south-west, but also occurred on the old land. Despite the difficult circumstances in the last quarter of the sixteenth century, it appears that prosperous farmers were the driving force behind the transformation of agriculture in central Holland. Very probably they profited from the fact that many of their poorer colleagues were forced to give up their land.

Creating such large holdings with numerous herds of dairy cows, however, would have been impossible without proper drainage of the land. The investments that had been made in the erection of drainage mills and the creation of mill polders from the sixteenth century onwards now paid off. Although supported by credit from town dwellers, most of these investments were made by farmers. Again, it appears to be the rural population itself – or at least prosperous groups within that population – that was able to bring about change.

It is sometimes assumed that the rise of relatively large dairy farms in Holland led to a larger group of landless, proletarian labourers.[35] There is no evidence for that. Although more labour intensive than cattle breeding, dairy farming requires much less labour than arable farming. Evidence from the Delfland village of Maasland shows that the vast majority of dairy farmers could manage with a few live-in servants and that farm labourers

---

Abgrenzung. Systematischer Teil einer Agrarbetriebslehre (Stuttgart, 1964), pp. 138–9; Bieleman, 'De ossen', p. 88.

34 'Register van de verboeking van landen onder Oegstgeest 1588', transcribed by A. van der Tuijn, retrieved from www.hogenda.nl, 12 July 2019.

35 F. Van Roosbroeck and A. Sundberg, 'Culling the Herds? Regional Divergences in Rinderpest Mortality in Flanders and South Holland, 1769–1785', *TSEG/ The Low Countries Journal of Social and Economic History* 14/4 (2017), p. 48.

were very scarce in this village with large dairy farms. Only for haymaking were some seasonal labourers hired, usually from Westphalia.[36] In central Holland, the emergence of a class of large, specialised farmers did not lead to proletarianisation, but to an increase of the number of workers in the non-agrarian sector, such as merchants, craftsmen, transporters, and so on.[37] There was room for such non-agrarian occupations because farmers concentrated on purely agricultural activities and outsourced other tasks such as the maintenance of buildings and implements to specialists such as carpenters and cartwrights.

Guelders and Overijssel

In the Guelders river area the development towards large-scale commercial farming had already taken place in the sixteenth century and it continued over the following centuries. In the Over-Betuwe district, in 1650 large farmers with 40 *morgen* (34 hectares) or more comprised only 14 per cent of the population and yet used 56 per cent of all agricultural land. More to the west, in the Tielerwaard region, the largest farms held 100 *morgen* or more.[38] Farms were usually mixed over here, combining arable farming with fruit growing, stock raising and horse breeding. The orientation of the farm could be changed, depending on the relative prices of the crops and animals that were produced. In the late seventeenth century, horse breeding for the army became very important in the Tielerwaard due to the almost continuous warfare the Dutch Republic was involved in. This pursuit could reach a very large scale. In 1694, for instance, ten farmers from Tielerwaard formed a consortium that agreed to deliver 578 artillery horses to the army, a transaction amounting to the sum of 85,000 guilders. This shows that these were very commercially oriented farmers who ran capital-intensive enterprises. It is thus not surprising that farmers in this area already kept account books in the seventeenth century.[39]

The farms of most of the Guelders river area were at least as large and commercial as those of the Flemish coastal plain and Zeeland. Crop yields, however, were much lower than those in Zeeland. Wheat yields were less than 10 hl/ha, barley yielded 18–20 hl/ha, less than on the coastal plain of Flanders and less than half of the yields in Zeeland (see p. 170).[40] Farmers in Zeeland were able to raise crop yields through intensive weeding and

---

36 D. J. Noordam, *Leven in Maasland: Een hoogontwikkelde plattelandssamenleving in de achttiende en het begin van de negentiende eeuw* (Hilversum, 1986), pp. 46–7, 55.
37 De Vries, 'The Transition', p. 78.
38 Brusse, *Overleven door ondernemen*, pp. 49–50; E. de Bruijn, *De hoeve en het hart: Een boerenfamilie in de Gouden Eeuw* (Amsterdam, 2019), pp. 176–7.
39 De Bruijn, *De hoeve*, pp. 182–6, 227–30.
40 Brusse, *Overleven door ondernemen*, p. 242.

ploughing. The resulting costs were more than compensated for by the high yields of the fertile marine clay soils. In the river area, however, where farmers struggled with seepage of river water making the land wet and abundant with weeds, intensive cultivation as in Zeeland was hardly feasible. It would require enormous wage costs for the large farmers which would not be repaid by much bigger harvests. Therefore, the large farmers of the river area preferred more labour-extensive farming. Yields were low, but then so were costs. Zeeland farmers had high yields, but also high costs. The quality of the soil and the hydrological circumstances determined this difference between the choices made by the large farmers in these two regions.[41]

Smallholders in the Guelders river area made different choices. In the Bommelerwaard district, small and family holdings of less than 15 *morgen* (c. 13 hectares) were predominant, while farms of more than 40 *morgen* were completely absent. The main source of income of these small farmers was the labour-intensive cultivation of hops. In one village two thirds of the arable land was under hops. Hundreds of small farmers lived off hop-growing.[42] Therefore, apart from the soil, farm size was also an important variable determining farmers' choices. Smallholders, who could use cheap family labour, could afford to grow labour-intensive crops. Large farmers could only do so if the soil was fertile enough to recompense for high labour costs. One question remains: why were farms in the Bommelerwaard district that much smaller than those in other parts of the river area? It appears the smallholders there were mostly owner-occupiers, whereas the large farmers in the Over-Betuwe were predominantly tenants who leased their farms in large units from mostly urban landlords.[43] Thus, in the Guelders river area large farms were mostly created by landlords, whereas in Holland, the Flemish coastal plain and Zeeland the farmers themselves formed large holdings by accumulating owner-occupied and lease land.

The problem of seepage confronting farmers in the river area began with riverbeds silting up due to the rivers having been enclosed between continuous dikes. Meanwhile, the land within those dikes compacted and the level of this land became lower than the riverbed, allowing riverwater to seep under the dikes to the agricultural land. This was the inevitable consequence of the construction of dike rings. These dike rings created safety, but eventually also problems for agriculture. It is good to remember that there were alternatives and that these were still also practised in the seventeenth century and later.

[41] Van Cruyningen, 'From Disaster', p. 256.
[42] P. Brusse, 'Property, Power and Participation in Local Administration in the Dutch Delta in the Early Modern Period', *Continuity and Change* 33 (2018), pp. 69–70.
[43] *Ibid.*, p. 82.

Agriculture after the Second Reclamation, c. 1550–1700

The islands in the IJssel delta, the Kampereiland, were only protected from the sea and the river by low summer dikes. Kampereiland proved that farming under such circumstances could be profitable. The number of farms situated on artificial mounds doubled from 26 in 1627 to 52 in 1682. Data from 1572 shows that on the higher parts of the area spring-sown crops such as barley and oats were grown. Most of Kampereiland was dedicated to dairy farming, however, and large amounts of high-quality hay were exported. The winter floods deposited sediment which increased the fertility of the land. The downside of this was that early floods could destroy any grass that had remained as grazing after the first cut.[44] Such a drawback could be averted by controlled flooding of the land. This occurred on the Ooijpolder near Nijmegen, where each autumn sluices were opened to let the polder flood. In this way, the river could deposit fertile sediment without causing damage. On this polder, the farmers, who lived on artificial mounds, could dedicate themselves to lucrative fattening of oxen to provide beef to urban consumers.[45] The positive effects of sediment deposition were widely known and sometimes attempts were made to introduce it elsewhere. In 1626, for instance, a plan was launched to let the southern part of the Gelderse Vallei near Wageningen flood during winter to fertilise the land. The plan was never implemented though due to resistance from landowners in other parts of the valley who were afraid their land would suffer from the flooding.[46] It appears there were alternatives to enclosing the land within dikes, but once the choice was made to construct dike rings it was difficult to return to the former situation because the land had become more densely populated, valuable buildings had been erected on it and farming had become more intensive, making the choice to let the land flood in winter costly.

The Wadden Coast

In the clay areas of the Frisian lands, farming intensified during the sixteenth and seventeenth centuries. In the province of Friesland, as in Holland, improved drainage contributed to a rise in yields. In the *grietenij* (rural district) of Idaarderadeel, straddling the border between the clay area and the central fenlands, and discussed extensively by Jan de Vries, farmers dug new drainage ditches, erected drainage mills, and increased the area of agricultural land from 3800 hectares in 1511 to 5500 hectares in 1700. Investors from Leeuwarden and Amsterdam also contributed to

---

[44] J. Dirkx, P. Hommel and J. A. J. Vervloet, *Kampereiland: Een wereld op de grens van zout en zoet* (Utrecht, 1996), pp. 36, 41, 44.
[45] J. van Eck, *Historische atlas van Ooijpolder & Duffelt: Een rivierengebied in woord en beeld* (Amsterdam, 2005), pp. 15, 20.
[46] Stol, *De veenkolonie Veenendaal*, pp. 203–5.

this by draining a lake. As in Holland, a class of large specialised dairy farmers emerged. The number of large farms with more than 100 *pondematen* (c. 37 hectares) of land increased almost fourfold from seven in 1511 to 27 in 1700. In the 1690s average herd size was 26.5 head of cattle, even more than in Delfland. These large farmers owned only seven per cent of the land; the remainder was leased from the nobility and from urbanites.[47] As in other lowland areas a lively land market contributed to the engrossment of farms. Farmers leased land from different owners.[48] Those smallholders who quit farming often engaged in non-agricultural activities such as transport, retailing and crafts, but there was also a considerable group of landless labourers.[49]

De Vries attributes the changes in the Frisian countryside to specialisation and investment. His conclusions are confirmed by more recent research by Merijn Knibbe, although Knibbe places more emphasis on the importance of investment and technological change than on specialisation. Non-agrarian specialists were already present in the Frisian countryside at the beginning of the sixteenth century according to Knibbe.[50] Productivity of the land was raised by manuring it more intensively. In the first half of the sixteenth century in Friesland, cattle dung was often dried and cut into sods that could be used as fuel. During that century farmers started buying turf for fuel, using the dung to manure their pastures, together with the ashes of the turf that was now burned in the hearths.[51] This investment was worthwhile due to the strongly improved drainage resulting from the introduction of drainage mills. Therefore, improvement of drainage by digging ditches and erecting drainage mills was crucial.[52] This may also have contributed to the increasing importance of large farms, because even small drainage mills required a considerable investment in construction and maintenance. The most important contribution to the increase of productivity, according to Knibbe, was made by the more intensive cultivation of marginal land, which also required improvement of water management.[53]

Most marginal agricultural land in Friesland was situated in the central peat lands. At the end of the Middle Ages, this area had been mostly abandoned due to the shrinkage and wastage of the soil. Only part of the land was used as meadow by farmers living on the higher sandy soils to the east. From the mid-sixteenth century onwards, improved drainage made it possible to settle here again and create viable farms (see also pp.

---

[47] De Vries, *The Dutch Rural Economy*, pp. 121–4.
[48] Knibbe, *Lokkich Fryslân*, p. 61.
[49] De Vries, *The Dutch Rural Economy*, pp. 126–7.
[50] Knibbe, *Lokkich Fryslân*, p. 202.
[51] O. Postma, *De Fryske boerkerij en it boerelibben yn de 16e en 17e iuw* (Snits, 1937), p. 27; Knibbe, *Lokkich Fryslân*, p. 189.
[52] Knibbe, *Lokkich Fryslân*, pp. 177, 185, 192.
[53] *Ibid.*, p. 249.

Agriculture after the Second Reclamation, c. 1550–1700

159–60). As on the clay soils, hundreds of small drainage mills owned by individual farmers or landowners pumped excess water to the lakes and canals of the 'Frisian bosom'. The mills only drained a small plot of land around the farm buildings, where some cows could graze. Farms here were islands amidst an extremely watery landscape of lakes, rivers, canals and large areas of undrained land that could be used for haymaking providing there were not too many floods in spring. By concentrating on relatively extensive cattle breeding, farmers could make a living in this area.[54] The impact of farming on the landscape was limited because only a small part of the land was drained. Farming was well adapted to the environment in the central fenlands of Friesland, but was dependent on the functioning of the Frisian reservoir system. When canals and sluices were not maintained properly, the low peat lands suffered most. In the seventeenth century, any problems due to insufficient means to drain excess water appear to have been limited, however.

In all areas discussed above, from the Flemish coastal plain up to and including Friesland, holding size was increasing during the late sixteenth and seventeenth centuries, and commercial and specialised farms became predominant. This cannot be concluded for the province of Groningen, immediately to the east of Friesland. In the nineteenth century, Groningen became renowned for its huge arable farms, but at the beginning of the eighteenth century there still were many smallholdings. In the district of Fivelingo in 1721 more than half of the farms were smaller than 20 *grazen* (c. 8 hectares) and less than three per cent were larger than 40 hectares. Holding size was increasing here, however, and already by 1630 two thirds of the land was held by farmers cultivating 25 hectares or more. The number of larger holdings increased, at the cost of the middling holdings of c. 8–24 hectares. In the north-east of the province, by 1691, large farms were not unusual, and subsistence farming was exceptional here.[55]

In the district of Oldambt holdings were mostly small by 1721. The share of farms smaller than six hectares varied from 50 to 75 per cent of all holdings in most villages of this district. In only one village was their share just 15 per cent. That village was also the only one where there was a considerable number of larger holdings with more than 20 hectares, a quarter of the total.[56] The territory of that village, Midwolda, was

---

54 Van Cruyningen, 'Water Management', pp. 75–7.
55 De Vries, *The Dutch Rural Economy*, p. 135; E. Karel and R. Paping, 'In the Shadow of the Nobility: Local Farmer Elites in the Northern Netherlands from the 17th to the 19th Century', in Freist and Schmekel, eds., *Hinter dem Horizont*, pp. 44–5; D. R. Curtis, 'Danger and Displacement in the Dollard: The 1509 Flooding of the Dollard Sea (Groningen) and its Impact on Long-Term Inequality in the Distribution of Property', *Environment and History* 22 (2016), pp. 130–1.
56 P. C. M. Hoppenbrouwers, 'Grondgebruik en agrarische bedrijfsstructuur in het Oldambt na de vroegste inpolderingen (1630–1720)', in J. N. H. Elerie

composed of fertile young marine clay, embanked during the seventeenth century. It appears that like in other regions, the emergence of large holdings began in the recently reclaimed areas. Farming in Oldambt was mixed, with farmers both owning cattle and tilling arable land.[57]

In the German part of the coastal zone of the Wadden Sea there also was a tendency towards increasing holding size. As mentioned in chapter four, this was already in progress around the mid-sixteenth century in Ostfriesland, driven by the farming population. This development continued in the seventeenth century in areas such as Ostfriesland, Dithmarschen and Nordfriesland. In some areas the rural population was busy engrossing their farms, but in Nordfriesland many large farms were created by the capitalists who invested in the reclamation of new polders.[58] One of the reasons why farms were being engrossed was the increase of arable farming, especially on the younger polders.

On older, less well drained polders animal husbandry remained predominant. It was even extended, by Dutch settlers in particular, who introduced drainage mills and improved dairy production in regions such as Eiderstedt and the Holsteinische Elbmarschen.[59] In the sixteenth century, butter, cheese and cattle were still the most important export products of the German coastal area, but the cultivation of barley and oats increased in importance. For large-scale arable farming, consolidated plots were required. Although arable farming became more important, all farmers continued to breed oxen and horses.[60] The farmers of the Schleswig-Holstein coast in particular participated in the large-scale Danish exports of oxen to the Low Countries, which reached its zenith around 1610. Farmers from Eiderstedt, for instance, annually exported thousands of oxen, and also large amounts of cheese. In 1610 three million pounds of cheese destined for export were weighed at the port of Tönning.[61]

An exception to the rule of increasing holding size was the Alte Land near Hamburg. In 1657, of 422 holdings in six villages in the Alte

---

and P. C. M. Hoppenbrouwers, eds., *Het Oldambt, deel 2: Nieuwe visies op geschiedenis en actuele problemen* (Groningen, 1991), pp. 77–80.

[57] Ibid., pp. 87-90.

[58] Swart, *Zur friesischen Agrargeschichte*, pp. 229–39; C. P. Rasmussen, 'Yeoman Capitalism and Smallholder Liberalism: Property Rights and Social Realities of Early Modern Schleswig Marshland Societies', in L. Egberts and M. Schroor, eds., *Waddenland Outstanding: History, Landscape and Cultural Heritage of the Wadden Sea Region* (Amsterdam, 2018), pp. 230–1.

[59] Lorenzen-Schmidt, 'Reiche Bauern', p. 33.

[60] Swart, *Zur friesischen Agrargeschichte*, pp. 211–12; Knottnerus, 'Yeomen and Farmers', pp. 153, 156–8.

[61] W. Gijsbers, *Kapitale ossen: De internationale handel in slachtvee in Noordwest-Europa (1330–1750)* (Hilversum, 1999), pp. 47, 97; Müller and Fischer, *Das Wasserwesen*, vol. 3/3, p. 156.

Land, only 14 were larger than 24 hectares; no less than 323 holdings were smaller than eight hectares and 185 even smaller than one hectare. Apart from income from non-agricultural activities and the growing and processing of flax, very probably an important cause of this small farm size was the rise of fruit cultivation. On the smallest holdings, of less than one hectare, about a quarter of the land was planted with orchards. Fruit cultivation provided a higher income per hectare than arable or animal husbandry and thus could make smaller holdings more viable. A farmer cultivating flax and fruit earned about one half more than a farmer who did not do so. The fruit was mostly sold in Hamburg, but also exported to Bremen, Jever and Bergen in Norway. Data from around 1690 confirms the relatively small size of the holdings. The Alte Land remained a territory characterised by smallholdings and family farms.[62]

In all regions discussed thus far, agriculture was performing well until at least the mid-seventeenth century, with both production and holding size expanding, mostly due to a dynamic farming population. An exception to this were the districts of Butjadingen and Stadland. These former free republics had been conquered by the count of Oldenburg in 1514. The count had enlarged the land area of the districts by reclaiming marshes. As a result, he owned about two fifths of the land. Part of this was rented to tenant farmers and part belonged to large estates exploited by the count himself, where cattle and horses were bred. Tenants had to pay rents and perform services for the count, and during winter all farmers, whether tenant or owner-occupier, had to feed cattle belonging to the count, without compensation.[63] This by itself might not have had particularly negative consequences. However, in combination with the fact, as mentioned in chapter five, that many privileged landowners were exempt from dike maintenance, and the burden of maintenance had to be borne by the farming population, the effects on the rural population could be disastrous. The recurrent floods in particular could and did lead to indebtedness of farmers, for instance after the series of floods in 1625, 1626 and 1628. Since only the villages immediately behind the dikes were liable for maintenance and those more inland were not, the coastal villages suffered most.[64]

Livestock farming remained predominant in most of the Wadden area in the seventeenth century. In the fenlands and on the heavy clays

[62] P. Brümmel, *Die Dienste und Abgaben bäuerlicher Betriebe im ehemaligen Herzogtum Bremen-Verden während des 18. Jahrhunderts* (Göttingen, 1975), pp. 67, 109, 114, 126.
[63] Swart, *Zur friesischen Agrargeschichte*, pp. 279–81; Norden, *Eine Bevölkerung*, pp. 226–31; J. Bölts, 'Die Rindviehhaltung im oldenburgisch-ostfriesischen Raum vom Ausgang des 16. Bis zum Beginn des 19. Jahrhunderts', in H. Wiese and J. Bölts, *Rinderhandel and Rinderhaltung im nordwesteuropäischen Küstengebiet vom 15. Bis zum 19. Jahrhundert* (Stuttgart, 1966), pp. 171–2, 182–3.
[64] Norden, *Eine Bevölkerung*, pp. 222, 255–6.

adjoining them, insufficient drainage made other forms of agriculture impossible. On the new polders with their well-draining young marine clay soils, however, arable farming was feasible. These were the 'young' marshes abutting the Wadden Sea, and there indeed mixed farming took root, especially after prices of arable products began to increase strongly after c. 1570.[65] Cereal production became very important in the northern part of the province of Friesland. Farmer Dirck Jansz, for instance, who exploited a 60-hectare farm on the polder Het Bildt (reclaimed in 1505), had in the early seventeenth century 60 per cent of his land sown with crops and the remaining 40 per cent in fallow and under grass. According to his account book, wheat yields varied from 14.4 hectolitres per hectare in 1617 to 31.8 hl/ha in 1604. The former was considered by him to be very low.[66] This shift to mixed farming remained mostly limited to a narrow band of recently reclaimed polders along the Wadden coast. Further inland arable farming remained relatively insignificant. That is not surprising, considering the frequent low crop yields. On the fields of a monastery in the Alte Land near Hamburg, for instance, average yield ratios over the period 1596/7 to 1617/18 were 1:4 for rye, 1:7 for barley and only 1:2 for oats.[67] Therefore mixed farming only became predominant where soil quality and drainage permitted it.

On the *Halligen*, the islands in Nordfriesland without dikes, where people lived on artificial dwelling mounds, extensive cattle and sheep husbandry remained predominant. The pasture, which was flooded regularly, was held in common per *Warft* (dwelling mound), and each household had the right to graze a number of animals in proportion to its share in the common. The land was re-distributed among the inhabitants on a yearly basis. Due to regular flooding, manuring was not possible. Instead, dung was gathered, dried and then cut into sods to be used as fuel. On Ketelswarf on the Hallig Nordmarsch-Langeneß, six farms held on average five cows and 20 sheep. Such small and extensive farming operations could not provide sufficient income for a family. The people of the *Halligen* could survive only because most men worked as sailors, many of them on Dutch whalers. While they were away at sea, the women took care of the farm.[68]

In the coastal zone from Flanders to Friesland the occupiers of the land were either short-term tenants or owner-occupiers. In Groningen and the territories further east, forms of hereditary lease were predominant. I will

---

[65] Swart, *Zur friesischen Agrargeschichte*, p. 211; Knibbe, *Lokkich Fryslân*, p. 206.
[66] J. A. Faber, 'Bildtboer met ploeg en pen', in P. Gerbenzon, ed., *Het aantekeningenboek van Dirck Jansz (1604–1636)* (Hilversum, 1993), pp. 23–5.
[67] Hauschildt, *Zur Geschichte*, p. 168, n. 26.
[68] Postma, *De Friesche kleihoeve*, pp. 136–43; Meier, Kühn and Borger, *Der Küstenatlas*, p. 168; Meier, *Die Halligen*, pp. 25–9.

focus here on Groningen and adjacent Ostfriesland. In these territories the tenant owned the buildings of the farm; the landlord only owned the land. This land was attached to the farm and could not be split off from the holding. As discussed in chapter four, in Ostfriesland this often concerned land belonging to different owners. This legal arrangement, known as *Beherdischheit* in Ostfriesland and as *beklemming* in Groningen, naturally strengthened the position of the tenant. In Ostfriesland, *Beherdischheit* was recognised as a hereditary lease by the Estates and the count in 1611. During the seventeenth and eighteenth centuries, the position of the tenants was further strengthened, especially as the rents tended to become fixed and therefore to decrease in real terms.[69]

The strengthened position of the tenants vis-à-vis their landlords also had a drawback. Faced with hereditary leases and fixed rents, landlords had no incentives to invest in dike maintenance. It was left to the tenants to carry that burden. The owners did not engage in the maintenance or improvement of dikes.[70] In combination with the sale of offices in water authorities to people who considered these offices as pure investments, as noted in the previous chapter, this could lead to suboptimal maintenance. The flood disasters of 1686 and 1717, which created thousands of victims in these areas, can be partly attributed to this.

The English Marshes and Fens

Of the English marsh areas, Romney Marsh is the most well-documented for the period under discussion in this chapter, owing to the work of Stephen Hipkin. Sheep farming was by far the most important branch of agriculture on Romney Marsh. It was booming in the first quarter of the seventeenth century due to high demand for wool from the cloth industry. When this collapsed in the second quarter of the century, most of the gentry families that had established large holdings in the area withdrew from direct farming and tenant farmers became predominant. In this period of depression, it was the largest farmers with holdings of over 200 acres (81 hectares) who performed best. They doubled their share in the land on the level from one fifth at the beginning of the century to two fifths around 1700. The number of middling holdings with 20 to 100 acres (c. 8–40 hectares) decreased.[71]

The largest tenant farmers were best equipped to deal with the depressed circumstances for sheep farming in most of the seventeenth century. They were able to increase output of wool to compensate for the lower prices. They could also shift towards production of meat for

---

[69] Swart, *Zur friesischen Agrargeschichte*, pp. 270, 284–9; Knottnerus, 'Yeomen and Farmers', pp. 164–6.
[70] Soens, 'Resilient Societies', pp. 171–2.
[71] Hipkin, 'Tenant Farming', pp. 661–2, 666, 669.

which prices appear to have been better than for wool. This required substantial amounts of capital to buy and fatten stock, and larger tenants had more of that than middling tenants. Moreover, scots – the local taxes for maintenance of dikes and drains – were also raised during the seventeenth century, for instance on dike repairs in the 1640s and 1650s. Since on Romney Marsh the scots had to be paid by the occupiers rather than the owners of the land, the farmers had to shoulder this burden, and again the larger tenants were better able to do so than middling farmers.[72] All the large tenant farms on Romney Marsh were composed of several plots leased from different landlords. According to Hipkin the initiative lay predominantly with the tenants, who could profit from the active land market on the level. The increase of large tenant farms occurred independently from the consolidation of landownership in the region.[73]

The marshes in the Thames estuary continued to be used for extensive grazing by sheep, with one exception, Canvey Island, which had been protected by a seawall financed with Dutch capital in 1622. The Dutch investors had been granted one third of the land on the island to recompense for their expenditure. This land was leased out to tenants from the Netherlands who practised arable agriculture. In 1628, the Dutch community of some 200 persons were granted the right to hold services in their own language in a chapel on the island. Conflicts over water management arose between the Dutch settlers and the English population who owned the remaining two thirds of the land and continued to practise sheep farming. The Dutch drained the land thoroughly to make it dry enough to grow crops, but that meant that there was not enough water in the ditches for the livestock of the English farmers.[74]

The rural economy of the Lincolnshire marshes appears to have stagnated during much of the seventeenth century. That was partially caused by coastal erosion, which hit some townships heavily, although others still managed to reclaim new land. The land that was lost on one part of the coast often silted up again on another part. Developments on the Lincolnshire marshes show clear similarities with those on Romney Marsh. Farming mostly concentrated on the grazing of sheep and cattle, and the larger yeoman farmers and members of the gentry were engrossing farms at the cost of the middling farms. The same economic pressures were probably at play here.[75]

Whereas on several marsh areas on the opposite shore of the North Sea, in the south-west of the Netherlands in particular, a shift took place from livestock to mixed farming, in England the economy of the marshes remained firmly pastoral. The only place we know of where arable

[72] *Ibid.*, pp. 669–70.
[73] *Ibid.*, pp. 654–5; Hipkin, 'The Structure of Landownership in the Romney Marsh Region, 1646–1834', *Agricultural History Review* 51 (2003), p. 71.
[74] Cracknell, 'Canvey Island', pp. 20–5.
[75] Thirsk, *English Peasant Farming*, pp. 144–56.

Agriculture after the Second Reclamation, c. 1550–1700

farming was introduced was Canvey Island, and there it was introduced by Dutch landowners and farmers. In England, much of the coastal marshes were owned or leased by graziers from the uplands.[76] Even in the period of high grain prices in the first quarter of the seventeenth century they did not consider changing marshland into arable, although much of this soil would have been suitable for growing crops. For them, to shift to arable husbandry would have meant a shift to much more labour-intensive farming, requiring more personnel and more monitoring of the workers. It must have seemed preferable to stick to livestock farming and increase its profitability by upscaling the operation instead of investing in time and energy to run arable farms. Therefore, although soil type and hydrology were important factors in determining the social agrosystem of a region, so were property relations.

The most interesting question regarding farming in the English lowlands in the seventeenth century is, of course, what happened in the fens where large drainage projects had been implemented? Did this result in a change in the farming system in these areas? For several fen areas, such as the East and West Fens and the Lindsey level in Lincolnshire, the answer is simple. During the years of political turmoil in the 1640s the inhabitants regained their commons, and traditional pastoral farming was restored. Where the drainage schemes were more or less maintained, such as in the Great Level and Hatfield Chase, some change did occur. The adventurers tended to put their land under the plough, and oats and rapeseed became popular and well-yielding crops. Oats also became popular among the inhabitants. By 1690 one in two farmers were growing this crop.[77]

Hatfield Chase appears to be one of the more successful projects, even after allowing for the conflicts with the commoners of the Isle of Axholme. Already in 1632 considerable amounts of rapeseed were being exported to Rotterdam, and in 1661 it was still being grown in Hatfield Chase, with four windmills in the level crushing rapeseed to produce oil. Introducing arable farming was not easy, however. In 1631 one of the participants was complaining that of his 600 acres (243 hectares) only 30 (12 hectares) were under the plough.[78] He was not the only participant who had reason for complaint. By 1635 only 105 hectares of newly drained land were under crops, 1064 hectares were used as pasture, 304 hectares were meadow, and 405 hectares were laying waste. According to Daniel Byford, agriculture in Hatfield Chase was already mixed before Vermuyden's drainage scheme, with arable farming on higher ground and grazing in the wetlands. Drainage appears not to

---

[76] Hipkin, 'Tenant Farming', pp. 662–6; Thirsk, *English Peasant Farming*, pp. 148–50.
[77] Thirsk, *English Peasant Farming*, pp. 125–8.
[78] Van Cruyningen, 'Dutch Investors', p. 34; W. Dugdale, *The History of Imbanking and Draining of Divers Fens and Marshes in Foreign Parts and this Kingdom and of the Improvements thereby* (2nd edn., London, 1772), p. 145.

have wrought much change in this mixed farming system. Flooding also continued to remain a recurrent problem.[79] Luckily, Hatfield Chase was an area that was mostly composed of silt fens that were less liable to soil compaction and wastage than peat fens such as those of the Great Level. Notwithstanding all problems, some tenants managed to do well. Abraham Venny, a descendant of Walloon settlers, owned 32 hectares of land and leased another 133 hectares in 1717. When he died four years later, he left investments worth £1629.[80] The improvement of productivity the drainers had hoped for, however, would only be realised after 1750. Outside of Hatfield Chase the rural economy of the Lincolnshire fenlands remained firmly pastoral, depending on the production of cattle, sheep and horses, and in the wetter areas also on fishing and fowling.[81]

On the Great or Bedford Level the adventurers also put their share in the land under the plough after the completion of drainage in 1653. Initially, this appeared to be a brilliant success. From 1655, crops such as rapeseed, onions, peas, hemp, flax and wheat were sown in the fens and bumper harvests were recorded. Although major parts of the fens remained very wet, arable agriculture seemed to be introduced here successfully and animal husbandry was also performing well. The Thorney estate of the earl of Bedford not only produced more rapeseed, oats and hemp, but cattle breeding also improved. The number of sheep that were grazed in the fens increased as well.[82] In 1670 the fenland village of March wanted to set up a market to trade in the increased amounts of crops, cattle and dairy that were being produced in the area.[83] Farming in the drained fens appeared to be booming. By 1700, however, 'disaster was abroad everywhere', according to H. C. Darby.[84] Vast stretches of the fens were flooded time and again and became all but useless for agriculture. Two problems were at the root of this: the outfalls of the fenland rivers were silting up and, as to be expected, the surface level of the land became lower, thus impeding the evacuation of excess water. The latter was of course the consequence of the draining of the peat soils, which caused shrinkage and wastage. What had happened in Holland in the Middle Ages now occurred in England, and as in Holland the only solution was the introduction of devices to lift water: drainage mills.[85] In the late

---

[79] D. Byford, 'Agricultural Change in the Lowlands of South Yorkshire with Special Reference to the Manor of Hatfield 1600–c. 1875' (unpublished PhD dissertation, University of Sheffield, 2005), pp. 18, 42–4, 103; Robson, 'Improvement', p. 189.
[80] Byford, 'Agricultural Change', p. 129.
[81] Thirsk, *English Peasant Farming*, p. 129.
[82] Darby, *The Draining*, pp. 86–91; Summers, *The Great Level*, p. 93.
[83] Bowring, 'Between the Corporation', p. 254.
[84] Darby, *The Draining*, p. 113.
[85] Ibid., pp. 94–115.

Agriculture after the Second Reclamation, c. 1550–1700

seventeenth century the effect of windmill drainage appears to have been limited, however.

It is not surprising that the fenland economy remained pastoral. The economy of the fen-edge villages studied by J. R. Ravensdale was based on cattle and sheep in the years 1660–1710, and cheese production was very important here.[86] In Willingham less sheep were held but cattle were very important and here too cheese was produced.[87] People in the villages on the edge of the Great Level continued to use the resources of the fens in the same way they had been used before drainage. This can also explain the fact that contrary to what we have seen in just about all regions discussed above, farm size in Willingham decreased. The number of holdings in Willingham almost doubled between 1603 and the 1720s, from 58 to 105. This was probably caused by fathers who wanted to establish their younger sons on a plot of land, and the resources of the fens made it possible to survive on a small holding.[88] Willingham is just one village in the vast English fenlands, but it seems that elsewhere in the fens there was no shift towards large-scale farming either. In the Lincolnshire fens farm size hardly changed during the seventeenth century. Some gentry landlords created large holdings, but small and middling farms remained predominant.[89]

Therefore, the drainage projects of the seventeenth century failed to transform agriculture in the fens, but neither did they change much in fenland society. This remained an area dominated by small and middling farmers. On the Great Level, even the land that had been granted to the adventurers did not produce much social change. That may have had to do with the fact that the 83,000 acres (c. 33,600 hectares) that had been granted to them were soon split up among an increasing number of owners. There were already 469 owners of adventurers' lands in 1697; by 1734 that had increased to 501.[90] With the exception of the earl (later duke) of Bedford's 18,000-acre (c. 7300 hectares) Thorney estate there were hardly any large estates on the Level and therefore also few landlords who might have wished to create large holdings. Above we have seen, however, that in many cases elsewhere farmers themselves created large holdings by accumulating tenancies, sometimes combined with owner-occupied land. However, this did not occur in the fens. It appears the need was not felt to do so. A judicious use of the multiple resources of the fens made a sober existence possible for small and middling farmers and that was deemed sufficient. Moreover, so long as drainage was not efficiently performed and frequent floods continued to hit the fens, even

---

[86] Ravensdale, *Liable to Floods*, pp. 60–1.
[87] Spufford, *Contrasting Communities*, pp. 129, 131–2.
[88] Ibid., pp. 156–8, 160.
[89] Thirsk, *English Peasant Farming*, p. 134.
[90] Summers, *The Great Level*, p. 188.

enterprising farmers may not have been interested in investing in the formation of large commercial farms.

Conclusion

With the notable exceptions of the English fenlands and the Dutch province of Groningen a transformation occurred in the regions of the North Sea Lowlands discussed in this chapter. Holding size increased and farmers became specialised entrepreneurs, employing wage labourers to produce cereals, industrial crops, dairy products and meat for urban markets. This has often been described as the rise of 'agrarian capitalism'.[91]

It is difficult to identify society in the lowlands with the three-tiered society of classical agrarian capitalism. Indeed, there were landlords and tenant farmers, but many farmers were also owner-occupiers or leased land out to other farmers. More problematic is the fact that there were not that many wage labourers, also because the farmers in the Lowlands could employ seasonal labourers from the uplands. As became abundantly clear in this chapter, in most of the lowlands livestock farming was predominant, and that type of farming is relatively labour extensive. Even in central Holland, where dairy farming was predominant, and more labour intensive than stock breeding or fattening, the number of agricultural labourers was small. Only in the areas where arable farming became prevalent, such as the south-west of the Netherlands and parts of the Wadden coast, were there large numbers of labourers. Many of these labourers were not proletarians, however, because they had their own smallholdings and kept some cows.

In one respect, however, society in the North Sea lowlands fits well with Brenner's definition: that farming takes place on the basis of capital and improvements.[92] There are few regions imaginable where improvements – in the form of water management infrastructure and land reclamation – were as important as in these lowlands, and where as much capital was required to implement those improvements. Rural society itself was not able to provide all the capital required to finance the investments in infrastructure. As demonstrated in the previous chapter, urban investors had to step in. They financed reclamation of hundreds of thousands of hectares of land and provided credit for the improvement of fenlands through the construction of drainage mills. Institutions such as government patents for reclamation projects and drainage companies made investing in land attractive for them. Where the urbanites' interventions failed, such as in much of the English fenlands, society was not transformed.

Ultimately, however, the rural population was responsible for the transformation of economy and society of the lowlands. In Romney

---

[91] For instance by Knottnerus, 'Yeomen and Farmers', pp. 149–50.
[92] Brenner, 'Agrarian Class Structure', p. 25.

Marsh, the Flemish coastal plain, Zeeland, Holland, the Guelders river area and much of the Frisian lands we can observe a group of farmers that accumulated tenancies and created large holdings. Often, but not always, they also owned some land themselves, which could either be used as collateral in difficult times or leased to other farmers in exchange for plots situated closer to the farmstead. These active, relatively prosperous people could also profit from their less fortunate neighbours who had to leave farming, by adding their smallholdings to the farms. Those who had to give up farming seldom became real proletarians. Many became non-agricultural specialists, serving the farming population as artisans, traders or transporters. The large, capital-intensive farms that emerged in the early modern period could make a profit because they could sell their cereals and industrial crops such as flax, madder and rapeseed, and dairy products and meat, on the markets of the burgeoning cities, those of the Dutch Republic in particular. Again, it should be stressed that it was rural people who decided to cultivate these cash crops or produce dairy. Urban landowners leased out their land to farmers and it was those farmers who decided the way in which the land was to be cultivated.

With some reservations concerning the three-tiered structure of society, one might define the farming system in most of the lowlands around 1700 as agrarian capitalism, even if one does not adhere to the Marxist concept of the transition from feudalism to capitalism.[93] It cannot be denied that large-scale, capital-intensive, commercial farming emerged in the North Sea Lowlands in the early modern era. It should be kept in mind, however, that in the North Sea Lowlands, as scholars such as De Vries, Lorenzen-Schmidt and Hallam have pointed out, feudalism was never that strong.[94] It was precisely the non-feudal character of coastal society, with its free rural population and free land and lease markets that made the transformation to a new farming system and new rural society possible. Freedom, acquired in the Middle Ages, gave prosperous and enterprising country dwellers the opportunity to engross their farms and transform rural society.

---

[93] See, for instance, Aston and Philpin, *The Brenner Debate*; Hoppenbrouwers and Van Zanden, *Peasants into Farmers?*; T. J. Byres, 'Differentiation of the Peasantry under Feudalism and the Transition to Capitalism: In Defence of Rodney Hilton', *Journal of Agrarian Change* 6 (2006), pp. 17–68; S. Ghosh, 'Rural Economies and Transitions to Capitalism: Germany and England Compared (c. 1200–c. 1800)', *Journal of Agrarian Change* 16 (2016), pp. 255–90.

[94] De Vries, 'Transition to Capitalism', p. 75; K.-J. Lorenzen-Schmidt, 'Northwest Germany, 1000–1750', in Thoen and Soens, eds., *Struggling with the Environment*, pp. 320, 334; Hallam, *The Agrarian History*, p. 507.

7

State and Steam: Water Management, c. 1700–1880

Introduction

During the eighteenth and nineteenth centuries tendencies that were already visible in earlier times continued: the role of oligarchies within the water authorities was strengthened; there was trust within the water authorities, but not between them. Localism, as Dorothy Summers termed it, was rampant.[1] This tended to lead to stagnation in institutional development. At the beginning of the eighteenth century, for instance, Friesland was one of the last regions to introduce monetisation of dike maintenance while other regions stuck to maintenance in kind. Although some changes occurred in the eighteenth century, real change came mostly in the nineteenth. That change came largely from outside through both state intervention and, as a result of the Industrial Revolution, the introduction of the steam pump, which would cause a revolution in water management.

Following a general introduction and a survey of land reclamation in the eighteenth and nineteenth centuries this chapter will discuss six case studies that will illuminate the problems and opportunities confronting the North Sea Lowlands in this period: the slow and tortuous introduction of monetisation of water management in parts of the Netherlands and north-west Germany; the introduction of Internal Drainage Boards, and later the steam engine, in the English fenlands; the struggle against deteriorating drainage in the Frisian fenlands; the consequences of and response to the devastating floods of 1715 and 1717; and the development of water management in the Dutch province of Groningen. These case studies confirm S. R. Epstein's view that pre-modern states did not possess undivided sovereignty as they were confronted with groups and bodies claiming to have legal privileges, which were defended fiercely against the state and any other party.[2] In this case, water authorities

---

[1] Summers, *The Great Level*, p. 115.
[2] S. R. Epstein, *Freedom and Growth: The Rise of States and Markets in Europe, 1300–1750* (London and New York, 2000), pp. 13–15.

defended the privileges their charters had granted them, even to the detriment of the safety of the reclaimed land. The last section before the conclusion will discuss how during the nineteenth century several states along the southern North Sea managed to overcome resistance and reform water management.

The North Sea Area in the Eighteenth and Nineteenth Centuries

The Industrial Revolution was the most incisive event of the 1700–1900 period. Maybe it is better to call it a process, because it took a long time to unfold. It started in the North Sea area, in England, around 1760. In the early nineteenth century it reached Belgium, later Germany, and finally, at the end of the nineteenth century, the Netherlands. This is not the place to discuss the Industrial Revolution, about which an enormous literature exists. Let it suffice to say that life in western Europe was revolutionised through the use of fossil fuels, the introduction of the steam engine and steam transport, industrialisation, and unprecedented urbanisation.[3]

The North Sea Lowlands remained mostly agricultural and rural. Although they were often situated within the orbit of major cities, industrialisation and urbanisation occurred elsewhere. The only parts of the Lowlands that had ever been urbanised were Holland and Zeeland, but there urbanisation ground to a halt and was even reversed in the eighteenth century.[4] The cities of Holland and Zeeland fell into a deep sleep, from which some would never awake. Indirectly, the Industrial Revolution did of course also influence the Lowlands. The strongly increasing urban and non-agricultural population led to a growing demand for agricultural products, and later in the century, as purchasing power increased, also for more 'luxury' products such as meat and dairy. The Industrial Revolution also influenced lowland farming by putting means at its disposal to improve productivity of land and labour: first steam pumps, which meant an enormous improvement in drainage, from c. 1850 onwards mechanical reapers, seed drills and steam-driven threshing machines, and at a later stage artificial fertilisers and pesticides.

Political revolutions also had consequences for the lowlands, at least on the continent. The French Revolution of 1789 and the Batavian Revolution of 1795, followed by the revolutionary and Napoleonic wars, had far-reaching consequences for the states on the eastern shore of the North Sea. The federal Dutch Republic emerged from this period

---

[3] E. A. Wrigley, *The Path to Sustained Growth: England's Transition from an Organic Economy to an Industrial Revolution* (Cambridge, 2016).

[4] De Vries and Van der Woude, *The First Modern Economy*, pp. 52–3; P. Brusse and W. W. Mijnhardt, *Towards a New Template for Dutch History* (Zwolle, 2011), pp. 20–4.

as the unitary Kingdom of the Netherlands. The Southern Netherlands were part of that kingdom from 1815 until 1830, when they seceded and became the Kingdom of Belgium. In north-west Germany, chunks of the coast had fallen into the hands of the Kingdom of Hannover at the Congress of Vienna, but the biggest changes occurred here when Bismarck enforced German unity in the 1860s. Denmark, which had regained control over Schleswig and Holstein in 1713, was forced to give up the duchies to Prussia in 1864 and two years later that kingdom also gobbled up Hannover. Most of the eastern shore was by then in the hands of the strong, unified states of Belgium, the Netherlands and Prussia. Of the old principalities only the grand duchy of Oldenburg still controlled a stretch of the coast as an autonomous part of the German *Reich*. The consequences of these revolutions and wars reached further than just the shifting of borders. As Tocqueville remarked, the restoration princes of Europe rejected all innovations of the French Revolution except centralisation.[5] As a result, the new nation-states were more prepared to intervene in water management and limit the autonomy of water authorities than their predecessors. Those predecessors had already tried to intervene in the eighteenth century, but had in most cases shown they were not able or willing to overrule local elites. This changed in the nineteenth century, when legislation was introduced that was intended to make the organisation of water management more uniform, in the Netherlands and Germany in particular. The Dutch state also began to implement drainage schemes itself, culminating in the drainage of lake Haarlemmermeer with steam pumps between 1839 and 1852.

Economically, for the Lowlands the 1700–1880 period started with depression. Economic growth halted around 1650 – in some areas even earlier – and problems were compounded by disastrous floods and a pandemic or *panzootic*: rinderpest, which hit western Europe several times during the eighteenth century. The Lowlands, most of which were dependent on cattle farming, suffered severely from this disease (see pp. 226–7, 229). There was nonetheless a ray of light during these dark years: the climate began to improve. After the extremely cold 1690s the Little Ice Age became less cold. There still occurred some very cold winters, such as those of 1709 and 1740, but in general the climate became somewhat more mild.[6] The 1810s again brought cold weather, which reached its nadir in 1816 – the 'Year without a Summer' – caused by the ashes thrown into the atmosphere by the explosion of the Tambora volcano in Indonesia in the previous year.[7] This was followed by a period of depressed cereal

---

5   A. de Tocqueville, *De la démocratie en Amérique* (2 vols., Paris 1835 and 1840, reprint 1981), vol. 2, p. 370.
6   Le Roy Ladurie, *Histoire humaine et comparée*, pp. 531–612; De Vries and Van der Woude, *The First Modern Economy*, pp. 610–19.
7   W. Behringer, *Tambora und das Jahr ohne Sommer: Wie ein Vulkan die Welt in die*

prices, caused by massive imports from Russia. Apart from these years, however, the 1750–1880 period was characterised by prosperity. The many large, imposing farms that can still be found in the North Sea Lowlands mostly date from this period. During this prosperous period, the population also increased. All the regions for which data is available for the 1750–1850 period (Zeeland, Friesland, Groningen, the coastal marshes of Oldenburg) show population growth.[8]

Land Reclamation and Steam Drainage

As soon as the rural economy recovered during the eighteenth century, land reclamation gathered speed. Gaining new land was once again lucrative and attracted investors. In Holland, many of the peat lakes that had come into existence due to destructive peat mining during the sixteenth and seventeenth centuries were pumped dry and turned into land during the eighteenth and nineteenth centuries. This resulted in agricultural land that was situated several metres below sea level. In the south-west, reclamation of marshes was resumed around 1750. In the three provinces of Holland, Friesland and Zeeland alone, about 128,000 hectares of land were reclaimed between 1750 and 1900.[9] Along the German coast, in Ostfriesland, over 10,000 hectares were reclaimed during the eighteenth and nineteenth centuries.[10]

In this period, land reclamation was no longer entirely left to private investors. Lake drainage was sometimes done by the state, because it was very expensive and technically challenging, and the prospects of making any profit on it were therefore bleak. The reason the state was prepared to engage in such difficult and risky projects was the fear that the many peat lakes that were slowly eroding the surrounding land might at some moment change central Holland into one large inland sea.[11] At first, the province of Holland took the initiative by draining lakes in southern

---

Krise stürzte (Munich, 2015), pp. 19, 43–51.

[8] Priester, *Geschiedenis*, pp. 49–52; Faber, *Drie eeuwen*, p. 413; R. F. J. Paping, *Voor een handvol stuivers: Werken, verdienen en besteden: de levensstandaard van boeren, arbeiders en middenstanders op de Groninger klei, 1770–1860* (Groningen, 1995), p. 52; E. Hinrichs, R. Krämer and C. Reinders, *Die Wirtschaft des Landes Oldenburg in vorindustrieller Zeit: Eine regionalgeschichtliche Dokumentation für die Zeit von 1700 bis 1850* (Oldenburg, 1988), pp. 27–9.

[9] P. van Cruyningen, 'Régulation des eaux, investissement urbain et croissance agricole: l'agriculture dans les provinces littorals des Pays-Bas (1400–1900)', in L. Herment, ed., *Histoire rurale de l'Europe, XVI–XXe siècle* (Paris, 2019), p. 63.

[10] J. Kramer, 'Küstenschutz und Binnenentwässerung zwischen Ems und Weser', in Kramer and Rohde, eds., *Historischer Küstenschutz*, p. 213.

[11] W. van der Ham, *Hollandse polders* (Amsterdam, 2009), pp. 72–5.

Holland in the period between 1772 and 1796.¹² From 1797 the new centralised Dutch state took over, and between 1828 and 1839 it drained the 4000 hectare Zuidplas lake between Rotterdam and Gouda. This more-than-six-metres-deep lake was pumped dry by a combination of 30 windmills and two steam pumps. The steam engines had greater capacity than the windmills; together the two steam pumps removed 40 per cent of the water.¹³ The first time a steam engine was used to drain land in the North Sea Lowlands was in 1787, when a Boulton & Watt engine was installed in the polder Blijdorp near Rotterdam.¹⁴ Subsequent experiments with steam pumps were not very successful. Fuel was the main problem. Coal had to be imported and therefore was too expensive, and when turf was used as an alternative fuel, as in the drainage of Zuidplas, the machines did not function properly.¹⁵

Steam drainage became successful in the Netherlands in the second quarter of the nineteenth century, as coal prices decreased and steam pumps were successfully used in the largest drainage project of that time, that of lake Haarlemmermeer between 1839 and 1852. With its size of 18,000 hectares and its location in between the cities of Leiden, Haarlem and Amsterdam, lake Haarlemmermeer was a serious threat to the safety of the centre of Holland. Plans had been made for its drainage from the seventeenth century onwards, but the sheer size and enormous cost of such a project discouraged potential investors. Therefore, eventually the state decided to intervene. While in the seventeenth century Dutch capital and expertise had helped to drain parts of the English fenlands, now English technology and expertise was crucial for draining this huge lake. Haarlemmermeer was pumped dry by three huge steam-driven pumping stations. The pumps used were enlarged versions of Cornwall mine pumps, and the technology of the stations was designed by the English engineers Joseph Gibbs and Arthur Dean. The most complex parts of the stations were made by English factories.¹⁶ Drainage of lake Haarlemmermeer had cost the Dutch state the enormous sum of 14 million guilders, whereas the proceeds from the sale of the new land were only eight million guilders. However, due to the drainage of this and other lakes, safety in central Holland had increased strongly, while money no longer had to be spent on trying to stop erosion of the shores of those lakes.

The success of the Haarlemmermeer project gave a boost to the use of steam pumps in the Netherlands after 1850. Many mill polders replaced

---

12 *Ibid.*, pp. 97–100.
13 Van de Ven, *Man-Made Lowlands*, pp. 302–3; Van der Ham, *Hollandse polders*, p. 119.
14 Van der Ham, *Hollandse polders*, pp. 100, 117.
15 *Ibid.*, pp. 119–20.
16 Van de Ven, *Man-Made Lowlands*, pp. 305–10; Van der Ham, *Hollandse polders*, pp. 121–2.

Water Management, c. 1700–1880

FIGURE 6. Cruquius steam pumping station near Haarlem, Noord-Holland, 1974. Photo L. M. Tangel. Collection Rijksdienst voor het Cultureel Erfgoed. The huge steam pumps of this station, designed and manufactured in England, were used to pump the water from the 18,500 hectare Haarlemmermeer lake.

their drainage mills with steam engines. The number of steam-driven pumping stations increased from 29 in 1861 to 418 in 1884. There was considerable variation between the provinces. Holland had 273 steam pumps in 1884, Friesland only ten.[17] This was the result of the difference in acreage of the polders in Holland and Friesland. As demonstrated in chapter five, in Holland relatively large mill polders were formed during the sixteenth and seventeenth centuries. The scale of these polders was large enough to make investment in steam pumps profitable. In Friesland, however, thousands of small mills drained the land of individual farms. Investment in a costly and powerful steam pump for such a small area was simply not an option, since in the 1870s the erection of a steam pumping station cost 70,000 guilders or more.[18] Most improvement in

[17] J. L. van Zanden, *De economische ontwikkeling van de Nederlandse landbouw in de negentiende eeuw 1800–1914* (Wageningen, 1985), p. 236.
[18] M. J. E. Blauw, *Van Friese grond: Agrarische eigendoms- en gebruiksverhoudingen en de ontwikkelingen in de Friese landbouw in de negentiende eeuw* (Leeuwarden, 1995), p. 222; A. Bijl, *Het Gelderse water: Waterstaatkundige en sociaal-economische ontwikkelingen in de polders van de westelijke Tielerwaard* (Leiden, 1997), pp. 178–80.

drainage that was achieved in Friesland was the result of investment by the province in improvement of canals and sluices.[19]

Steam drainage made it possible to control the water level during all seasons; the fenlands were no longer flooded during autumn and winter. This resulted in a longer growing season for the grass and thus in higher productivity of the land. The drawback was that because the land was no longer flooded during part of the year, the processes of shrinkage and wastage could now continue all through the year. Therefore, land subsidence in the fens sped up.[20] Looking at it with the benefit of hindsight, one can conclude that this was a fatal step, strengthening the vicious cycle that had started when the fens were reclaimed in the Middle Ages.

The Persistence of Maintenance in Kind

The slow shift from maintenance in kind to monetisation has already been discussed in chapters three and five and that discussion is continued here. By 1700, most water authorities in England, Flanders, Brabant, Zeeland and Holland had monetised dike maintenance and other tasks. These tasks were no longer performed in kind by landowners or their tenants. Instead, the water authorities levied rates from the landowners, and let contractors do the maintenance work under their supervision. Originating in Flanders, monetisation slowly spread northwards and reached part of Friesland in the seventeenth century.[21] Then diffusion of monetisation ground to a halt. Along the Wadden coast, in the Dutch river area and here and there in Holland and Zeeland, maintenance in kind held its own until well into the nineteenth century and in some cases even into the twentieth. This has surprised experts, who consider maintenance in kind to be inferior to maintenance by professional contractors under the supervision of the water authority. Contemporary reports confirm this view. In 1771, for instance, a report on the maintenance of the dikes of the polder Schouwen, the only polder in Zeeland that still had maintenance in kind, stated that the quality of the work done by the several landowners liable for maintenance showed considerable variation.[22] A German observer noted in 1780 that it was impossible to find a straight stretch of dike along the lower Elbe due to *Kabeldeichung* (maintenance in kind). Stretches of no more than four feet were sometimes maintained by five *Interessenten*

---

[19] G. ter Haar and P. L. Polhuis, *De loop van het Friese water: Geschiedenis van het waterbeheer en de waterschappen in Friesland* (Franeker, 2004), pp. 100–3.

[20] Duyverman, *De landbouwschekundige basis*, pp. 158–60, 190.

[21] Van Tielhof, *Consensus en conflict*, pp. 86–91; Beekman, *Het dijk- en waterschapsrecht*, pp. 504–5.

[22] J. A. Schorer, *De geschiedenis der calamiteuse polders in Zeeland tot het reglement van 20 januari 1791* (Leiden, 1897), p. 18.

(landowners liable for maintenance).²³ In Groningen those liable for maintenance of very small plots often shirked their duties.²⁴ At the places where the stretches of dike maintained by different owners touched, the resulting uneven work could cause problems: a dike is only as strong as its weakest spot, and seams in a dike are weak spots. Another problem arising from maintenance in kind was the variation in costs of the upkeep of different stretches of dike. Dikes that were protected by a broad foreshore were much cheaper to maintain than dikes that lacked such protection. Therefore, some landowners faced crippling costs, whereas others only had to spend a little.²⁵

Chapter five showed that the rural population was convinced they had good reasons to stick to maintenance in kind, based on distrust of rapacious elites – not entirely unfounded – and the wish to keep the cost of dike maintenance under control. For the Alte Land, Brümmel has demonstrated that the system of maintenance in kind was beneficent for both larger and smaller farmers. Calculated in *Reichsthaler*, the financial burden of dike maintenance was heaviest for the small farmer. When that farmer performed the work himself, or allowed family members to do it, he could avoid spending money on wages, and could also source a lot of material from his own holding. Larger farmers, on the other hand, had to pay wages, but they needed to perform relatively less maintenance. Therefore, both groups profited from the system. Moreover, they could perform the work during slack seasons on the farm. Farm work was always prioritised over dike maintenance. Whether or not that was conducive to good maintenance is questionable.[26] Landowners also opposed change because a new system would undermine existing rights. Land with a high burden of dike maintenance was sold at low prices, while land with a low burden was much more costly to purchase.[27] Introducing monetisation of dike maintenance and making landowners pay proportionally for maintenance would therefore harm the interests of those who had purchased expensive land with a low burden of maintenance. The system had its own perverted logic that was an obstacle to change. Moreover, in the Land Kehdingen on the lower Elbe river, landowners perceived their dike stretches as private property, and of course any interference with private property was considered unacceptable.[28]

For a mix of these reasons, water authorities, especially on the Wadden coast and in the Dutch river area, stuck to maintenance in kind until

---

[23] Ehrhardt, *Ein Guldten Bandt*, p. 195.
[24] Schroor, *Wotter*, p. 52.
[25] Ehrhardt, *Ein Guldten Bandt*, p. 197.
[26] Brümmel, *Die Dienste und Abgaben*, pp. 102–3; Ehrhardt, 'Kabel und Kommunion – Formen des individuellen und genossenschaftlichen Deichens im Elbe-Weser-Raum', in Fischer, ed., *Zwischen Wattenmeer*, p. 139.
[27] Ehrhardt, *Ein Guldten Bandt*, p. 203.
[28] Fischer, *Wassersnot und Marschengesellschaft*, pp. 17, 165.

well into the nineteenth century, or even into the twentieth. That did not mean they were unaware of the drawbacks of this form of maintenance. Landowners and water authorities thought of ways to mitigate the consequences. Many water authorities that had preserved maintenance in kind differentiated between ordinary and extraordinary maintenance. Ordinary maintenance, keeping the dike at the required height and strengthening the outward slope with straw mats, was done by the individual landowners. All more complicated and expensive work, such as large-scale repairs after storm floods, construction of groynes and sluices, sinking of osier mats on the foot of the dike under water, was monetised and performed by professional contractors.[29] Moreover, that maintenance in kind persisted did not necessarily mean that each spring every farmer trotted to the dike with a spade and wheelbarrow. On the polder of Schouwen, for instance, the landowners liable for maintenance often left the work to *waardijkers*, professional contractors.[30] Farmers in the Dutch river area also often made use of contractors. This even led to the emergence of groups of specialised contractors for the construction of dikes and fortifications.[31] Moreover, in the course of the eighteenth century more and more of the stretches of dike that were most expensive to maintain became the responsibility of villages or districts instead of individual landowners.[32]

Water authorities also found ways to reduce the differences in the burden of maintenance. On the polder of Schouwen, for instance, every seven years the stretches of dike that had to be maintained were re-distributed between the landowners. That way, each owner was certain that he or she never had to maintain a 'bad' (costly) stretch of dike for more than seven years on end.[33] On the north Frisian island of Nordstrand, in 1758 yet another solution was found for this problem. Maintenance was performed in kind by the individual owners, but they were reimbursed from a communal fund if their costs were higher than average. Should their costs be lower, then they had to pay a sum into the fund. In this way, dike maintenance costs were distributed equitably.[34]

The Land Hadeln, on the lower Elbe, had a very sophisticated maintenance system. The parishes that were situated far from the dike had monetised dike maintenance, whereas in the parishes close to the river the maintenance was still performed by individual landowners. For

[29] Van Tielhof, *Consensus en conflict*, p. 87; P. H. Gallé, *Beveiligd bestaan: Grondtrekken van het middeleeuwse waterstaatsrecht en hoofdlijnen van het dijksbeheer in dit gebied (1200–1963)* (Delft, 1963), pp. 147, 160.
[30] Gallé, *Beveiligd bestaan*, p. 156.
[31] De Bruijn, *De hoeve en het hart*, pp. 262–5.
[32] E. J. Bosch van Rosenthal, *De ontwikkeling der waterschappen in Gelderland* ('s-Gravenhage, 1930), p. 217.
[33] Gallé, *Beveiligd bestaan*, pp. 153–4.
[34] Feikes, *Die geschichtliche Entwicklung*, pp. 99–101.

the inland parishes it was more practical to contract maintenance out, because then the farmers did not have to travel to the dike and lose a lot of time. The farmers in the parishes adjacent to the river did not have to travel, and therefore they could more easily perform maintenance in kind. Moreover, all parishes and individuals liable for maintenance were responsible for several stretches along the whole length of the dike. In this way, 'bad' and 'good' plots were distributed equitably over all landowners and villages.[35] The drawbacks of the large numbers of owners with very small dike plots could be avoided by making groups of small landowners collectively responsible for one longer stretch of dike. This was the case on the polder of Schouwen, and from 1773 on the island of Pellworm in Nordfriesland.[36]

There were, therefore, several ways to mitigate the consequences of maintenance in kind, and that may explain why this system was maintained for so long: in Gelderland, for instance, until 1880; in Groningen until 1853–73; and in the Land Hadeln until 1930.[37] However, there were also regions where the drawbacks of the system were not in any way reduced. One of those was the Alte Land. The consequences there were exacerbated by the tradition that the offices were held in turn by the more prosperous farmers of a community. In the early nineteenth century, the *Oberdeichgräfe* of the Alte Land observed that the office-holders viewing the dikes of their neighbours tended to ignore bad maintenance because they knew that next year their neighbours might be viewing *their* stretches of dike.[38] Catastrophic floods, leading to breached dikes and wide flooding occurred here at least once in every 25 years in the period from 1602 to 1825.[39] Every generation was confronted with a devastating flood, but in the perception of the inhabitants this may have been compensated for by the fact that the costs of dike maintenance were kept low. The price was paid by those living near the breach: they were often financially ruined.[40] It is too easy to describe this attitude as simply fatalistic. Greg Bankoff has described a similar attitude towards disasters in the Philippines. He demonstrated that this is not just a 'passive sense of acceptance of one's fate', but also 'a sense of finely calculated assessment of the odds'. It is a kind of gamble, but one based on knowledge of patterns in the past that are likely to reoccur in the future (see also pp. 209–10).[41]

---

35 Fischer, *Im Antlitz*, pp. 52–4.
36 Feikes, *Die geschichtliche Entwicklung*, pp. 82–3.
37 Bosch van Rosenthal, *De ontwikkeling*, p. 325; Schroor, *Wotter*, p. 141; Fischer, *Im Antlitz*, p. 368.
38 Ehrhardt, *Ein Guldten Bandt*, p. 166.
39 Ibid., pp. 436–64.
40 Ibid., pp. 519–20.
41 G. Bankoff, 'In the Eye of the Storm: The Social Construction of the Forces of Nature and the Climatic and Seismic Construction of God in the Philippines', *Journal of Southeast Asian Studies* 35 (2004), pp. 102–4.

Internal Drainage Boards and Steam Pumps in the Cambridgeshire Fens

This section will focus on the Great Level of the Fens, which became known as the Bedford Level, to the north of Cambridge, which was drained between 1630 and 1653 by the earl of Bedford, Cornelius Vermuyden, and associates. An Act of Parliament of 1663 installed the Bedford Level Corporation, a Dutch-style water authority with the obligation to maintain many kilometres of water ways, about 500 km of embankments, as well as sluices, bridges and causeways, and the right to enforce maintenance of private ditches. The Corporation had the right to levy rates from the owners of the adventurer's land – 95,000 acres or 38,500 hectares – in order to finance all of this, and to form a board and a professional staff to implement it. The composition of the board and access to the meeting of owners that supervised it were firmly based on oligarchic principles. The governor and six other board members had to own at least 400 acres (162 hectares); for access to the general meeting ownership of 100 acres (40.5 hectares) was required.[42] It began well, but board, staff and landowners must soon have felt they had landed in a nightmare.

Some causes of the problems confronting the Bedford Level Corporation were already discussed in the previous chapter: the continuous subsidence of the peat soil due to drainage and the silting-up of the outfalls of the fenland rivers. It became increasingly difficult to drain excess water – like in Holland in the fourteenth and fifteenth centuries – and fields flooded. The Corporation could not do much about this, because it was now in chronic financial difficulty. Drainage had cost enormous sums and the upkeep of the physical infrastructure was also very expensive. The Corporation ran into debt and was unable to keep all infrastructure in good repair, let alone pay for the erection of drainage mills that might help with keeping the Level dry. An additional problem was that only the lands that had belonged to the original adventurers could be taxed. Extending this to all land drained by the works of the Corporation was not feasible. An attempt to do so in 1777 failed miserably due to fierce resistance by the owners of that land.[43]

The governors of the Bedford Level Corporation, realising they did not have the means to redress the problems, decided to leave the solution to local landowners. As a result in 1728 and 1753 decisions were taken that effectively split off the northern part of the Level, which subsequently had its own budget and commissioners.[44] In 1727, the inhabitants of the Haddenham Level, claiming that their land had been inundated

---

[42] Summers, *The Great Level*, pp. 79–83.
[43] Ibid., pp. 85–7, 117, 128.
[44] Ibid., pp. 119–20, 124.

almost continually for seven years, managed to get an Act of Parliament passed that granted them the right to drain their own lands at their own expense, for which they could raise rates and appoint administrators. By the end of the eighteenth century there were 20 such Internal Drainage Boards (IDBs) in the southern part of the Level alone.[45] Within such an IDB a group of landowners cooperated to keep their land dry enough for farming. Within the territory of the IDB the water level could be controlled independently from the surrounding land, and drainage mills pumped excess water to the main canals and rivers that were still maintained by the Corporation. In fact, the IDBs were the same kind of organisation as the mill polders that had been formed in the Holland and Utrecht fenlands in the sixteenth and seventeenth centuries. There was one crucial, and fatal, difference with Holland, however. In Holland, the regional water authorities had to grant permission to form mill polders and they coordinated their functioning. The Bedford Level Corporation lacked such rights and the IDBs could damage each other's land and the infrastructure of the Corporation without risking sanctions. As a result, the fenland was rife with conflict between drainage authorities. 'There was everywhere a chaos of authorities and an absence of authority', as Darby has put it.[46]

Such conflicts were of course not conducive to optimal water management. Unsurprisingly, descriptions from the late eighteenth and early nineteenth centuries paint a dismal picture of the fenlands. 'Look which way you will, you will see nothing but misery and desolation' an observer noted in 1777.[47] Arthur Young, visiting the fenlands in 1799, saw an area of 10,000 hectares that was completely under water, in summer![48] This was not solely caused by the suboptimal organisation of water management, but also by the insufficient capacity of the drainage mills. Pumping up water by way of a windmill with a scoop wheel was not easy, as farmers on the opposite shore of the North Sea also knew. The height to which the mills could lift the water was at most 1.5 metres. In winter they could not pump away sufficient water. And they of course needed wind to function at all. It appears that in this respect it was more difficult to run a windmill in the English fens than in Holland. Windmills function best when the wind has a speed of 8 to 12 metres per second. In Holland the wind blows at that speed for 1339 hours per year on average, but in eastern England only 450 hours. Therefore, in England, windmills had much less time within which to function effectively than in Holland. Moreover, the fenland drainage boards tried to pump their excess water directly into rivers; they had no storage system for water like the Dutch

---

[45] Darby, *The Draining*, pp. 119–21; Summers, *The Great Level*, p. 119.
[46] Darby, *The Draining*, p. 123.
[47] Ibid., p. 130.
[48] Ibid., p. 172.

hoogheemraadschappen.[49] All this meant that it must have been much more difficult to keep the fenlands dry in England than in Holland.

The solution to the drainage problems of the Fens was of course, as Arthur Young already noted in 1805, 'the application of steam-engines'.[50] Steam engines had a much larger capacity than windmills, could lift water higher, and could run all year because they were not dependent on the wind. As mentioned above, the first steam engine used to drain land was applied in Holland, near Rotterdam, from 1787 onwards, but it took until after 1850 for steam drainage to spread widely. The first area in England where steam pumps were installed appears to be Hatfield Chase, where an Act of 1813 authorised the construction of a steam pump.[51] In the Fens the first steam engine with a scoop wheel was functioning near Wisbech shortly afterwards, in 1817. From that year onward steam pumps spread rapidly. The benefits of steam engines soon became apparent and by 1850 most of the Fens were drained by steam pumps. Steam engines were not only more reliable, but they also required less labour input. One steam engine with its machinist and stoker could replace dozens of windmills, each with its own miller. In Littleport Fen, 75 windmills were replaced by two steam engines in 1848. When the scoop wheels were replaced by pumps their capacity increased further.[52] Of course, as in the Dutch fens, steam drainage had the drawback that it sped up shrinkage and wastage of the fen soil.

In the first half of the seventeenth century, drainage promoters such as Vermuyden had noticed the possibility of transforming the Fens into intensively farmed land. It was the steam engine, product of the Industrial Revolution, that 200 years later made it finally possible to realise this potential. However, although drainage had improved enormously, there were still problems and floods now and then still occurred. This was not due to technological issues but rather to institutional failings. By 1920 there were 83 drainage authorities in the drainage area of the river Great Ouse and 12 channel and outfall authorities.[53] These authorities spent a lot of energy, time and money fighting each other. Darby reports that authorities along the river Nene spent £100,000 on legal and parliamentary battles in the 50 years preceding 1882.[54] That enormous sum was thus not spent on keeping the land dry and fit for farming, but instead on settling conflicts with neighbours. There clearly existed no trust between the drainage authorities of the Fens, but the rest of this chapter shows that the Dutch and Germans did not do any better in this respect.

---

[49] R. L. Hills, *Machines, Mills and Uncountable Costly Necessities: A Short History of the Drainage of the Fens* (Norwich, 1967), pp. 32–3.
[50] Darby, *The Draining*, pp. 220–1; Summers, *The Great Level*, p. 164.
[51] Byford, 'Agricultural Change', pp. 273–4.
[52] Summers, *The Great Level*, pp. 167–74.
[53] *Ibid.*, p. 215.
[54] Darby, *The Draining*, p. 217.

Deteriorating Drainage in the Frisian Fenlands

Around 1800, farmers in the fens of central Friesland struggled with the same kind of problems as their counterparts in the English Fens. Each farmer drained a plot of land around his farmstead with a small drainage mill that pumped excess water to the rivers and lakes in the area. Most land was left undrained and was only used for haymaking. For two centuries at least this was a satisfactory arrangement. From the second half of the eighteenth century, however, complaints were being raised. This was probably partly due to farmers in the Frisian fens being no longer satisfied with their extensive farming system and wanting to profit from increasing dairy prices by shifting towards more intensive dairy farming. That was impossible under the existing system of water management. Yet, there are also indications that the drainage of the fenlands was deteriorating. Estimates of hay production from 1810 are illustrative. In the districts of Sneek and Heerenveen, where there were many undrained fenlands, undrained land yielded about a third fewer kilograms of hay than drained land. Not only did this undrained land yield less hay, but that hay was also of inferior quality, since its price was a third lower than that of hay from drained land.[55] Since, in the fenlands, drained land was mainly used for pasture while hay was made on undrained land, this meant that farmers in the fens may indeed not have produced enough hay, which must have strongly limited their opportunities to intensify farming.

The deterioration of the drainage of the Frisian fenlands had several causes. Firstly, after 1750 more land was reclaimed in the uplands, which drained towards the low central part of Friesland. Secondly, at the same time peat dredging for fuel production began in the fenlands. This was mostly done in the undrained lands. They were surrounded with a dike and then drained to make peat dredging possible. In this way the capacity of the Frisian *boezem* – the area of lakes, rivers and undrained land used to store and transport excess water – was reduced, at a time when it also had to store and transport more water. Thirdly, most sluices were situated in the south-west of the province, and south-westerly winds were predominant, making it necessary for water to be drained against both the wind and the sea water that was pushed up by the wind, which often meant that the sluices could not be opened. Fourthly, there was a lot of ambiguity about who was responsible for the upkeep of canals and sluices. In principle, this responsibility was shared by cities and rural districts, but they often shirked their duties. Again, we can perceive a lack of trust between communities.[56]

---

[55] Knibbe, *Lokkich Fryslân*, p. 80.
[56] Van Cruyningen, 'Water Management', pp. 75–7; J. P. A. Louman, *Fries*

The province of Friesland tried to support farmers and landowners in the fenlands as much as possible. It took over responsibility for the maintenance of part of the infrastructure. By 1771, for example, it maintained 11 major sluices in the province. Proposals to make the province liable for the maintenance of the whole *boezem* failed, however, due to a lack of funds and the problem of having to reach an agreement between the 11 cities and 30 rural districts that formed the Provincial States.[57] More successful was legislation introduced in 1774 making it easier to found mill polders. Until then, founding of mill polders was only possible if all landowners in the area concerned gave their permission. This unanimity was all but impossible to achieve, so new legislation was introduced which made founding a mill polder possible when two thirds of the owners agreed. The minority that disagreed had to acquiesce or sell their land to the majority at an agreed price. It appears that the 1774 regulation had the expected effect, and the number of mill polders began to increase rapidly. Until 1810, some 90 to 100 mill polders were founded in Friesland. The majority of the more than 2000 mills that dotted the Frisian landscape in that year were still small mills draining individual farms.[58] The drainage problems of the Frisian fens were only solved in the nineteenth century, when the province took the whole drainage system under its care and invested millions of guilders in it, culminating in the erection of a huge steam pumping station on the Zuider Zee coast in 1920.[59]

The Flood of 3 March 1715 in the South-Western Netherlands and its Aftermath

The storm that hit the south-west of the Netherlands in early March 1715 is not among the well-known disastrous storms of the North Sea area such as those of 1570, 1634, 1717, 1825 or 1953. This is probably due to the fact that, although dozens of polders were flooded, very few humans or animals were killed. Presumably, this was mainly because the dikes breached in the afternoon, giving people enough time to reach higher ground and to save themselves and their animals.[60] From a historical viewpoint, however, the 1715 flood is interesting because of the way in

---

waterstaatsbestuur: Een geschiedenis van de waterbeheersing in Friesland vanaf het midden van de achttiende eeuw tot omstreeks 1970 (Amsterdam, 2007), pp. 89, 91.
[57] Van Cruyningen, 'Water Management', p. 78.
[58] Ibid., pp. 76, 79; Louman, *Fries waterstaatsbestuur*, pp. 312–20.
[59] Ter Haar and Polhuis, *De loop*, p. 101.
[60] Buisman, *Duizend jaar*, vol. 5, pp. 414–15; A. M. J. de Kraker, 'Two Floods Compared: Perception of and Response to the 1682 and 1715 Flooding Disasters in the Low Countries', in K. Pfeifer and N. Pfeifer, eds., *Forces of Nature and Cultural Responses* (Dordrecht, 2013), pp. 186, 193–7.

which water authorities and government responded to it. The dikes in this region had been suffering from erosion by the deep gullies in the tidal rivers for decades. Fighting this erosion by constructing groynes and reinforcing the foot of the dike with osier mats was very costly and had probably been neglected due to the economic depression that set in around 1660. The damage caused by the flood of 3 March 1715 and smaller floods in the previous year appears to have served as a wake-up call. Water authorities raised their budgets and started investing in improvement of their sea defences.[61]

To finance these investments, water authorities increased rates, borrowed money, and looked to the provincial or federal government for support. These governments traditionally supported destitute polders by granting them exemptions from taxation. From 1715 the number of requests to the States of Zeeland and the States-General – which ruled what is present-day Zeeland Flanders – for tax exemption from 'calamitous' polders increased enormously. Therefore, the tax revenues of the province of Zeeland and the federal government decreased strongly and politicians started to look for means to mitigate this. The States of Holland suffered less from these problems. Only a few islands in the utmost south-west of Holland were in need of support. For one island, Goeree, a drastic decision was taken in 1717: the province itself became responsible for financing maintenance of the sea defences. This was not because the Holland *regenten* felt that much compassion for the islanders, but because Goeree was strategically important: it was situated next to the sea inlets that provided access to the port of Rotterdam and the naval base of Hellevoetsluis. Thus, it had to be preserved so as to prevent silting-up of those sea inlets.

Meanwhile, the States of Zeeland and the States-General tried to extend the area that was liable for dike maintenance. The south-west of the Netherlands was at that time an archipelago of small islands, each composed of several small polders. All of those polders were administratively autonomous, with only the ones immediately behind the flood defences being responsible for their particular maintenance. These polders faced extremely high costs, whilst the ones situated more inland contributed nothing. Only after major disasters were they sometimes, very reluctantly, prepared to provide some support. By 1720 dignitaries in both Zeeland and The Hague had reached the conclusion that it would not be unreasonable to ask the inland polders to pay a modest annual subsidy to the polders that were suffering from high maintenance costs. After all, they profited from the work and expenditure of the landowners

---

[61] Unless stated otherwise, this section is based on P. J. van Cruyningen, 'Sharing the Cost of Dike Maintenance in the South-Western Netherlands: Comparing "Calamitous Polders" in three "states" 1715–1795', *Environment and History* 23 (2017), pp. 363–83.

in the outer polders, and if those outer polders were flooded then they themselves would have to finance costly dikes. However, eighteenth-century landlords and farmers did not see it that way. Providing such support would be a violation of their privileges, which was unacceptable. Faced with the choice between maintaining either their rights or their safety, they chose the former.

Landowners were not always granted the freedom to make this choice, however. In the territory controlled by the States-General, present-day Zeeland Flanders, technical and financial experts ruled on requests for subsidies for destitute polders. In the 1720s, these experts proposed to unite all polders of an island in western Zeeland Flanders into one water authority that would be responsible for all its sea defences, thus spreading the costs of dike maintenance over the whole island. Such mergers of polders had been successful on the islands of Cadzand (1537) and Voorne (1630). In the 1720s, however, these plans were opposed so fiercely by local officials that the States-General did not even attempt to break the resistance. The federal government was prepared to support less far-reaching plans of the experts, in which adjoining polders also had to contribute to dike maintenance. Local powerholders could not change these decisions of the experts because the latter had the support of the federal government. Local interests were not represented in the States-General and exerted very little influence in The Hague. The inland polders may not have liked having to contribute, but they were left with no choice.

In the province of Zeeland, however, the situation was different. Zeeland was ruled by its States, composed of representatives of the six major cities. The representatives of those cities also perceived themselves to be representatives of the islands on which they were situated, and they were prepared to defend the interests of the inland polders. When in 1721 the Zeeland Audit Office proposed to make the inland polders contribute a small sum to the subsidies for destitute polders on the same island, the cities of Zierikzee, Goes and Tholen voted against it. These three cities were situated on islands with many inland polders, within which the city elites owned land. It was in their interest to shift the burden to the outer polders and the general taxpayer, and that was what they did. For 70 years they were able to prevent any regulation that would make the inland polders contribute. It took until 1791, when the financial situation of the province itself had become calamitous, to reach a compromise that forced the inland polders to contribute to the costs of the flood defences of the outer polders.

Both the federal authorities and the provincial authorities of Zeeland tried to introduce a more equitable way of distributing the costs of dike maintenance. Both were confronted with fierce opposition from the local polder boards, refusing to contribute anything that would undermine their privileges. In Zeeland Flanders, the federal government judged more ambitious plans unfeasible, but was able to push through subsidies by inland polders for the outer polders. In Zeeland, local interests had representatives in the Provincial States, and they were able to prevent contributions by inland polders for 70 years by shifting the burden to the general taxpayer. In both areas, as in the Cambridgeshire Fens, local water authorities were primarily aiming at defending their parochial interests at the cost of efficient water management. Trust was inward-looking; outsiders were not to be trusted, let alone supported. The States-General were better able to overcome this parochialism than the States of Zeeland, but both were successful in defending the land against the encroaching sea and rivers. Both did this by spending enormous sums on dike maintenance. The difference was that in Zeeland the inland polders could continue their free riding until 1791 at the cost of the general taxpayer and in the territories of the States-General they could not. It was wise to bar regional interest groups from influencing water management policy. This was also realised in the adjoining county of Flanders, where a department of public works was made responsible for flood protection in 1755, thereby terminating the influence of regional oligarchies on decision-making.[62]

In Zeeland and Zeeland Flanders enormous sums were invested to prevent flooding. This investment was successful; most polders in the area were flooded only once or twice during the seventeenth and eighteenth centuries. The Watering Cadzand, for instance, situated on an exposed spot in the south-western delta, was only flooded in 1682. In 1715 its dikes and dunes were damaged, but, due to huge investments, flooding in that year and later was prevented. Landowners in the Alte Land, as demonstrated above, followed another strategy: they maintained relatively inexpensive maintenance in kind and were willing to accept a flood once in each generation, for which only a small part of the community had to pay the price. They ran more risk, but very probably had to pay less, unless a very disastrous storm did extensive damage, as in 1825. Their approach may have resulted from the difference between a more fatalistic and a more optimistic attitude towards dike construction and flooding. The former is exemplified by the Groninger labourers who refused to work on dike repairs after the Christmas Flood of 1717, because they believed that 'floods came once every 30 years, and nobody could construct dikes against them'. Early seventeenth-century inhabitants of Nordstrand held the same view, only they estimated that floods came every 40 years.[63] For people acting from such a perspective,

[62] M. W. Serruys, 'The Societal Effects of the Eighteenth-Century Shipworm Epidemic in the Austrian Netherlands (c. 1730–1760)', *Journal for the History of Environment and Society* 6 (2021), pp. 116–18.
[63] Quoted in A. Sundberg, *Natural Disasters at the Closing of the Dutch Golden Age:*

it was perfectly logical to stick to old ways and not to invest too much in dike construction. And, as noted above, there was also an element of calculation in this 'fatalism' (see p. 201).

In the south-western delta huge sums were invested in dike maintenance, but much less in drainage. Since most polders here were situated above sea level, generally drainage was not very problematic, but it was also far from optimal. Drainage of the lower parts of polders in particular remained problematic.[64] Investment in steam pumping stations came late here. The first one was built on the island of Schouwen in 1877 and there were only six in the whole province by 1884.[65]

The Christmas Flood of 1717 and its Aftermath

On 23 December 1717 a strong south-westerly wind blew in the Wadden area. The following day the wind changed direction to come from the north-west, but seemed to abate somewhat before midnight. Since no spring tide was expected, the inhabitants of the coastal region were not worried and went to bed. Then suddenly and dramatically, in the early morning of Christmas Day, the power of the north-westerly increased enormously and the tide rose at an extraordinary speed, sometimes up to nine feet within half an hour. The dikes along the Wadden coast from Friesland to southern Denmark could not withstand this onslaught and breached at many places.[66] This was the beginning of probably the worst storm disaster in the southern North Sea area after 1570. Between 11,800 and 13,300 people lost their lives and c. 100,000 head of cattle and horses were killed.[67] Because the storm hit during the early morning and the water rose very rapidly, people had no chance to get themselves and their animals to higher and safer places in time. Many people saved themselves on the roofs of their houses, wearing nothing but their soaked night shirts in the severe winter weather.[68] Therefore, many people must have died of hypothermia. How relevant the timing of the disaster was is shown by the case of the Alte Land. As this region was situated quite far upstream on

---

    Floods, Worms, and Cattle Plague (Cambridge, 2022), p. 117; for Nordstrand: Mauelshagen, 'Disaster and Political Culture', p. 49.

64 A. P. de Klerk, 'Water naar de zee dragen II: Last en bestrijding van het binnenwater op Walcheren na 1795', in A. P. de Klerk, *Het Nederlandse landschap, de dorpen in Zeeland en het water op Walcheren: Historisch-geografische en waterstaatshistorische bijdragen* (Utrecht, 2003), pp. 246–7.

65 De Klerk, 'Water naar de zee', p. 253; Van Zanden, *De economische ontwikkeling*, p. 236.

66 M. Jakubowski-Tiessen, *Sturmflut 1717: Die Bewältigung einer Naturkatastrophe in der frühen Neuzeit* (Munich, 1992), pp. 13–14.

67 Buisman, *Duizend jaar*, vol. 5, pp. 451–3; Jakubowski-Tiessen, *Sturmflut 1717*, pp. 57–68.

68 Jakubowski-Tiessen, *Sturmflut 1717*, pp. 20–4, 40.

the Elbe river, the flood only reached it between five and six o clock in the morning, when people were already awake. Consequently, they were able to save themselves and their cattle on higher ground. Only four people drowned and the number of animals that died was limited as well.[69]

Human suffering was extreme during the disaster and the following months, particularly because it was difficult to provide the survivors with food, fresh water and shelter. Repair of the breached dikes was almost impossible: there was a lack of capital to finance repairs, the inhabitants of the coastal zone were discouraged by the disaster, and the situation was exacerbated by another heavy storm in February 1718.[70] Obviously, local communities were not able to repair the damage. For this, they were in need of support from the state. The state was willing to support the devastated areas, but it wanted something in return: more control over local water management and modernisation of maintenance, especially monetisation. State officials had observed that territories that had introduced better dike maintenance or institutional reforms suffered less damage, such as the province of Friesland or the Groningen region of Oldambt, or were better able to cope with the damage after any disaster, such as the Land Wursten (Wursten was one of the German areas where monetisation had been introduced successfully in 1625).[71] It seemed not unreasonable to demand reform in return for generous financial support.

The area most heavily hit by the Christmas Flood was Butjadingen, where almost 2500 people were killed from a population of just over 8000, and almost all cattle had drowned.[72] Reconstruction was only possible with generous support from the Danish Crown, and after some initial reluctance the Crown was prepared to give such support. More than 700,000 *Reichstaler* were spent by the Danish state on the reconstruction of Butjadingen, of which only some 50,000 was eventually repaid.[73] In return, the state required the introduction of more efficient dike maintenance. It actively promoted the introduction of monetisation and in 1729 succeeded in convincing representatives of the coastal villages to accept monetisation at the level of the *Vogtei* (rural district). Between 1752 and 1762 it was finally introduced in the coastal districts of Butjadingen.[74] The safety of the land was greatly enhanced. Yet monetisation was only a partial success. Although Butjadingen had been hit by devastating floods

[69] Ehrhardt, *Ein Guldten Bandt*, pp. 443, 447.
[70] Jakubowski-Tiessen, *Sturmflut 1717*, p. 40.
[71] Van Tielhof, *Consensus en conflict*, pp. 245–6; M. Ehrhardt, 'Zur Geschichte der Deiche an der Unterweser', in H. Bickelmann et al., eds., *Fluss, Land, Stadt: Beiträge zur Regionalgeschichte der Unterweser* (Stade, 2011), p. 308
[72] Jakubowski-Tiessen, *Sturmflut 1717*, p. 273; Norden, *Eine Bevölkerung*, pp. 78, 330.
[73] Norden, *Eine Bevölkerung*, p. 224.
[74] Jakubowski-Tiessen, *Sturmflut 1717*, pp. 234–5; Norden, *Eine Bevölkerung*, pp. 224–5.

just about every decade in the seventeenth century, during the period 1717 to 1825 there occurred no such floods. The downside of this was that all required investments still had to be made by the coastal districts, and in the second half of the eighteenth century they could no longer count on state subsidies. Therefore, the rates for dike maintenance in the coastal districts were very high, while the more inland districts contributed nothing. This problem was only solved in 1855, when all districts of Butjadingen became liable for maintenance of the seawalls.[75]

In the principality of Ostfriesland, landowners and local water authorities were also unable to finance repairs. Here too, the state had to intervene, but since it did not have the money either, it had to borrow. Eventually, it found the money in the Netherlands. In April 1720, the Dutch States-General were prepared to grant Ostfriesland a loan of 600,000 guilders.[76] The prince of Ostfriesland also wanted to strengthen his grip on the water authorities, and he succeeded in doing so by winning a power struggle with the States of the principality. In 1728 a *Deichdirektorium* was installed that would supervise the water authorities in the western part of the principality. In Harlingerland, the eastern part, the prince already had a strong grip on water management, and he was able to strengthen it even more. The administrators of water authorities in this area became state officials and the water authorities lost all autonomy.[77]

Outside Ostfriesland and Butjadingen, however, there was very little change in dike maintenance in north-west Germany after 1717. By protesting, referring to ancient privileges, and using delaying tactics, landowners managed to prevent most change. In the small territory of Jever, for instance, the dikes had been destroyed so thoroughly that landowners could not identify the stretches of dike they were responsible for. However, when the authorities tried to compile new dike books, the landowners did not cooperate and by 1735 there were still no new books. Landowners preferred continuation of the existing, dangerous chaos over the risk of maybe having to contribute more to maintenance.[78] On the left bank of the Lower Elbe, the kingdom of Hannover also tried to introduce reforms in dike maintenance, and also to no avail. The plans became stranded on the stubborn resistance of local communities, defending their ancient privileges against the state.[79]

From the late seventeenth century, brick and stone were substituted for wood in the construction of sluices in north-west Germany, beginning in Greetsiel in Ostfriesland in 1674. This was two centuries later than in the Low Countries. Why were the Germans so reluctant to introduce this

---

[75] Norden, *Eine Bevölkerung*, p. 226.
[76] Jakubowski-Tiessen, *Sturmflut 1717*, p. 184.
[77] Ibid., pp. 229–31.
[78] Ibid., pp. 231–2.
[79] Ibid., pp. 236–9; Fischer, *Wassersnot und Marschengesellschaft*, pp. 163–6.

innovation? Did it have to do with the fact that dike maintenance was still mostly performed in kind? For large wooden sluices, however, most work was done by specially trained carpenters. Farmers only assisted them in the simple digging and moving of earth.[80] It appears that farmers and landowners were simply reluctant to invest large sums in the construction of brick and stone sluices.

Water Management in Groningen

As noted in the previous chapter, the Dutch province of Groningen was one of the few regions in the North Sea Lowlands where agrarian capitalism had not made much headway by 1700 (see pp. 181–2). This was to change rapidly over the following two centuries. Groningen's swift catching up was, for the most part, caused by changes in the way in which water management was organised in this province. Groningen water authorities were mostly ruled by absentee elites that had purchased their offices and saw them purely as money-making investments (see p. 161). Farmers, whether they were owner-occupiers or tenants, had no influence. In this respect Groningen was different from the other coastal provinces of Holland, Zeeland and Friesland. In those provinces, even though water authorities were also ruled by elites, those elites owned substantial tracts of land in the jurisdiction of the water authorities, and they often included prosperous farmers. They were interested in the safety of that land and prepared to invest in it. In the two centuries under discussion in this chapter, Groningen was the only Dutch province to be hit twice by devastating floods, with thousands of casualties. The flood of November 1686 killed 1600 people and the Christmas Flood of 1717 took the lives of almost 2300 Groningers and 37,000 head of cattle.[81]

As in other areas hit by the Christmas Flood, the rural population was not able to repair the extensive damage. After some initial bickering over the costs, the States of Groningen took responsibility for the repairs and put the work out to contractors. The average height of the dikes was raised from 2.5 to 3.6 metres. Around 1730, new problems emerged: the wood that reinforced the seaward slopes of the dikes was destroyed by pileworm (*Teredo navalis*), a mollusk that bored holes into the woodwork. This tiny animal wreaked havoc in all wooden structures that were exposed to sea water, and the water authorities that used wooden palisades were hit particularly heavily.[82] The most sustainable solution for this problem was to reinforce the slopes of the dikes with stone. However, this was also a very expensive solution that could not be

---

[80] Krabath, 'Mittelalterlicher Deich- und Landesausbau', p. 92; Kramer, 'Binnenentwässerung und Sielbau', pp. 113, 123, 125.
[81] Buisman, *Duizend jaar*, vol. 5, pp. 137, 453.
[82] Schroor, *Wotter*, pp. 52–3; Sundberg, *Natural Disasters*, p. 126.

financed by the individual landowners or farmers. So, government, in the Dutch Republic the provinces, needed to either finance or subsidise: the provinces of Holland (mostly for the dikes around the Zuider Zee) and Groningen decided to do so.[83] In 1755 the province of Groningen also introduced new regulations that put an end to the worst abuses within the water authorities. Board members (*scheppers*) had to reside within the territory they represented, appeal of decisions of the *scheppers* became possible, the water authorities had to submit their accounts to the High Court of Groningen and opportunities to impose unjustified fines to raise income were reduced.[84] In this way, water management was organised more efficiently, and rapacious absentee elites had less opportunities to feed off the countryside.

By the end of the eighteenth century, individual farmers in the low central part of the province began to erect wind-driven drainage mills to enable more intensive farming on their land. Initially, this occurred in a rather haphazard way, resulting in higher water levels in the drainage canals and forcing farmers living on higher ground to install drainage mills as well. In 1807 the province introduced regulations to prevent such coordination problems. In 1857 some 68,000 hectares were drained in this way.[85] By that time, drainage mills were beginning to be replaced by steam pumps, the first of which was erected in Groningen in 1845. After that first pump it took a long time for steam drainage in Groningen to expand. In 1870 there was still only one pump, but between 1878 and 1884 the number suddenly jumped from seven to 38.[86] That sudden increase may have been caused by the reform of water management in Groningen by the States of the province. Between 1854 and 1873 the old water authorities were abolished and replaced by new organisations, whose limits were based – as much as possible – on hydrological requirements. This new, rational organisation of water management may have stimulated the construction of steam-driven pumping stations.[87] As a result of all these changes in water management, in which the provincial administration played a prominent part, Groningen agriculture would be in the vanguard of lowland farming by the third quarter of the nineteenth century.

---

[83] Z. Y. van der Meer, *Het opkomen van den waterstaat als taak van het landsbestuur in de Republiek der Vereenigde Provinciën* (Delft, 1939), pp. 75–7, 106; Schroor, *Wotter*, p. 76.

[84] Schroor, *Wotter*, p. 56.

[85] *Ibid.*, pp. 57–8; P. R. Priester, *De economische ontwikkeling van de landbouw in Groningen 1800–1910: Een kwalitatieve en kwantitatieve analyse* (Wageningen, 1991), pp. 274–7.

[86] Schroor, *Wotter*, p. 95; Van Zanden, *De economische ontwikkeling*, p. 236.

[87] Schroor, *Wotter*, pp. 84–104.

State Intervention in Water Management during the Nineteenth Century

In the previous sections we have seen that during the eighteenth century, states in the North Sea area attempted to intervene in water management to make it more uniform and efficient. The results were less than satisfactory because both local and regional water authorities, and their ruling elites, fiercely and often successfully opposed any change. National and provincial governments could either not overcome this resistance or preferred to avoid the social unrest that might emerge if they pursued their policies. This changed during the nineteenth century, when stronger unified states on the continent began to systematically reorganise water management. The prime mover was a state that has not been discussed in the previous chapters, because it controlled only a small part of the North Sea coast: France.

As a result of the revolutionary wars, in 1795 France was able to annex the whole Flemish coastal plain, including the Dutch part (Zeeland Flanders). The most western part of the coastal plain, near Calais and Dunkirk, had already been in French hands for much longer, and there the water authorities had been integrated into the machinery of the French state.[88] That did not prevent the new revolutionary leaders of the *département du Nord* from abolishing the water authorities in 1790 and making the department and the state responsible for maintenance of the water infrastructure. The results were extremely disappointing, and from 1801 to 1809 the water authorities were reintroduced in the departments of Nord and Pas-de-Calais, albeit somewhat reorganised and under intensified control from the state.[89] In 1797 the administration of the *département de l'Escaut* also decided to abolish the water authorities in former Zeeland Flanders and the adjoining parts of former Austrian Flanders, and hand over the task of maintaining drainage and flood defences to the newly created municipalities. The result was administrative chaos. The municipalities had no expertise in water management, and in many cases it was unclear regarding which municipality a water authority belonged to. Within two months the decision was rescinded.[90]

After this false start, Guillaume Charles Faipoult, appointed *préfet* of the *département* in 1799, made a serious study of water management in Zeeland Flanders and its problems. Maintenance of the flood defences had been neglected during the final years of the eighteenth century.

---

[88] R. Morera, 'Conquest and Incorporation: Merging French-Style Government Practices with Local Water Management in Seventeenth-Century Maritime Flanders', *Environment and History* 23 (2017), pp. 409–30.

[89] Blanchard, *La Flandre*, pp. 272–3.

[90] J. Maenhout, *Het polderbeleid in het Scheldedepartement (1794–1814): Juridische – technische – financiële problematiek: De publieke opinie tegenover dit beleid* (Eeklo, 2021), pp. 74–5.

That was not the fault of the French government, but of the Zeeland Flanders' water authorities. They had made use of the collapse of the Ancien Régime to stop supporting the outer polders that were liable for dike maintenance. The States-General and the States of Zeeland were abolished and were no longer able to force the inland polders to contribute to the maintenance of the sea defences. The outer polders, many of which were 'calamitous' (destitute), faced enormous financial problems and in several cases stopped maintaining their dikes. Faipoult realised he had to revive the old subsidy regulations to save the coastal zone from flooding. He was confronted with the same stubborn resistance as his predecessors: water authorities preferred maintaining their ancient privileges over safety. Parochialism was still rampant.[91] The administrators of the inland polders speculated that the outer polders would somehow continue to maintain their dikes, even without sufficient funds. Around 1800, that speculation proved to be mistaken. From 1797 to 1802 several coastal polders flooded and were abandoned by their owners, forcing the inland polders to make high investments to improve their own sea defences.[92]

The reasons the inland polders gave for refusing to support the outer polders were always the same: they claimed the problems of the outer polders were caused by insufficient maintenance. Moreover, if the dikes of the outer polders breached, that would not be a problem for the inland polders because they had their own inland *slaperdijken* (sleeper dikes) that could serve as sea defences in such circumstances.[93] How hollow such arguments were is proved by a letter written by the *dijkgraaf* of the Watering Cadzand on 22 January 1802, on the day a high flood destroyed the dikes of an outer polder near Cadzand. In this letter he describes how large numbers of people were having to work desperately to reinforce the sleeper dikes, as those dikes had not in fact been maintained well enough to withstand the onslaught of the waves.[94]

By protesting with the central government in Paris, the inland polders managed to postpone Faipoult's new regulations. After the devastating flood of January 1802, however, new legislation was introduced to make the entire hinterland of the dikes contribute to their maintenance. Again, this elicited loud howls of woe from the inland polders. This time, though, they had bad luck; First Consul Napoleon Bonaparte was

---

[91] K. J. J. Brand, 'Van een poging tot opheffen van polders en waterschappen in 1797, tot handhaving van deze publiekrechtelijke instellingen door de keizerlijke decreten van 1811', in A. M. J. de Kraker et al., eds., *Over den Vier Ambachtenz: 750 jaar Keure 500 jaar Graaf Jansdijk* (Kloosterzande, 1993), p. 588.

[92] M. H. Wilderom, *Tussen afsluitdammen en deltadijken* (4 vols., Vlissingen, 1961–73), vol. 4, pp. 230, 261–3; Maenhout, *Het polderbeleid*, pp. 102–5.

[93] Gallé, *Beveiligd bestaan*, p. 208.

[94] Wilderom, *Tussen afsluitdammen*, vol. 4, pp. 262–3.

not impressed by their protests and prefect Faipoult was able to press this new legislation through. The inland polders were now obliged to subsidise dike maintenance of the outer polders. This measure was made slightly more palatable by a subsidy of 500,000 francs by the French government.[95] Napoleon stayed interested in dike maintenance during the remainder of his rule. In 1811 several imperial decrees introduced regulations for water authorities for the Low Countries south of the river Waal, including the Flemish coastal plain, Zeeland, western Brabant and the southern part of the Guelders river area. These decrees had two goals. The first was intended to broaden the financial basis for dike maintenance by grouping all water authorities defended by the same flood defences into polder *arrondissements*. All polders within these *arrondissements* were liable for dike maintenance, but were divided into classes depending on their distance from the seawalls. Those closest to the dike had to pay the highest contributions, those further away paid less.

More revolutionary was the decree that introduced general rules for the organisation of water authorities. Until 1811 each water authority had its own unique bylaws. Napoleon's decree put an end to that by promulgating a regulation that contained basic rules valid for all water authorities.[96] In this way, water management became more uniform and, more importantly, water authorities lost much of their autonomy because now the central government could change their bylaws. Their privileges were no longer sacrosanct, and that did not alter after Napoleon's fall. In Belgium the Napoleonic decrees remained the basis of water management organisation until the 1950s. The province of Zeeland introduced a new regulation in 1841, based, however, on the Napoleonic decrees.[97]

In 1795 France also occupied the Dutch Republic, but it permitted the Dutch to implement their own, 'Batavian', Revolution. This resulted in a new constitution in 1798 that changed the federal republic into a unitary state and made that state responsible for water management.[98] This did not mean that the tasks, rights and organisation of the local and regional water authorities were changed. It only meant that the state from then on supervised maintenance of the most important physical infrastructure, such as sea and river dikes. A national water management service was founded in 1798 to implement this and was maintained after the French left in 1813. The influence of this service on the water authorities was

---

[95] Brand, 'Van een poging', pp. 589–90; Maenhout, *Het polderbeleid*, pp. 108–12.
[96] Brand, 'Van een poging', pp. 594–5; M. de Vleesschauwer, *Van water tot land: Polders en waterschappen in midden Zeeuws-Vlaanderen 1600–1999* (Utrecht, 2013), p. 76; Maenhout, *Het polderbeleid*, pp. 175–9.
[97] Gallé, *Beveiligd bestaan*, p. 222; A. Pauwels, *Polders en wateringen* (Brussels, 1935), p. 37.
[98] B. Toussaint, 'Eerbiedwaardig of uit de tijd? De positie van de waterschappen tussen 1795 en 1870', *Tijdschrift voor Waterstaatsgeschiedenis* 18 (2009), pp. 43–4.

limited. It was mostly involved in king Willem I's ambitious programme of canal and road construction, and lake drainage.[99] There were many changes in national policy regarding water management in the 1795–1813 period, most of which were rescinded within a few years and had no lasting effect.[100]

One idea survived the Batavian-French period: that the state had the right to intervene in the organisation of water authorities. Both the autocratic constitution of 1815 and the democratic constitution of 1848 granted the provinces, under supervision of central government, the right to promulgate general regulations for water authorities, change their bylaws, merge water authorities or found new water ones.[101] About this, the autocratic government and the Liberal opposition were in agreement. Between 1835 and 1841 the provinces of Overijssel, Gelderland, Noord-Brabant and Zeeland introduced general regulations for the water authorities under their jurisdiction. When the Gelderland landowner Anne Gerrit Brouwer protested in 1842, claiming that water authorities were organisations based on private contracts between landowners, which could not be changed by the province, his arguments were refuted by the Liberal statesman Johan Rudolph Thorbecke. The future prime minister claimed that in the modern state public interest had to prevail and thus the constitution of 1815 was justified in granting provinces the right to supervise and regulate water authorities. In 1845, in advice to the *Hoogheemraadschap* of Rijnland, Thorbecke claimed that the rights that had been granted to it in the Middle Ages were delegated public authority. Water authorities were lower organisations of public administration and therefore subject to supervision by higher authorities such as the provinces.[102] From the 1850s, Thorbecke's view was generally accepted and water authorities became cogs in the machinery of Dutch public administration. They remained largely autonomous, but subject to provincial supervision, and the provinces used their legislative power to make water management more efficient and uniform.

North-west Germany and England were less, or not at all, exposed to the influence of Napoleonic legislation. In Germany new regulations were introduced in Ostfriesland (1853) and Oldenburg (1855).[103] Interesting is the new regulation for dike maintenance introduced in Schleswig-Holstein in 1800/1805. This addressed the same problem as

---

[99] T. Bosch, *Om de macht over het water: De nationale waterstaatsdienst tussen staat en samenleving 1798–1849* (Zaltbommel, 2000), pp. 49, 164–75.
[100] Gallé, *Beveiligd bestaan*, pp. 215–19; Toussaint, 'Eerbiedwaardig', pp. 40–6.
[101] Bosch van Rosenthal, *De ontwikkeling*, p. 235; Bosch, *Om de macht*, pp. 125–6.
[102] J. Korf, 'Het tijdvak van 1838 tot 1954', in Moorman van Kappen, Korf and Van Verschuer, *Tieler- en Bommelerwaarden 1327–1977*, pp. 259–70; R. Aerts, *Thorbecke wil het: Biografie van een staatsman* (Amsterdam, 2018), pp. 333–7.
[103] Peters, 'Entwicklung', p. 193.

the Napoleonic legislation for the south-western Netherlands: how to support the outer polder with heavy maintenance costs. The Danish Crown applied the same solution as Napoleon: create districts composed of outer polders and those inland polders that were protected by the dikes of the former. The latter were now obliged to contribute to the maintenance of the flood defences.[104]

The squabbling water authorities in the English fenlands had to wait until 1930, when the Land Drainage Act of that year finally introduced a more efficient and less conflict-prone organisation of water management (see p. 255).[105] This drawback brought unexpected notoriety: the English water authorities are the only ones that were ever mentioned in a work of detective fiction. In Dorothy L. Sayers' 1934 novel *The Nine Tailors*, her sleuth Lord Peter Wimsey has a chat with the keeper of a dilapidated sluice in the Fens. The keeper explains the cause of the sluice's neglect:

> Nobody knows whose job this here sluice is, seemin'ly. The Fen Drainage Board, now – they say as it did oughter be done by the Wale Conservancy Board. And *they* say the Fen Drainage Board did oughter see to it. And now they've agreed to refer it, like, to the East Level Waterways Commission. But they ain't made their report yet.[106]

Later in the novel, the sluice breaches, causing part of the fenlands to flood. Obviously, Sayers, who grew up in a rectory near the Fens, knew about the institutional causes of insufficient maintenance of infrastructure in the Fens.

Conclusion

The development of water management in the North Sea Lowlands during the eighteenth and nineteenth centuries was mostly determined by two factors: increasing state intervention and the introduction of steam drainage. In the eighteenth century, state policy was mostly reactive. The state reacted to demands from the lowlands, for instance, for loans and subsidies after storm events, or for Acts of Parliament to found new Internal Drainage Boards. Now and then the state tried to introduce reforms in water management, but usually backed down when confronted with determined resistance. Groningen was one of the few territories where successful reforms were introduced, but there they had been long overdue. On the continent, change came during the Napoleonic interlude. In 1811 legislation was introduced in the area south of the river Waal that basically turned the water authorities in that area into cogs in the French bureaucratic machinery, while still leaving them

---

[104] Müller and Fischer, *Das Wasserwesen*, vol. 3/2, pp. 297–301.
[105] Summers, *The Great Level*, p. 218.
[106] D. L. Sayers, *The Nine Tailors* (London, 1982), p. 154.

some autonomy. In the Batavian Republic, water authorities were also subjected to government supervision. During the Restoration, old times did not return. Ancient privileges were no longer sacrosanct and water authorities became part of the state apparatus at the local level. This was one of the few things the authoritarian kings and the liberal opposition agreed on. In Belgium there was also no return to old practices: the Napoleonic legislation remained in place.

Another breakthrough was achieved with the introduction of steam drainage. Although the Dutch first experimented with it, it was in the English Fens that steam engines were first used and rapidly diffused from 1817 onwards. Not being dependent on the wind and with much more capacity, steam engines made it possible for the first time to control the water level in the lowlands, and in fenlands in particular, all year through. The growing season could be prolonged, and crop and hay yields enhanced. After the success of the drainage of lake Haarlemmermeer in the 1840s, the new technology was also rapidly diffused throughout the Netherlands. This breakthrough had one serious drawback: it exacerbated the vicious cycle the fenlands had got caught in once they were reclaimed and drained. The processes of shrinkage and oxidation were accelerated.

8

The Apogee of Lowland Farming, c. 1700–1880

Introduction

During the eighteenth and nineteenth centuries farming in the North Sea Lowlands reached its zenith. Following the difficult years of the first half of the eighteenth century, characterised by economic depression, devastating floods and cattle disease, a period of prosperity began that extended until the 1870s, with only one interruption between c. 1817 and 1835. The numerous large, imposing farms that can still be found in the lowlands mostly date from this predominantly prosperous period. In the advanced regions, farming stabilised at a high level of productivity, holding size tended to increase further and as a result social inequality also increased. The regions that were relatively 'backward' started catching up. By 1870 Groningen and the English fenlands were among the most prosperous and productive areas in the Lowlands. The late nineteenth-century agricultural depression brought an end to this period of prosperity, prompting the beginning of a new period, in which the lowlands were no longer in the vanguard, but instead had trouble keeping up with the uplands. Signs of the beginning of stagnation could already be observed before 1880 in the coastal regions from Flanders up to and including Friesland, and it is these regions which will be the subject of the next section. This is followed by a section on regions that still had potential for further development and which were catching up (or not): the English Fens, Groningen, and the Frisian fens. The third section reveals how north-west Germany fared after the devastating Christmas Flood of 1717. Finally, the regions that had not been transformed will be discussed.

Flanders to Friesland

Mid-nineteenth-century observers were sceptical about the performance of farming in the Flemish coastal plain, the Dutch coastal provinces (Groningen excepted) and the central river area. Their comments on the perseverance of fallowing, small cattle herds causing a lack of manure, insufficient care for the soil, and resulting decreasing crop yields were often scathing.[1] They were also mostly wrong. For Zeeland, for instance, Priester has shown that farmers were able to maintain soil fertility and keep crop yields at a high level for over two centuries.[2] Much of the criticism of farming in the coastal zones and the river area was based on a lack of understanding of the ecological constraints confronting farmers in reclaimed wetlands. The nineteenth-century 'experts' also appear to have been unaware of the law of diminishing returns, especially when they pleaded for adoption of the labour intensive farming methods of areas such as inland Flanders in the coastal plain.[3] Such methods would have been extremely expensive to implement on the large farms along the coast, which were already employing large numbers of wage labourers. Those enormous extra expenses on wages would never have been compensated for by the increased yields. Still, there was a grain of truth in their criticism. Jan Luiten van Zanden has demonstrated that, with the exception of Groningen, land productivity in the Dutch coastal provinces was stagnating in the first half of the nineteenth century and that labour productivity was even decreasing.[4] This section will explore the causes of this stagnation and how farmers managed to overcome it, concentrating on four regions: the coastal zone from Calais to Rotterdam, mainland Holland, the Guelders river area, and the Frisian clay zone.

On the Flemish coastal plain and the archipelago in Zeeland and southern Holland, large-scale, commercial mixed farming was already predominant in the seventeenth century. During the following two centuries holding size tended to either stabilise at a high level, as in the districts of Dunkirk and Veurne, or to increase even further.[5] In the western part of Zeeland Flanders – the part of the Flemish coastal plain controlled by the Dutch Republic – in 1750 almost 80 per cent of agricultural land was part of holdings of at least 40 hectares and 93 per cent of holdings over 20 hectares.[6] Smallholdings and family farms were unusual in this region of mixed farming on relatively heavy clays. The need to have at least three to four horses for ploughing and harrowing required a farm size of at least 20 hectares. The smallholders in this region were mostly labourers or non-agriculturalists for whom the holding was a side-activity. In Zeeland Flanders and on the islands of Zeeland, farm size continued to increase until the mid-nineteenth century and even

[1] Priester, *Geschiedenis*, pp. 376–7; Brusse, *Overleven door ondernemen*, p. 4; E. De Laveleye, *L'agriculture Belge: rapport présenté au nom des sociétés agricoles de Belgique et sous les auspices du gouvernement* (Brussels, 1878), appendix 2, 21–3; for a nuanced discussion De Hoon, *Mémoire*, pp. 104–6.
[2] Priester, *Geschiedenis*, pp. 291–2.
[3] Blanchard, *La Flandre*, p. 305.
[4] Van Zanden, *De economische ontwikkeling*, p. 219.
[5] Vandewalle, *De geschiedenis*, p. 106; Vandewalle, *Quatre Siècles*, p. 121.
[6] P. J. van Cruyningen, *Behoudend maar buigzaam: Boeren in West-Zeeuws-Vlaanderen 1650–1850* (Wageningen, 2000), p. 99.

The Apogee of Lowland Farming, c. 1700–1880

beyond.[7] By 1800, in most of this coastal zone, farms of 40 to 80 hectares were predominant. Until recently, it has been assumed that large farms were successful here because they were more efficient than smaller ones: they needed less labour and draught power per hectare and were often able to sell their grain and other crops at higher prices because they did not have to sell immediately after the harvest.[8] Recently this view was criticised by Berghmans, who concluded that the main driving force behind engrossment of farms in the Flemish coastal plain was the damage caused by warfare and the rising burden of taxation required to finance war. According to him, the large tenant farmers were supported by their landlords who permitted them to run up rent arrears – more than small farmers – and they were given preferential treatment by tax collectors. The motivation behind this support was that it was difficult to find individuals wealthy enough to take over a large tenant farm.[9] The fact that holding size in Zeeland and Zeeland Flanders stabilised during the peaceful 1749–95 period, after years of strong growth in more belligerent times, may be an indication that warfare indeed played a part.[10]

A striking characteristic of agriculture in this region with its large farms was its intensity. Farming in Zeeland Flanders and the Zeeland–South Holland archipelago had from the Second Reclamation onwards been predominantly arable, as was demonstrated in chapter six. It was also labour intensive, with much weeding of the crops and preparing of the soil by abundant ploughing and harrowing. As holding size increased, farming did not become less labour intensive. If anything, it became more so, because on several islands the cultivation of madder was increased, which required much labour.[11] Other farmers in this region concentrated on different crops demanding a high labour input, such as wheat, flax and rape.[12] On the Flemish coastal plain, where stock farming had been more important during the sixteenth and seventeenth centuries, in the next centuries a shift took place towards more intensive mixed farming with an important arable branch. In the Castellany of Veurne, known for its dairy production, the share of arable in the area of agricultural land increased from 34 per cent in the second half of the sixteenth century to 54 per cent in 1846, and in the region of Dunkirk even from 40.6 per cent in 1615 to 80.2 per cent in 1893.[13] On the part of the

---

7   Van Cruyningen, *Behoudend*, pp. 100–2; Priester, *Geschiedenis*, pp. 184–90.
8   Van Cruyningen, *Behoudend*, pp. 175, 239–42.
9   S. Berghmans, 'War, Taxation and the Enlargement of Farms in Coastal Flanders (Seventeenth–Eighteenth Centuries', *Agricultural History Review* 70 (2022), pp. 193–218.
10 Priester, *Geschiedenis*, pp. 700–4; Van Cruyningen, *Behoudend*, p. 100.
11 Priester, *Geschiedenis*, pp. 354–8.
12 Van Cruyningen, *Behoudend*, pp. 131–6; Baars, *De geschiedenis*, pp. 135–6.
13 Vandewalle, *De geschiedenis*, p. 167; Vandewalle, *Quatre Siècles*, p. 175.

Flemish coastal plain situated in the province of West Flanders, of 76,000 hectares of agricultural land in 1846, 58 per cent was used as arable, 40 per cent as pasture or meadow and 2 per cent for market gardening. Only where the land was too wet, such as in the Calaisis, did extensive stock farming remained predominant. On the heavier soils of the Castellany of Veurne and on the low plain of the IJzer river around Diksmuide, dairy farming was practised.[14] Crop yields in the coastal plain were considerable. In the Castellany of Veurne, wheat yielded 21 hl/ha in 1846 and barley 44 hl/ha; in Cadzand in the western part of Zeeland Flanders, wheat yielded on average 25 hl/ha and barley 56 hl/ha over the years 1784–1825.[15]

For a century, the large farmers of the coastal plain and the islands to the north of it did well. They even began to purchase their land and farms. Whereas, on the younger polders in particular, in the seventeenth century farmers usually owned less than ten per cent of the land, by 1800 this may have increased fivefold or more. On ten polders and *wateringen* in Zeeland, Zeeland Flanders and the islands of South Holland, covering 16,700 hectares in total (7.5 per cent of the land area in the region), 62 per cent of all land was cultivated by owner-occupiers.[16] There were wide variations, from 15 per cent in border villages in Zeeland Flanders to 90 per cent or more in Zeeland, but a conservative estimate would be that farmers owned about half of all land in the province. The success of the large farmers was also expressed in the construction of impressive farm buildings with characteristic huge wooden barns for crop storage. On the part of the coastal plain that was to become Belgian territory, farmers may have acquired less land because the bourgeoisie of the Belgian cities was very interested in investing in land. That was also felt in Zeeland Flanders, where burghers of Ghent and Bruges had purchased up to 40 per cent of all land in the border villages.[17]

However successful the farmers of this coastal zone may have been, there were signs that farming was beginning to stagnate by the first half of the nineteenth century. Self-declared experts from the landowning but usually non-farming elites criticised the farmers for their conservatism, which would have led to declining yields of crops such as wheat and beans.[18] The critics were overdoing it; recent research has demonstrated that yields did not decrease but stagnated at a high level. However,

---

14 Blanchard, *La Flandre*, pp. 283–4, 304; *Statistique de la Belgique. Agriculture. Recensement général de 1846* (Brussels, 1846), vol. 2.
15 Vandewalle, *De geschiedenis*, pp. 194, 200; Van Cruyningen, *Behoudend*, pp. 183–4.
16 Baars, *De geschiedenis*, pp. 89, 95, 101; Priester, *Geschiedenis*, p. 695; Van Cruyningen, *Behoudend*, pp. 114–15.
17 Van Cruyningen, *Behoudend*, pp. 114–15.
18 Priester, *Geschiedenis*, pp. 8, 14.

there is no denying that productivity of farming in this coastal zone did stagnate in the first half of the nineteenth century.[19] To explain this I will focus in on Zeeland, the province that has been studied most intensively. 'Experts' reproached the farmers for keeping insufficient cattle to manure the fields. They advised the farmers to abolish fallow and keep more land under grass and fodder crops. In Zeeland, however, with its brackish groundwater, it was difficult to expand the cattle herd because of lack of fresh drinking water for the animals. Moreover, transforming arable into pasture would have meant extensification of farming and thus a reduction of income, which might have made it difficult to pay the high rents that were usual in this province.[20] The fallow year was required to fight weeds by repeated ploughing and harrowing, and it was only during the summer season that the land was dry enough to manure it. It appears the traditional farming system had reached a ceiling it could not break through.

From the third quarter of the nineteenth century the Zeeland farmers began to increase their cattle herds and to grow more fodder crops, mangelwurzels in particular. By more intense integration of arable and animal husbandry, productivity was beginning to increase again.[21] The real breakthrough came around 1870, both on the Flemish coastal plain and on the Zeeland and South Holland islands, and in the older literature this has been attributed to the introduction and diffusion of sugar beet cultivation.[22] More recent research has shown that imported seeds and artificial fertilisers also played an important part in this.[23] In Zeeland Flanders the first sugar beet were grown in the early 1850s, for Belgian sugar refineries. At first, diffusion was slow, because sugar beet tends to deplete the soil. Sugar beet cultivation is only possible when the land is well fertilised, and this only became possible after artificial fertilisers became available around 1870. About that time cultivation of madder, an important cash crop in Zeeland, collapsed because the dyestuff that was made from it could not compete with the newly invented chemical dyestuff, alizarine.

Sugar beet proved to be a good substitute for madder. It should be noted that this change in the farming system began in the 1860s, and thus it was not the case – as older literature suggested – that conservative, somnolent farmers were finally woken up by the late nineteenth-century depression. By 1880, they had already again improved both land productivity and labour productivity. Admittedly, they were stimulated by sugar manufacturers who offered them attractive advances for growing

---

[19] Ibid., p. 378; Van Zanden, *De economische ontwikkeling*, p. 204.
[20] Priester, *Geschiedenis*, p. 418.
[21] Ibid., p. 385.
[22] Blanchard, *La Flandre*, p. 299.
[23] Priester, *Geschiedenis*, pp. 407, 409–10.

sugar beet.[24] Sugar beet cultivation was only feasible when there was an industry nearby that could process the beet as there were high transport costs of the bulky product. In Belgium the first factories were founded before mid-century, in the south-western Netherlands from 1858. The factory-owners then had to find farmers willing to grow the crop by offering them advantageous conditions, and from the late 1860s they were helped by the decline of madder cultivation. What they also needed was government support, because beet sugar needed to compete with less expensive cane sugar. The countries on the continent supported their beet sugar industry by providing export bounties, which allowed beet sugar refineries to dump their product on foreign markets, the British market in particular.[25]

Agriculture in mainland Holland was dominated by dairy farming since most of the land was too wet for arable farming or market gardening. South of the IJ river both butter and cheese were produced, north of it mostly cheese. During the depression in the first half of the eighteenth century, butter prices did not decline much, but cheese prices decreased strongly.[26] Both butter and cheese producers were hit heavily by the rinderpest epidemics of 1714–20 and 1744–54 and to a lesser degree by the 1769–83 epidemic.[27] Due to this combination of low cheese prices and *epizoötics*, the dairy farmers in northern Holland suffered most. In 1744 alone, almost 85 per cent of all 20,000 cows in the region of eastern Westfriesland died of rinderpest. The disease did not end there, however, because the farmers who had bought new cattle often lost several of those animals to the disease in the following years. Many farmers had to purchase a new herd several times, and especially the smaller, poorer farmers often were not able to do so and had to give up farming.[28] This led to an increase in holding size of the remaining farmers, but also to heavily indebted farmers 'abandoning' their land: giving up all rights to the land because they were no longer able to pay the land tax and rates of the water authorities. Around the mid-eighteenth century in several regions of northern Holland, hundreds of hectares of land were abandoned, meaning that this land was no longer cultivated and

[24] Van Zanden, *De economische ontwikkeling*, p. 235; Priester, *Geschiedenis*, p. 394.
[25] U. Bosma, *The World of Sugar: How the Sweet Stuff Transformed our Politics, Health, and Environment over 2,000 Years* (Cambridge, MA and London, 2023), pp. 224–7.
[26] A. M. van der Woude, *Het Noorderkwartier: Een regionaal historisch onderzoek in de demografische en economische geschiedenis van westelijk Nederland van de late Middeleeuwen tot het begin van de negentiende eeuw* (Wageningen, 1972), pp. 575, 840–1.
[27] *Ibid.*, pp. 531–3.
[28] L. Schuijtemaker, 'Koeien, kaas, kentering: De agrarische geschiedenis van oostelijk Westfriesland tot 1811' (unpublished MA thesis, University of Amsterdam, 2018), pp. 83, 107.

no longer brought in any tax income.[29] The surviving farmers adapted to the circumstances by keeping more sheep and by expanding arable farming. The latter strategy was only feasible, however, on higher and better drained land.[30] When the cheese prices increased again from the 1750s onwards, farmers in northern Holland returned to dairy farming, although they also kept considerable flocks of sheep. They added wool production to their farming operations in order to spread risks and no longer be entirely dependent on cheese.[31]

By the 1780s, Holland dairy farming had overcome the agricultural depression and cattle epidemics, and remained a highly prosperous and highly productive sector of the economy. It also suffered less from the 1817–35 depression than did arable farming, because dairy prices were not as reduced during these years.[32] However, until 1850 productivity in the Holland dairy sector hardly increased. This was due to productivity being already very high, and to the sector's strong market orientation. To start with the latter, most produce of the dairy farmers was sold on the urban markets of Holland and Zeeland. The cities of these two provinces were in a deep demographic and economic crisis that reached its nadir in 1795–1815 and was only resolved after 1850. Therefore, pre-1850, the Holland dairy farmers lacked a market for the sale of their increased production.[33] They also struggled with the fact that productivity was already very high, and it was difficult with existing technology to find ways to raise it. Some increase in production per animal was possible by feeding more oil cakes and hay, but that led only to limited increases in milk production. As with the arable and mixed farmers in the Calais–Rotterdam coastal zone, the Holland dairy farmers became victims of their own success: they suffered from the high degree of efficiency and specialisation they had reached.[34]

Export provided the way out of the impasse for the Holland dairy farmers. After 1845 exports of butter and cheese, but also of live cattle, increased strongly. In 1838 only 8000 head of cattle were exported, but this rose to 153,000 in 1864. That peak was reached because of rinderpest in Great Britain, which provided opportunities for cattle exports. In the following year, however, the Netherlands itself was confronted with rinderpest and exports collapsed, although they did recover some years

---

[29] Schuijtemaker, 'Koeien', p. 92; K. van der Wiel, 'De boer als Assepoester van de Zaanse geschiedenis: Het boerenbedrijf van de Zaanstreek en de invloed van de industriële ontwikkeling', in E. Beukers and C. van Sijl, eds., *Geschiedenis van de Zaanstreek* (Zwolle, 2012), p. 227.
[30] Schuijtemaker, 'Koeien', pp. 113–17.
[31] Ibid., pp. 127–9.
[32] Van Zanden, *De economische ontwikkeling*, p. 214.
[33] Ibid., pp. 215–16.
[34] Ibid., pp. 214, 219.

later. The large exports of dairy products, cattle and later meat were made possible by raising the productivity of land and animals. The increase in land productivity was the result of better drainage owing to the introduction of steam pumping stations.[35] Increased income owing to increased exports made it feasible for farmers and landlords to invest in expensive pumping stations, leading to increased land productivity and thus to even more lucrative exports.

Around the mid-nineteenth century, farmers in the Guelders river area received a bad press. Experts such as the agronomist Winand Staring particularly chastised the large farmers for their assumed incompetence, lack of judgement and sloppiness. Ironically, he held the Zeeland farmers – who were severely criticised by their own experts – as a laudable example to them. The Zeeland farmers had their fields weeded more thoroughly than their counterparts in the river area; therefore their arable land was not as weed-infested. What experts like Staring forgot, however, was that the circumstances in the two regions were very different. The young marine clay in Zeeland was very fertile and relatively well-drained, and therefore it was worthwhile spending considerable sums on ploughing and weeding. This would be amply repaid by high crop yields. The Guelders river area, however, suffered from seepage of water from the silted-up riverbeds to the fields, which remained cold for a long time in spring because they were saturated with water. As a result, crop growth began late and had to compete with weeds such as corn flowers, and so in spring the fields looked like flower gardens. Fighting this luscious weed growth was extremely costly, and in the end a lost battle. Therefore the large farmers decided not to fight the weeds, which led to low yields, but also maintained low costs.[36] Under the circumstances, this was a rational choice.

That their choice was rational is evidenced by the fact that the large farms, which had emerged in the region in the late Middle Ages, managed to maintain themselves until the end of the nineteenth century. Their tenants also managed to purchase land. In 1650 the farmers in the Over-Betuwe district owned only 27 per cent of the land, but this had increased to 44 per cent in 1790 and 46 per cent around 1820. Many large farms were sold to their tenants, although some of these tenants were forced to sell them again during the depression of the 1820s.[37] The extensive farming system of the large farmers in the river area, based on a combination of cereal cultivation, fruit growing, fattening stock for

---

[35] Ibid., p. 221.
[36] H. K. Roessingh and A. H. G. Schaars, *De Gelderse landbouw beschreven omstreeks 1825: Een heruitgave van het landbouwkundig deel van de Statistieke beschrijving van Gelderland* (Wageningen, 1996), pp. 70–1; Brusse, *Overleven door ondernemen*, pp. 204, 237–8; Bieleman, 'Farming System Research', p. 246.
[37] Brusse, *Overleven door ondernemen*, pp. 123–4, 126–7, 130–1.

slaughter and horse breeding, which could be easily adapted to changing market prices, proved to be economically quite sustainable, although its yields were indeed modest. Smallholders, of which there were more in the western part of the river area, concentrated more on labour-intensive crops such as potatoes, sugar beet and tobacco, because they could work with cheap family labour.[38] In the last quarter of the nineteenth century their number increased also in the eastern part of the river area. The large farmers leased small plots of land to them for fruit cultivation or growing of labour-intensive crops such as sugar beet. The large farmers themselves continued their extensive way of farming, and since cereal cultivation was no longer profitable, they concentrated on horse breeding and stock fattening.[39]

Whereas other regions in the Lowlands were able to break through production ceilings after 1870 by growing new crops and applying artificial fertilisers, this appears not to have occurred in the Guelders river area. The main reason for this was that drainage remained problematic. Not only could the problem of seepage not be solved, it also remained impossible to cultivate the heavy impermeable clays that were present in much of the area. They could only be used for some hay production so long as the weather was not too wet. This was not the result of any lack of judgement or sloppiness on the part of the farmers. Just like farmers elsewhere, they were prepared to invest in steam pumping stations. They were even quite precocious; the first steam pumps were built here in 1846. These pumps were not sufficient to solve the drainage problems of the river area, however.

Agriculture in the province of Friesland suffered heavily from the eighteenth-century rinderpest *epizootics*. The second *epizootic* in particular hit the Frisian clay areas heavily; in the pasture areas 95 per cent of all cows died in 1745 and in the areas with mixed farming 64 per cent.[40] Moreover, like the farmers in Holland, the Frisians suffered from lower prices for dairy products. Frisian farmers were less dependent on dairy sales, however, than their counterparts in Holland. When dairy prices were low, they concentrated more on stock breeding and hay production for the Holland market. Where possible, arable farming was extended. This occurred mostly in the mixed farming region along the Wadden coast, but also further inland. The account books of a farm at Achlum show that the area under oats was extended here strongly during the first half of the eighteenth century, probably owing to export opportunities to Britain. Achlum is only some 5 kilometres from the port of Harlingen, so

---

38 P. Brusse, 'De Gelderse landbouw in de periode 1850–1920', in J. Bieleman et al., eds., *Anderhalve eeuw Gelderse landbouw: De geschiedenis van de Geldersche Maatschappij van Landbouw en het Gelderse platteland* (Groningen, 1995), p. 113.
39 Ibid., pp. 115–18.
40 Faber, *Drie eeuwen*, p. 469.

the oats could be easily exported. Yields of oats were high on this farm: 49 hectoliters per hectare in 1736. Frisian farming was more diverse than Holland farming: the Frisians exported butter, cheese, cattle, horses, rape, and cereals in large amounts.[41] Therefore it was easier for them to adapt to adverse circumstances. The farmers on the wettest and heaviest clays, which were entirely unsuited for arable farming, must have suffered more than the mixed farmers on the fertile silts, however. For the province as a whole, Knibbe has concluded that the agricultural economy was, in spite of all its problems, growing from 1720 onwards.[42]

This growth continued until the early nineteenth century. Then, as in Flanders, Holland, Zeeland and the Guelders river area, stagnation set in. Crop yields stagnated or even decreased. The cause was the same as in the regions discussed above: existing farming technology only permitted production growth through strongly increased labour input, which was economically unrealistic due to relatively high wages and the large holding size prevalent in this zone. Cattle farming fared better during this period. Butter prices were less affected by the 1817–35 depression, and by feeding the animals more cereals and oil cakes production per animal increased. The quality of the grass land was also improved by applying *terpaarde* (terp soil). The soil of the dwelling mounds that had protected the Frisians from floods before dikes were built was very rich in nutrients because it consisted partly of manure. Around 1840, entrepreneurs began to cut away parts of the *terpen* to sell to farmers in the clay pasture area to improve their land. In this way more than three quarters of c. 900 *terpen* in Friesland either disappeared or became much diminished in size. Frisian farmers suffered from the same problem as those in Holland, however: the urban markets in Holland and Zeeland for which they produced were shrinking due to economic and demographic decline. As in Holland, relief came when exports to Britain increased after 1850.[43]

Catching Up or Not: the English Fens, Central Friesland and Groningen

Three regions in the North Sea Lowlands can be said to have lagged behind in the development towards large-scale, specialised commercial farming: the English Fens, the central area of Friesland, and Groningen.

---

[41] Knibbe, *Lokkich Fryslân*, pp. 122, 169, 180; P. L. G. van der Meer, *Opkomst en ûndergong fan in boerenbedriuw ûnder Achlum: De famylje Hibma, 1697–1824* (Leeuwarden, 2001), pp. 19–23.
[42] Knibbe, *Lokkich Fryslân*, p. 250.
[43] Nicolay and Huisman, 'Ploughing', p. 57; Van Zanden, *De economische ontwikkeling*, pp. 205, 213–15; J. J. Spahr van der Hoek and O. Postma, *Geschiedenis van de Friese landbouw* (Leeuwarden, 1952), pp. 420–2.

The Apogee of Lowland Farming, c. 1700–1880

In this section I will try to answer the question of whether or not they were able to catch up after 1700.

I will first discuss the largest fenland area: the Great or Bedford Level, to the north of Cambridge. English historians have paid ample attention to the drainage of this c. 140,000 hectare area during the reign of Charles I and Cromwell, but have shown surprisingly little interest in what happened afterwards. Darby provided an impressionistic outline of farming in the Fens after drainage, and most later authors – excepting Richard Hills and Dorothy Summers – limited themselves to copying Darby.[44] The rather patchy overview that results from this literature can be complemented, however, with data from the archive of the Thorney estate and with the outcome of Joan Thirsk's extensive research on the Lincolsnhire Fens.[45]

Darby painted a picture of a region with a pastoral economy, where here and there small patches of spring-sown grain were grown, where smallholders kept dairy cattle to produce butter and cheese and complemented their incomes with fishing and fowling, and where cattle and sheep were fattened.[46] Records of the management of the Thorney estate in the northern part of the Level, owned by the Duke of Bedford, confirm and complement this picture for the second half of the eighteenth century. A 'Plan for the management of the Thorney estate', dating from 1771, showed that the estate agent deemed only 2727 acres (1104 hectares) on a total of 17,192 acres (6963 hectares) of agricultural land to be suitable for tillage.[47] The economy of the Thorney estate was indeed pastoral and not intensive. An inventory of the herds of the tenants from c. 1750 shows that they held 2605 head of cattle, of which 822 were cows.[48] The low number of cows indicates that the tenants of the estate concentrated on labour-extensive stock breeding and fattening rather than dairy farming. Considering that the tenants leased somewhat less than 7000 hectares, cattle density was low. In the fenland areas of Holland, cattle density was three times higher. However, the Thorney tenants probably held considerably more sheep than the fenland farmers of Holland, who hardly possessed sheep.[49] Sheep were not counted in the inventory of c. 1750, but of two tenants it is mentioned that they owned 70 and 80 sheep, respectively.

Herd size of the tenants is shown in table 8. As in Delfland in 1672 (table 7), middling and large cattle owners with 21 head of cattle or more were predominant, but there was also a group of smallholders.

---

[44] Darby, *The Draining*, pp. 153–62; Hills, *Machines, Mills*, pp. 37–46, 127–37; Summers, *The Great Level*, pp. 183–210.
[45] Thirsk, *English Peasant Farming*.
[46] Darby, *The Draining*, pp. 153–7.
[47] Bedfordshire Archives and Records, R4/4033.
[48] Bedfordshire Archives and Records, R4/4029.
[49] Bieleman, *Five Centuries*, pp. 126, 140.

These individuals with up to ten animals had 115 cows from a total of 197 head of cattle, which indicates that they concentrated more on dairy production than the larger farmers, as Darby had already pointed out.

TABLE 8. Herd size of the tenants of the Thorney estate, c. 1750. Source: Bedfordshire Archives and Records, R4/4029.

Herd size	Number of tenants	%
1–5 head	26	21
6–10 head	15	12
11–20 head	26	21
21–40 head	39	32
41–69 head	17	14
Total	123	100

Labour-extensive breeding and pasturing of cattle and sheep appears to have been the mainstay of farming on the Thorney estate. Another document from the Thorney archives shows why this was so. It mentions that the north bank of Morton's Leam was breached in 1763, 1764, 1767 and 1770 and the land of the estate was flooded in these years.[50] Under such circumstances it was wise to stick to extensive livestock farming. Change was in the air, however. A new cut, dug in 1773/4, improved the drainage of the northern part of the Level, and it appears the estate was no longer flooded, until the end of the century. Also more land seems to have been used as arable: estate agent John Wing mentions 7100 acres (2875 hectares) of arable in a letter from 1800, considerably more than in 1771.[51] Therefore, the increase in rental of the estate from £5457 in 1749 to £11,949 in 1797 was not just caused by the general increase in rents during the Napoleonic years, but also reflects improvements in drainage and farming.[52]

Elsewhere in the Fens the chronology appears to be the same. A landlord from Waterbeach in the southern Level, for instance, complained about the regular flooding of his land in the early 1770s.[53] Later, however, the increasing grain prices appear to have given landowners renewed courage and inspired them to new drainage projects. Already in 1762 landowners had procured an Act of Parliament to drain the land along

---

[50] Bedfordshire Archives and Records, R4/4034.
[51] Bedfordshire Archives and Records, R4/4034, Wing to the Duke of Bedford, 6 December 1800.
[52] Bedfordshire Archives and Records, R4/4034.
[53] Hills, *Machines, Mills*, pp. 38–9.

The Apogee of Lowland Farming, c. 1700–1880

the river Witham.[54] Eventually, all remaining unimproved silt fens were drained in this period. The naturalist Sir Joseph Banks, for example, was the driving force behind the drainage project of East, West and Wildmore Fens near Boston, implemented by John Rennie.[55] Farming the Fens stood or fell with sufficient drainage, and for as long as that could not be guaranteed, risks remained high. Wheat could yield a respectable 22 hl/ha in the Fens, and oats even 42 hl/ha, but one flood could destroy the whole harvest. In 1799 oats sown in the Fens near Boston had to be reaped by men in boats. On the Boston market 25 shillings per last were paid for these oats, whereas good oats were sold at £10 per last.[56]

A transformation of fenland farming could only come about after the introduction of steam pumping stations around 1820. At that point drainage was finally reliable enough for the Fens to become the English breadbasket. Wheat now became the most important product of fenland farming. It is somewhat surprising that the fenland farmers, who had always known and appreciated a pastoral economy, shifted to arable farming as soon as it was possible. In the 1870s, 75 per cent of the land in both peat and silt fens was arable.[57] Their colleagues in the Holland/Utrecht plain, the same kind of fenland environment, stuck to animal husbandry after the introduction of steam drainage. Moreover, prices cannot have stimulated this change in the Fens, because grain prices were low in the first two decades after the introduction of steam pumps.[58] Maybe farmers were inspired by the high grain prices of the Napoleonic period to install the first steam engines and then continued growing wheat in the hope of better prices, and around 1840 they were finally rewarded. Prices increased again and wheat yields had increased to 25–27 hl/ha in normal years, and to over 30 hl/ha in very good years, while the quality of the fenland wheat had also improved.[59] Not only did crop yields increase, but also the cropping pattern changed. In the Lincolnshire Fens between 1801 and 1870 the area under wheat and potatoes increased strongly at the expense of oats. In Deeping Fen in 1801, 300 acres (c. 120 hectares) of wheat were grown; by 1870 that had increased to 7500 acres (c. 3000 hectares).[60] These changes were possible because farmers had invested in soil improvement. They had started claying: digging up deeper layers of clay and mixing it with the peaty topsoil to give

---

[54] Thirsk, *English Peasant Farming*, p. 208.
[55] Hills, *Machines, Mills*, pp. 39–40.
[56] Ibid., p. 43; Darby, *The Draining*, p. 169.
[57] T. Williamson, *The Transformation of Rural England: Farming and the Landscape 1700–1870* (Exeter, 2002), p. 146.
[58] Hills, *Machines, Mills*, pp. 133–4.
[59] Ibid., p. 133.
[60] Thirsk, *English Peasant Farming*, pp. 224–5.

it more weight and substance and to neutralise the acid in the soil.[61] This expensive and labour-intensive form of soil improvement was only worthwhile when the level of the water could be adequately controlled, in other words after the introduction of steam pumping. The same goes for underdrainage with tile drains, which was underway around 1850.[62]

Not much is known about holding size on the Great Level. The records of the Bedford Level Corporation only mention the owners of the land. Their number doubled in the early nineteenth century, but that does not necessarily mean that farm size decreased, because many new owners were non-farmers who did not live in or near the Fens. As in many other regions, farmers may have combined owner-occupied land with tenancies to create substantial holdings. One William Wells from Upwell, for example, owned some land, but also leased 80 hectares from four different owners. Therefore, the large number of smallholdings under 20 hectares, which the ledgers of the Bedford Level Corporation appear to indicate, may well be due to the fact that we only know how much land farmers owned, not how much they leased. What is beyond doubt is that large estates, so characteristic for most of England, were almost entirely absent from the Fens, with the exception of the Duke of Bedford's Thorney estate.[63] For the Lincolnshire Fens, Joan Thirsk has found that holdings there were indeed often small as late as 1870. On the Isle of Axholme many holdings were even smaller than two hectares, and in the entire Lincolnshire Fens only 11 per cent of all holdings were larger than 40 hectares. Only in a recently drained parish such as Deeping St. Nicholas in Deeping Fen were 53 out of 73 holdings over 40 hectares.[64]

The numerous small farms in the Fens could be quite profitable, as even Arthur Young admitted.[65] On the Isle of Axholme farmers specialised in the cultivation of potatoes for the London market. Much of the land on the Isle and the adjoining Hatfield Chase was 'warped'. This meant that silt-rich water from the tidal rivers was fed through a canal to the fields, which were surrounded by six-foot-high banks. The canal was provided with a sluice, which was opened when the tide was high to let the water on to the fields. It stayed there until the sluice gates were opened again at low tide. All through summer this was repeated twice daily. The procedure was costly, but it resulted in the deposition of a 15 to 40 centimetre thick and very fertile silt layer. Some low-lying land was raised in this way by 90 to 150 centimetres, thus also reducing flood risk. After warping, wheat and potatoes could be alternated on this

---

[61] Hills, *Machines, Mills*, pp. 129–30; Summers, *The Great Level*, p. 194; Thirsk, *English Peasant Farming*, p. 223.
[62] Williamson, *The Transformation*, p. 147.
[63] Summers, *The Great Level*, pp. 186–90.
[64] Thirsk, *English Peasant Farming*, pp. 215–16.
[65] *Ibid.*, p. 215.

The Apogee of Lowland Farming, c. 1700–1880

fertile land for more than a decade. It appears warping was introduced in the mid-eighteenth century and was widely practised by 1800.[66] A smallholding could easily reproduce itself because it was profitable, and because land in the Fens was regularly sold at death. Selling the land was the easiest way to divide it among the heirs. This meant that small plots often came on the market, and therefore it was relatively easy to start a small farm.[67] Owing to steam drainage and soil improvement, fenland farmers experienced a period of great prosperity from c. 1840 to the 1870s, although this came at the price of increasing shrinkage and oxidation of the peat layers.

In the fenlands of central Friesland, farming had been adapted to the extremely wet environment. Part of the land was drained by small drainage mills, but most of it was undrained and used for hay production to support extensive cattle farming. From the late eighteenth century, however, more and more complaints were voiced about the functioning of the *boezem* – the system of lakes, rivers and canals required to store and transport excess water to the open water of the Zuider Zee – and about the quality of the hay harvests. The main complaint was that the undrained land flooded too often, also in spring and early summer, resulting in smaller and lower quality hay yields. That something was indeed wrong was demonstrated by estimates of the size and quality of the hay harvest in Friesland from 1810. In the district of Sneek – in which much of the fen area was situated – hay yields from undrained land were about 1000 kg/ha lower than those from drained land, and hay from undrained land was also sold at a considerably lower price, which strongly indicates lower quality.[68] This was caused partly by land reclamation in the upland sandy part of Friesland, which increased the amount of water the reservoir system had to deal with, and partly by peat mining in the fenlands themselves, which reduced the size of the reservoir system; then there was the centuries-old problem of insufficient maintenance of canals and sluices by the cities and rural districts.[69]

Until recently, the general verdict on agriculture in the central Frisian fenlands was that it failed to grow until the province of Friesland built a huge steam pumping station on the Zuider Zee coast in 1920. The slow increase in the number of polders and steam pumps in Friesland, mentioned in the previous chapter, is perceived as evidence for this lack

---

[66] S. R. Haresign, 'Agricultural Change and Rural Society in the Lincolnshire Fenlands and the Isle of Axholme, 1870–1914' (unpublished PhD dissertation, University of East Anglia, 1980), pp. 20–1; Thirsk, *English Peasant Farming*, pp. 230–1; Byford, 'Agricultural Change', pp. 276–82.
[67] Thirsk, *English Peasant Farming*, p. 313.
[68] Knibbe, *Lokkich Fryslân*, p. 80.
[69] Van Cruyningen, 'Water Management', pp. 77–8; Blauw, *Van Friese grond*, p. 223.

of development.[70] Recent research throws doubt on this view. It shows that while the area of agricultural land in the Frisian fenlands remained stable at about 56,000 hectares between 1832 and 1914, the number of cattle almost doubled from somewhat less than 36,000 in 1811 to almost 61,000 in 1910. There was a modest but almost continuous increase of the cattle herd in the area over the nineteenth century.[71] It appears that despite all the complaints of the agricultural sector, the provincial investments in canals and sluices were bearing fruit in the course of the nineteenth century.

The English Fens and Groningen were the parts of the North Sea Lowlands that developed most spectacularly during the eighteenth and nineteenth centuries. At the beginning of the eighteenth century, farms were still relatively small in Groningen. Mixed farming was practised, and as late as 1810 land productivity was considerably lower there in comparison to Zeeland. By the 1890s, however, wheat, barley, oats, and potatoes all yielded considerably more per hectare in Groningen than in Zeeland.[72] At the root of this remarkable achievement lay two interacting developments: the consolidation of the tenants of most Groningen farms as *de facto* owners of the land and the reorganisation of water management from 1755 onwards (see pp. 213–14).

During the eighteenth century the right of the tenants to the land (*beklemrecht*) developed to the further detriment of the landowners. When the province decided to sell its 25,000 hectares of land in 1764–74 this resulted in litigation between the purchasers and the tenants, the *meiers*. The latter won, meaning the tenancies became perpetual and hereditary and the rents fixed. The owners were left with *bloot eigendom* (bare property): the right to receive the fixed rent and a 'present' when the tenancy was sold or inherited. The *meiers* were completely free to farm the land according to their own preferences and reap the fruits of all improvements, and could sell, mortgage and bequeath the tenancy. Since the regulations concerning the 'provincial lands' were followed by other owners of farms that were *beklemd*, the tenants of those farms also profited from the outcome of these lawsuits. Unsurprisingly, nineteenth-century statistics considered these 'tenants' as owners. In 1862, 90 per cent of the land in the province was property of owner-occupiers, of which 63 per cent was held in *beklemming*.[73] In the long run, this shift in landownership also resulted in a shift in political power. After the constitution of 1848, provincial politics was

---

[70] A. van der Woud, *Het landschap, de mensen: Nederland 1850–1940* (Amsterdam, 2020), pp. 183, 212; Blauw, *Van Friese grond*, p. 225.
[71] D. Vellema, 'De landbouw in het Lage Midden van Friesland, 1800–2020' (unpublished BSc thesis, Wageningen University, 2021), pp. 25, 31.
[72] Van Zanden, *De economische ontwikkeling*, pp. 204, 230.
[73] Priester, *De economische ontwikkeling*, pp. 107–11.

dominated by large farmers, many of them *meiers*.[74] It was in this period that the farmer-dominated Provincial States completely reformed the organisation of water management in the province, as described in the previous chapter, and that may have been the main cause of the diffusion of steam pumping from the 1870s onwards.

These changes provided incentives for farmers to invest in improvement of their land. The most conspicuous result of this was an increase in arable land. Whereas in 1807, in the part of the province with clay soils, 42 per cent of the agricultural land was arable, by 1910 that had increased to 72 per cent. In Oldambt this shift towards arable had already begun in the eighteenth century.[75] In the past this was interpreted as a shift from a more pastoral to an arable farming system. Priester has demonstrated that this was not the case; farming in Groningen was and remained mixed. What occurred was that arable and animal husbandry became more integrated. Convertible husbandry, which was already known in Groningen in the sixteenth century, became predominant in the province in the first decades of the nineteenth century. By ploughing grassland, the farmers could make use of the nitrogen that had been stored up in the land. Because the *meiers* had acquired *de facto* ownership of the land, there were no landlords who could prevent them from doing so. Later, grass and clover were introduced into the crop rotation, thereby improving soil fertility and providing more fodder for the stock. In this way both animal and arable husbandry could be improved. Another way in which soil fertility was improved was by *kleigraven* (clay digging), which involved deeper layers of calcium-rich clay being dug up and spread over the land. It is mentioned in archival sources for the first time in 1800.[76] Investing in such expensive improvements was worthwhile for the *meiers* because they were certain that all benefits would accrue to them, and because improved water management guaranteed such improvements would be effective. These changes resulted in yields of wheat increasing from 18.5 hl/ha around 1832 to 22.1 hl/ha in 1862 and 30.5 hl/ha in 1888/90.[77]

Groningen was the only region in the North Sea Lowlands where underdrainage was introduced on a large scale before 1900. Underdrainage was a land-saving technology because pipes below the surface replaced the surface network of ditches and drains. By 1900 some 39,000 hectares of mostly arable land in Groningen was provided with sub-surface drains. Again, this very costly investment was possible because the farmers were the *de facto* owners of the land, and because the water management system in the province had been improved. In many places, the results of underdrainage were disappointing, and farmers returned to the trusted system of

---

[74] IJ. Botke, *Boer en heer: 'De Groninger boer' 1760–1960* (Assen, 2002), p. 359.
[75] Priester, *De economische ontwikkeling*, p. 79; Curtis, *Coping with Crisis*, p. 190.
[76] Priester, *De economische ontwikkeling*, pp. 216–17, 246–7.
[77] Ibid., p. 351.

FIGURE 7. Farm at Finsterwolde, Groningen, 1976. Photo P. van Galen. Collection Rijksdienst voor het Cultureel Erfgoed. During the boom period of the nineteenth century, Groningen farmers demonstrated their prosperity by building huge, monumental farmhouses.

surface drainage. Often the level in the drainage reservoirs was too high, and farmers had failed to understand that the drains had to be maintained regularly or even replaced after seven to 15 years.[78]

Both the expansion of the arable and the introduction of clay digging meant that farming became more labour intensive; clay digging especially required a huge amount of heavy labour. Underdrainage also required a large labour input to construct the drains, but afterwards less labour was required to clean out the ditches. In general, however, labour intensity increased.[79] This intensification occurred during a period in which farm size in Groningen increased. There were 585 holdings over 40 hectares in the clay area in 1755, and that number had increased to 912 in 1862. The share of the land cultivated by these large farms increased from about one third to one half over the same period.[80]

[78] Ibid., pp. 278–81.
[79] Curtis, Coping with Crisis, pp. 193–4.
[80] Paping, Voor een handvol stuivers, pp. 71–3.

After the Christmas Flood: North-West Germany

The 1717 Christmas Flood was probably the most catastrophic flood to hit a part of the North Sea Lowlands during the last millennium (see pp. 210–13). One would expect this to have had far reaching consequences, but in a recent article Tim Soens concluded that the economy of the lowland areas was remarkably resilient, and even this 1717 flood caused surprisingly little long-term disruption, with coastal societies never showing signs of collapse.[81] This was also valid for agriculture in the north-west German coastal region after 1717. Some farmers and many labourers lost their lives, and many farmers went bankrupt, but farming recovered after the dikes were repaired. Around 1740, agriculture was ready to enter a new phase of expansion and prosperity. The only exceptions were the villages of Butjadingen that were liable for dike maintenance, which entered a phase of decline that would last until 1790. This was mainly because the burden of dike maintenance was borne exclusively by the villages immediately behind the seawalls. The villages situated further inland did not have to contribute until 1855. In conjunction with the consequences of the cattle plague epidemics, this caused the bankruptcy of half of all farms in the district of Eckwarden between 1772 and 1781. Many farms were also sold to absentee landlords; as a result, 60–80 per cent of farms were leased during the 1750–1850 period.[82] Apart from Butjadingen, however, farming on the German Wadden coast showed remarkable resilience after the Christmas Flood, sometimes also at the individual level. An interesting case is the Eylerts family from the village of Wiemsdorf on the Lower Weser. After the death of Sebbe Eylerts, who died in the aftermath of the 1717 flood, the family went bankrupt and had to part with a farm of over 100 *Jück* (c. 56 hectares). In 1750, however, a grandson, also named Sebbe Eylerts, managed to re-purchase the farm with 48 *Jück* of land, and by 1771 he had enlarged this holding to its original size.[83] Such a happy ending was probably unusual, but it does demonstrate that even losses due to the 1717 flood did not necessarily reduce farming families to poverty for ever.

Until c. 1870 the development of farming in the north-west German Lowlands had two characteristics: holding size increased; and farming became more oriented on arable husbandry and more labour intensive. Farm sizes had been increasing since the late sixteenth century and this process continued during the eighteenth and nineteenth centuries.[84] Small and middling farmers in particular had to give up their holdings under the pressure of storm damage and high costs of dike maintenance, and the larger farmers were able to purchase or lease their

---

[81] Soens, 'Resilient Societies', p. 160.
[82] Norden, *Eine Bevölkerung*, pp. 225, 247.
[83] Cronshagen, *Einfach vornehm*, pp. 211–12.
[84] Knottnerus, 'Yeomen and Farmers', pp. 165–8.

holdings.[85] In territories such as Butjadingen and Jever the number of farms decreased, middling farms disappeared and large holdings became predominant.[86] At the same time, the area of land that was owned by absentee – often urban – landlords increased. Butjadingen was not the only area where tenant farmers became more numerous; it also occurred in Ostfriesland and the Jeverland.[87] The Land Hadeln and Dithmarschen, where the large farmers remained owners of their holdings, appear to have been exceptions.[88] Jessica Cronshagen has rightly stated, however, that for the *Hausleute* – the farming village elites – the difference between ownership, long-term lease and short-term lease was not that relevant. What mattered was that they had access to the land market. Only this top layer of village society possessed the reputation and the collateral to access the land market and engross its holdings, excluding the rest of the population.[89] Therefore, the major part of the land remained in the hands of the village elites, either as owner-occupiers or as some kind of tenant. By accumulating own land and tenancies they remained able to create large holdings. Otto Knottnerus has shown that at the end of the nineteenth century in the diverse regions of the north-west German Lowlands, between 65 and 91 per cent of all agricultural land was held by farmers cultivating at least 20 hectares.[90]

It seems unlikely that an increase in holding size was accompanied by increasing labour intensity. Intuitively one would expect large farmers to try to cut as much as possible from labour costs, because they needed to hire large numbers of wage labourers to run their farms. The outcome would be, as Bas van Bavel stated, high labour productivity in combination with relatively low land productivity due to insufficient care for soil and crops.[91] We have already seen, however, that in the south-western Netherlands and in Groningen an increase in the size of farms was not accompanied with a decrease, but with an *increase* of labour intensity (see pp. 223, 237). The same thing happened in many parts of north-western Germany from c. 1700 to 1870. In Ostfriesland, Jever and Butjadingen arable farming was extended, and in Hadeln, where it was already predominant, it was intensified.[92] In Hadeln, farmers concentrated on the

---

85 Cronshagen, *Einfach vornehm*, pp. 209–14.
86 Hinrichs, Krämer and Reinders, *Die Wirtschaft*, pp. 50–1, 55.
87 Cronshagen, *Einfach vornehm*, p. 228.
88 L. Bierwirth, *Siedlung und Wirtschaft in Lande Hadeln: Eine kulturgeographische Untersuchung* (Bad Godesberg, 1966), p. 56; Knottnerus, 'Yeomen and Farmers', p. 168.
89 Cronshagen, *Einfach vornehm*, p. 230.
90 Knottnerus, 'Yeomen and Farmers', p. 151.
91 Van Bavel, *Manors and Markets*, p. 332.
92 C. Reinders-Düselder, 'Zur Landwirtschaft in Ostfriesland um 1800', in O. S. Knottnerus et al., eds., *Rondom Eems en Dollard/Rund um Ems und Dollart: Historische verkenningen in het grensgebied van Noordoost-Nederland en*

The Apogee of Lowland Farming, c. 1700–1880

cultivation of demanding crops such as rapeseed and wheat. Rape was introduced here around 1740 and became a prominent part of the crop rotation by 1770.[93] In Jeverland and Ostfriesland wheat and rape were also important cash crops around 1800. By the 1830s, when Jever had become a part of Oldenburg, this small region alone accounted for half of all wheat exports of the grand duchy.[94] Even in a traditional cattle farming area such as Butjadingen farmers shifted towards cereal cultivation. Here with less success, however. The heavy clays of many Butjadingen polders were not very suitable for cereals and the farmers lacked experience in arable farming. As well as insufficient ploughing and harrowing, their maintenance of drains and ditches was inadequate. This contributed to the long-lasting depression of Butjadingen agriculture until 1790.[95]

In north-west Germany the same methods to improve soil fertility were adopted as in Groningen. Pasture was ploughed up and grass land, now often sown with clover, became part of the crop rotation. As in Groningen and in the English Fens, deeper nutrient-rich layers of clay were dug up and spread over the surface of the fields to improve soil fertility; here this was called *wühlen* or *kuhlen*.[96] And, as in those other regions, this expensive way of soil improvement would not have been worthwhile if it had not been possible to guarantee sufficient drainage. In Ostfriesland these investments led to a strong increase in corn yields from the second half of the eighteenth century onwards. Another means to improve soil fertility was controlled flooding. Along the river Oste, for instance, wooden constructions were built in the dikes, which were opened during winter. This resulted in the depositing of layers of fertile sediment of up to 73 centimetres.[97] As in Groningen, the increasing orientation towards arable farming did not lead to neglect of animal husbandry. In Ostfriesland, for instance, the cattle herd increased from 87,000 head in the early eighteenth century to 130,000 in 1867.[98] This increase in herd size led to the production of more manure and thus to increased crop yields. It was possible to sustain the increased herd size because grass and clover were inserted in the crop rotation. The

---

Noordwest-Duitsland (Groningen and Leer, 1992), p. 404; Hinrichs, Krämer and Reinders, *Die Wirtschaft*, pp. 102–5; Bierwirth, *Siedlung*, pp. 42–52; Norden, *Eine Bevölkerung*, p. 281.

[93] Bierwirth, *Siedlung*, p. 42; Schürmann, *Die Inventare*, pp. 112–15, 119.
[94] Hinrichs, Krämer and Reinders, *Die Wirtschaft*, pp. 102–4; Reinders-Düselder, 'Zur Landwirtschaft', pp. 404–5.
[95] Norden, *Eine Bevölkerung*, p. 281.
[96] L. Leemhuis, 'De landbouw op de klei in de 19e eeuw: Een vergelijking tussen ontwikkelingen in Groningen en Oost-Friesland', in Knottnerus et al., eds., *Rondom Eems en Dollard*, pp. 422, 424; Bierwirth, *Siedlung*, pp. 47, 52.
[97] N. Fischer, 'Deich und Marschengesellschaft: Fallstudien zur Mentalitäts-, Technik- und Landschaftsgeschichte des Wasserbaus', in N. Fischer, ed., *Zwischen Wattenmeer und Marschenland*, pp. 110–15.
[98] Reinders-Düselder, 'Zur Landwirtschaft', pp. 404, 410.

farming system remained mixed and was able to increase both vegetable and animal production by integrating the two branches of agriculture.

Wheat Yields, Seventeenth–Nineteenth Centuries

Although wheat was just one of the crops grown on the lowland fields, the productivity of this crop can be perceived as indicative of overall farm productivity, at least for the mixed farms; measuring productivity of dairy farms is much more difficult. Wheat is a demanding crop, and therefore, if its yields are stable or increasing, probably overall arable production will develop positively as well. Moreover, in this period before the introduction of artificial fertilisers, maintaining or increasing yields unavoidably also meant increasing production of manure, and therefore expanding or intensifying animal husbandry. Table 9 presents data for the Groningen clay area, for the Cambridgeshire Fens, and for three areas on the Flemish coastal plain.

TABLE 9. Wheat yields in five areas in the North Sea Lowlands, 1615–1862 (in hl/ha). Sources: Vandewalle, *De geschiedenis*, pp. 193–4; Vandewalle, *Quatre siècles*, pp. 261, 464; Hills, *Mills, Machines*, pp. 43, 133; Van Cruyningen, *Behoudend maar buigzaam*, pp. 181–4, 426; Priester, *De economische ontwikkeling*, p. 351.

	Cambridgeshire Fens	Groningen Clay	Dunkirk District	Veurne District	Western Zeeland Flanders
1615–44			12	11	
1698					16
1740–42					18
1760–84			16		
1784–1825					25
1790s	22				
1830s	25–27	18.5			
1843					25
1846			21	21	
1862		22			

The table shows that by the first half of the nineteenth century, yields were relatively high and the trend was upward, in particular in the regions of Dunkirk and Veurne. Statistics can be deceiving, however, as the case of Western Zeeland Flanders demonstrates. There, the 1698 harvest partly failed due to the severe winter of 1697/8. The data for 1740–2 includes one harvest year, 1740, in which the wheat harvest completely failed due to the extremely harsh winter of 1739/40. In the following two years the yields were 25 and 29 hl/ha, respectively. Therefore, we may assume that, apart from years with extreme weather conditions, yields in this area usually were above 20 hl/ha in the first half of the eighteenth century. In Western Zeeland Flanders, where yields were already high, they remained stable. This is confirmed by Priester's results for Zeeland, where yields also remained stable during the seventeenth and eighteenth centuries.[99] Where yields were already high, they stayed high; where they were lower, they increased. On the eve of the introduction of artificial fertilisers, land productivity in the North Sea Lowlands appears to have reached a high level.

Conclusion

Between 1750 and 1870 the North Sea Lowlands went through the most prosperous phase of their history. Holding size increased in most regions, and in several areas tenant farmers were able to purchase a considerable share of the agricultural land and erect impressive farm buildings. In the first half of the nineteenth century, however, there were indications that the most advanced areas, Flanders, Zeeland, Holland and Guelders, were beginning to stagnate. They appeared to have reached a productivity ceiling they found difficult to break through, although they managed to increase productivity somewhat in the third quarter of the century by more integration of arable and animal husbandry. A real breakthrough occurred here only after 1870 through the introduction of new crops such as potatoes and sugar beet, and the application of artificial fertilisers. Good drainage was crucial for these improvements. In the Guelders river area drainage problems could not be solved and, there, productivity stagnated.

Most of the areas that had lagged behind in the development towards large-scale, specialised commercial farming caught up in the nineteenth century. The most impressive cases are the English Fens and the Dutch province of Groningen. In the English Fens the introduction of steam drainage made it finally possible to realise the potential which seventeenth-century drainage projectors had perceived in the area. Wheat yields increased significantly and the Fens became the breadbasket of England. Farmers shifted from animal husbandry to arable farming and realised high crop yields. Technological change was at the basis of this because the organisation of water management remained far from optimal. Many authorities were active in relatively small areas and spent more time and money on fighting each other than on improving water

---

[99] Priester, *Geschiedenis*, p. 288.

management. This shows that agriculture can also develop well even when encumbered with less than efficient institutional arrangements.

The impressive development in Groningen was the result of both technological and institutional developments. Drainage was improved by the erection of drainage mills, and later steam pumping stations. This enabled farmers to enlarge the arable area and introduced convertible husbandry in places where earlier only extensive stock farming had been possible. It was also the result, however, of the fact that in the late eighteenth century the tenant farmers in Groningen, the *meiers*, had become the *de facto* owners of their farms. This gave them the freedom to plough the land and to reap all the benefits from their investments in improvement. Importantly, it also provided them with political clout. After 1848 they became the dominant group in the Groningen Provincial States, and this provided them with the power to push through a complete overhaul of the organisation of water management between 1854 and 1873, thus creating favourable circumstances for the introduction of steam drainage.

A remarkable conclusion that can be drawn from this chapter is that engrossment of farms in most cases did not lead to more labour-extensive farming. In fact, in many regions – Zeeland, the Flemish coastal plain, Groningen, several north-west German regions and very probably also the English Fens – engrossment of holdings was accompanied by a shift towards more intensive arable farming, also with more demanding crops like madder, wheat and rape, and labour-intensive improvements such as claying. It appears that farmers looked less to the expenditure of their enterprise than to the financial yields of their crops. That attitude sometimes paid off, like in Zeeland and Groningen. Yet it could also be a miscalculation, as in Butjadingen, where the old impermeable clay was badly suited to arable farming and the farmers were not used to the cultivation of arable crops. The farmers of the Guelders river area were therefore wise when they stuck to their extensive mixed farming operations. Soil type and drainage still determined the limits of the viable options for farming.

9

Nature Tamed? Water Management, c. 1880–1980

Introduction

During the century between 1880 and 1980 most local communities in the North Sea Lowlands relinquished control over water management to the state. Therefore, this chapter is almost entirely dedicated to the increasing grip of the state on water management. This was achieved through reorganisations of water authorities and through the implementation of enormous infrastructural projects that went far beyond the capacities of the traditional water authorities, but also through smaller yet still substantial improvements in drainage. Because of the ever-increasing role of the state, the regional approach of the previous chapter will be abandoned here. In this period it was the state which determined the development of water management and its organisation. First we will take a short look at the general history of the North Sea area during this period. The period ends in 1980, when doubts about the environmental sustainability of modern agriculture began to influence government policy, leading to the introduction of environmental legislation. Then a period of transformation began, which will be addressed in the epilogue.

The North Sea Area 1880–1980

The episodes from this period that left the deepest impressions on collective memory were, of course, the two World Wars (1914–18 and 1939–45). During the first of those wars, the Flemish coastal plain was hit heavily when part of it was inundated in October 1914 to halt the German advance. Afterwards the war raged for four years in the western part of Flanders, destroying much of the region. The remainder of the North Sea Lowlands survived the war unscathed. After the war, the German *Reich* had to cede the most northern part of the duchy of Schleswig, with a predominantly Danish population, to Denmark; Holstein and the southern part of Schleswig remained German. At the end of the Second World War, both the Allied forces and the Germans used inundation of parts of the Lowlands to hinder the enemy's operations. In the south-west of the Netherlands and the Guelders river area large areas were flooded. After the Second World War, the borders between the nation states that controlled the North Sea coasts were not changed. A reorganisation did take place within (West-)Germany, however. The state of Prussia was abolished, and the former Prussian provinces between the Dutch–German border and the Elbe river were merged with the grand duchy of Oldenburg along with some other territories into the *Land* (state) of Lower Saxony. This state became responsible for the supervision of water management along most of the German coast, together with the state of Schleswig-Holstein (also a former Prussian province), while limited stretches of coast remained under the control of the city-states of Hamburg and Bremen.

Economically, the period started with a deep agrarian depression that had set in during the 1870s. It was caused by massive cereal imports from the United States, enabled by improved and cheap transport by steam train and steam ship. As a result, domestic cereal prices collapsed. The depression, which lasted until the mid-1890s, was therefore mostly a crisis for arable farming, although in the Netherlands the dairy sector suffered as well. This, however, was not due to imports but rather to Dutch butter being outcompeted on the important British market by the better-quality Danish product. Germany and France reacted to the depression by introducing import tariffs on cereals, while most other states in the North Sea area stuck to free trade. While Belgium, the Netherlands and Denmark supported horticulture, and especially animal husbandry, by financing research, education and extension, and encouraging the founding of cooperatives, the United Kingdom did little to support its agricultural sector. The result was that British agriculture, once admired by agronomists in neighbouring countries, now began to lag behind the continent and appears to have profited less from the recovery after 1895. Whereas in other west European countries production and productivity were substantially raised, in Britain they stagnated.[1]

During the period from c. 1895 to the early 1920s the farming sector was doing well; particularly during and shortly after World War I prices were high. From 1921 prices fell, and after 1929 the agricultural sector fell victim to the Great Depression. Now all countries, even Great Britain, reverted to protectionism and regulation of production. The grip governments now had on farming was not to be released after World War II, and from 1964 was continued in the Common Agricultural Policy (CAP)

---

[1] Bieleman, *Five Centuries*, pp. 154–5; Y. Segers and L. Van Molle, *Leven van het land: Boeren in België 1750–2000* (Leuven, 2004), pp. 50–1; M. Tracy, *Agriculture in Western Europe: Challenge and Response* (London, 1982), pp. 45–52, 116–19; J. L. van Zanden, 'The First Green Revolution: The Growth of Production and Productivity in European Agriculture, 1870–1914', *Economic History Review* 44 (1991), pp. 228–9.

of the European Economic Community (EEC). By 1973 all countries in the southern North Sea Area were members of the EEC. The post-1945 period was characterised by fast economic growth and increasing prosperity for broad layers of the population of the western European countries. Farm incomes, however, kept trailing behind despite measures by government and the EEC to support the sector, especially through very costly price support.[2]

Meanwhile, the farming population of the countries in the North Sea area had been declining. In both Great Britain and Belgium it was declining both in absolute numbers and as a percentage of the total labour force from the nineteenth century onwards. Throughout the Netherlands and Denmark, the absolute number of workers in agriculture continued to increase during the nineteenth century, and in the Netherlands even until 1947. In all countries, however, the share of the farming sector in total employment decreased. By 1950, only five per cent of the total workforce in Britain was employed in agriculture, 12 per cent in Belgium, 19 per cent in the Netherlands and 23 per cent in Germany. Nonetheless, this did not lead to diminished influence of the sector on agricultural policy. Owing to the strong connections between farming organisations and political parties, and to the fact that politicians still had the food shortages of the war on their mind, governments continued to support agriculture. Post-War policy led to an enormous increase in production and productivity and overproduction of food instead of shortages.[3] That eventually forced the EEC to change course and reduce support for agriculture.

The first three decades after World War II were characterised by trust in science and technology, which were assumed to be able to solve most problems. But nature cannot be completely controlled. Oxidation and shrinkage of peat soils continued, and stronger pumps had to be constructed, which only strengthened the vicious cycle that had started when the fens were drained in the Middle Ages. Nor were disastrous flood events phenomena of the past. More than 2000 people in the Netherlands and England fell victim to the flood of February 1953, and the February 1962 flood killed over 300 people in Germany.[4] Those disasters functioned as catalysts. They opened the minds of the public and politicians to plans which they would not have contemplated under normal circumstances. This resulted in enormous public works and drastic reorganisations of water management.

---

[2] Tracy, *Agriculture*, pp. 159–64, 231–53; Bieleman, *Five Centuries*, p. 250.
[3] Tracy, *Agriculture*, p. 236; Bieleman, *Five Centuries*, p. 149.
[4] Van de Ven, *Man-Made Lowlands*, p. 399.

Great Works: State Projects in Water Management

The Industrial Revolution also had consequences for drainage and land reclamation. The Haarlemmermeer project, discussed in chapter seven, demonstrated that projects were now feasible on a scale that would have been unthinkable only a century earlier. Dreamers and visionaries came up with more, as well as less realistic, plans. For instance, during the nineteenth century, several groups in the Netherlands lobbied for plans to drain the Zuider Zee area and the Dutch part of the Wadden Sea. Due to their gargantuan scale and the enormous costs involved, such plans had never been taken very seriously. In 1916, however, a storm surge in the Zuider Zee area demonstrated the vulnerability of this region and the need for better flood protection. Minister for Public Works at that time was Cornelis Lely, who had already presented a plan to close off the Zuider Zee from the North Sea and change large parts of it into farmland in 1891. Struck by the disastrous storm and by the lack of land to feed the population during World War I, parliament in 1918 agreed to implement Lely's plan. This happened a few months before the end of his third and last term as minister, after he had promoted his plans for 30 years. From the beginning it was obvious that only the state would be able to organise and finance these huge works. The *Dienst Zuiderzeewerken* (Zuider Zee Works Agency) was founded and began the works in 1920 by building a dam from the North Holland coast to the island of Wieringen. Due to budget cuts the work was slowed down in the early 1920s, but in 1932 a second dam from Wieringen to the coast of Friesland was completed. This 30-kilometre dam replaced 300 kilometres of dikes as primary flood defence. The Zuider Zee became the freshwater lake IJsselmeer. The first new polder, Wieringermeer, with a land area of 20,000 hectares, was already drained in 1930, the year after Lely's death. By 1968, 165,000 hectares of sea had been changed into land, most of it being good farmland with light marine clay soils. The safety of the land surrounding the former Zuider Zee was much enhanced, and flood defence expenditure was reduced due to the shortening of the coastline.[5]

From the 1930s, Johan van Veen, an engineer of the *Rijkswaterstaat* (the state agency for public works and water management), was worried about the safety of the south-west of the Netherlands. This was mostly an archipelago of small islands with fertile soils, in the delta of the Scheldt and Meuse rivers. In his opinion, the dikes in this region were too low and not able to withstand heavy storm surges. His warnings led to the establishment of the Storm Surge Commission in 1939 and the closing off of sea inlets in 1950 and 1952. For Van Veen this was not enough, and he

---

[5] Ibid., pp. 370–8; Van der Ham, *Hollandse polders*, pp. 196–203; W. van der Ham, *Verover mij dat land: Lely en de Zuiderzeewerken* (Amsterdam, 2007), pp. 50, 59, 145–50.

Figure 8. Completion of the Afsluitdijk, 28 May 1932. Photo L. Verbost. © Rijkswaterstaat. This 32-km dam between Noord-Holland and Friesland was the first of the huge state-led and -financed infrastructural works that shortened the Dutch coastline and reduced flood risk. Boats and cranes are celebrating the historic moment by blowing their steam whistles.

continued pressing for more measures. His often rather untactful reports made him unpopular with several of his superiors and earned him the nickname 'Doctor Cassandra'. In 1951, a newly appointed director general of the *Rijkswaterstaat* showed greater appreciation for Van Veen and offered him the opportunity to formulate a plan to prevent disastrous flooding. Van Veen submitted his report on 29 January 1953. Two days later, during the night of 31 January, due to a heavy north-westerly storm combined with a spring tide, the dikes in the south-west of the Netherlands were breached in many places: 129,000 hectares of land were flooded, 3000 houses destroyed and 43,000 damaged, and 1835 people, 20,000 cows, 2000 horses and 12,000 pigs were killed.[6] It was the most disastrous flood that had hit this region since 1570.

[6] W. van der Ham, *Johan van Veen, meester van de zee: Grondlegger van het Deltaplan* (Amsterdam, 2020), pp. 220–4; Van de Ven, *Man-Made Lowlands*, pp. 399–400; K. Slager, *De Ramp: Een reconstructie* (Goes, 1992), pp. 351–2; W. van der Ham, 'Veiligheid: Het Deltaplan: Nieuwe bescherming voor een

Within three weeks after the flood, a commission was appointed with Van Veen as its secretary, tasked with devising a plan to prevent such disasters in the future. In 1954 the commission presented its first reports and in the same year the work began. During the next two decades most sea inlets in the south-west were closed off from the sea with huge dams, dikes were strengthened, and movable storm barriers were erected in some of the major rivers. The costs were enormous. In 1954 an economist estimated them at 2.7 billion guilders, but any cost was considered acceptable that would help to prevent a repetition of the 1953 disaster.[7] The Netherlands was not the only country that was hit by the 1953 flood. Along the east coast of England 83,000 hectares of land were flooded, from Lincolnshire to the Thames estuary. Over 300 people were killed, 58 of them on Canvey Island, where thousands of people became homeless. The British government also took measures to prevent future disasters. Their most important project was the Thames Barrier, a retractable barrier able to be raised at high tides to protect central London from flooding. However, the English needed more time to reach a decision: construction only began in 1974.[8]

North-west Germany suffered much less from the 1953 storm, but it served as a wake-up call for German policymakers. They had been made aware that the same weaknesses in flood defences existed in Germany. Shortly afterwards, the government of the state of Lower Saxony presented a programme for flood protection for the years 1955–64. This was too late, however, to prevent the storm disaster of February 1962. Where the programme had already started, as in the Land Hadeln in 1960, it proved its worth, because the damage was limited there. But elsewhere thousands of hectares of land were flooded and over 300 people died. The 1962 disaster led to the same kind of measures as in the Netherlands and England: raising of dikes, embanking islands in river plains, and constructing storm barriers in tributaries of the major rivers. In this way, the stretch of primary flood defences that had to be maintained was reduced from 1200 to 700 km. Here too, local communities were not able to finance the expensive measures and the Federal government had to subsidise the works. Seven billion *Mark* was spent on dike construction from 1954 to about 1990. The government

---

kwetsbaar land', in W. van der Ham et al., *Modern wereldwonder: Geschiedenis van de Deltawerken* (Amsterdam, 2018), pp. 23, 30–33.

[7] Van der Ham, *Johan van Veen*, pp. 250–62; Van der Ham,'Veiligheid', p. 47; P. Brusse, 'Economie: De effecten van de Deltawerken en regionaal overheidsbeleid op de economische ontwikkeling van Zuidwest-Nederland', in Van der Ham et al., *Modern wereldwonder*, p. 262.

[8] J. A. Steers, 'The East Coast Floods', *Geographical Journal* 119 (1953), p. 280; D. Summers, *The East Coast Floods* (Newton Abbot, 1978), pp. 7, 87–90, 138–45; J. Purseglove, *Taming the Flood: A History and Natural History of Rivers and Wetlands* (Oxford, 1988), p. 72.

of Schleswig-Holstein, where the *Halligen* in particular had been hit heavily by the 1962 flood, introduced legislation in that same year, which resulted in strengthening of the sea defences. In addition, the length of the sea dikes was shortened from 564 to 358 km by the construction of new dams and barriers in river mouths.[9]

'Small' Works: Drainage

Disasters, great works and the triumphs and tragedies of great men like Lely and Van Veen speak to the imagination and have been preserved in the collective memory. Much less known are the drainage schemes that were implemented during the twentieth century, even though they had a substantial impact on farming and the environment. This was more piecemeal work by farmers improving underdrainage of their land, and water authorities great and small improving arterial drainage.

For a fenland region such as the Great or Bedford Level, the risk of flooding by rivers had always been greater than that of flooding from the sea. Rivers such as the Great Ouse carried the excess water of large catchment areas through the Level. The amounts of water increased as much of the uplands became better drained. As early as 1638 Cornelius Vermuyden had proposed the construction of a canal that would lead most of the river water around the fens and then to the sea. Probably for budgetary reasons this part of Vermuyden's plan was never implemented. By the 1930s the situation had become so critical that far-reaching measures appeared to be required. The River Great Ouse Catchment Board sought advice from Sir Murdoch MacDonald & Partners. In July 1940 they presented a plan to construct a cut-off channel to intercept the upland waters, as Vermuyden had proposed. Due to World War II and high costs, this plan was not implemented. Then, in March 1947, after a severe and wet winter, riverbanks breached and 23,000 hectares of fenland were flooded. Following this disaster, the Catchment Board was convinced of the necessity of Sir Murdoch MacDonald's plans, but now the King's Lynn Conservancy Board opposed them as it feared the plan might have negative effects on the navigability of the river. They even employed Johan van Veen as an external adviser to prevent implementation of the MacDonald plan. Agreement was only reached in 1953, and finally in 1964 the Great Ouse Flood Protection Scheme was completed

---

[9] Fischer, *Im Antlitz*, pp. 387–9; Fischer, *Wassersnot und Marschengesellschaft*, pp. 330–40; Ehrhardt, *Ein Guldten Bandt*, pp. 561–2; J. Kramer, 'Entwicklung der Deichbautechnik an der Nordseeküste', in Kramer and Rohde, eds., *Historischer Küstenschutz*, p. 105; Bantelmann et al., *Geschichte Nordfrieslands*, pp. 420–1.

at a cost of over ten million pounds. The result showed many similarities with Vermuyden's original scheme.[10]

The scheme that was implemented in the Great Level was meant to reduce flood risk, but that risk was caused by improved drainage in the uplands. The British government – and other governments in the region – discovered that improved field drainage led to larger amounts of water needing to be evacuated and therefore required improved arterial drainage. Moreover, after the long agricultural depression and the First World War, drainage works were often in bad repair. In 1927 it was estimated that of about 1.8 million hectares of land in England requiring drainage, over 700,000 hectares were in need of new drainage works. During the late 1920s and the 1930s, the government invested large sums in arterial drainage and some sea defence schemes. The aim of this drainage programme was technical: to raise crop yields and livestock output. The agricultural capacity of the land had to be fully exploited. This agronomic perspective remained predominant until well into the 1970s. From 1971 to 1980, 84,000 hectares of land were drained. British farming was expected to produce as much food as efficiently as possible. There was very little attention given to the possible environmental consequences.[11]

Britain was far from exceptional. All countries surrounding the North Sea followed the same, 'productivist', track. In the transformed wetlands, this required improved drainage. In the Netherlands, the Directorate of Agriculture in 1917 published a report on 'The influence of drainage on Dutch farming', in which insufficient drainage was identified as a cause of suboptimal land productivity. Here too, the aim was to raise land productivity through better drainage.[12] Before the Second World War, however, there was no state programme for improvement of drainage in the lowland part of the Netherlands. Improving drainage was left to the water authorities and the provinces. An important innovation in the first decades of the twentieth century was the replacement of steam pumps by electric pumps. Electric pumping stations could be easily connected and organised in networks, which improved the efficiency of drainage. They also needed less maintenance and could function automatically.

[10] Summers, *The Great Level*, pp. 227–42; Van der Ham, *Johan van Veen*, pp. 204–10.
[11] J. Sheail, 'Arterial Drainage in Inter-War England: The Legislative Perspective', *Agricultural History Review* 50 (2003), p. 254; J. Bowers, 'Inter-war land Drainage and Policy in England and Wales', *Agricultural History Review* 46 (1998), pp. 64–8; J. Sheail and J. O. Mountford, 'Changes in the Perception and Impact of Agricultural Land-Improvement: The Post-War Trends in the Romney Marsh', *Journal of the Royal Agricultural Society of England* 145 (1984), pp. 53–4; Purseglove, *Taming the Flood*, p. 17.
[12] Directie van den Landbouw, 'De invloed van den waterafvoer op het Nederlandsche landbouwbedrijf', *Verslagen en Mededeelingen van de Directie van den Landbouw* 1917 no. 1.

These pumping stations were provided with a float: when the water rose, the float moved a switch, which started the engine. In this way, one engineer could operate several pumping stations. In England, a chain of electric pumping stations installed on the river Trent around the middle of the twentieth century finally solved the remaining drainage problems in Hatfield Chase. One Dutch province followed a different course: Friesland. Here the draining of the central fenlands remained problematic. After long drawn-out discussions, in 1913 the province decided to build a steam pumping station on the Zuider Zee coast near Lemmer. This was the biggest steam pumping station ever built. It was completed in 1920 and is still functioning.[13]

After the Second World War improvement of drainage became part of land consolidation schemes. In the Netherlands, land consolidation not only meant exchange and amalgamation of plots, but also improvement of productivity through better roads, farm buildings and drainage. The countryside was completely reorganised. These activities were heavily subsidised by the national government, which, since the Hunger Winter of 1944/5, was motivated even more by the wish to enhance productivity. Between 1924 and 1985 about three quarters of all agricultural land in the Netherlands was part of 452 land consolidation schemes, the vast majority of them implemented after the Second World War.[14] Productivity was indeed much improved, although this came at some cost: the disappearance of historic landscapes and neglect of the environmental consequences. In the fenlands, improvement of drainage meant that the processes of shrinkage and oxidation were sped up. Until the late 1970s, however, policymakers did not perceive this as a serious problem. Britain and the Netherlands followed the same policy. Belgium followed a bit later. An Act of 1956 made land consolidation possible here, and as in the Netherlands until well into the 1980s, land consolidation primarily served to enhance productivity of agriculture. In north-west Germany as well, large-scale drainage schemes were implemented to improve land productivity.[15]

Private Reclamation Projects along the Wadden Coast and the Wash

Although huge state-financed water management projects were predominant in the post-1880 period, private initiative had not completely disappeared. In the province of Groningen, for instance, the owners of the farms adjoining the seawall had maintained their rights to the foreshore,

---

[13] Van der Woud, *Het land*, pp. 211–12; Byford, 'Agricultural Change', p. 274; Louman, *Fries waterstaatsbestuur*, p. 112.

[14] Bieleman, *Five Centuries*, pp. 243–5.

[15] Segers and Van Molle, *Leven van het land*, pp. 148–9; J. Kramer, 'Binnenentwässerung und Sielbau', pp. 130–1.

including the right to reclaim it. In this region, groups of farmers now and then cooperated to reclaim sizeable polders. When successful, they could add vast stretches of land to their farms or create new farms in the polder. However, as in the past, this was a risky business. In 1872, for instance, nine farmers agreed to reclaim the marshes that were to become the Westpolder in the north-west of Groningen. The cost of building a seawall and creating a drainage system for almost 500 hectares of land was 750,000 guilders, much more than had been estimated. That was already a very large sum, but it got worse. In January 1877 the new dike was breached at several places and the polder flooded: 27 people were killed and six newly built farms heavily damaged. Repairs cost another 300,000 guilders. Three of the original participants withdrew from the project but the others decided to continue.[16]

Those who persisted became owners of large plots of very fertile polder land, but their investments were huge, and by the time the Westpolder was finished, the depression of the late nineteenth century hit the mixed farmers of northern Groningen. The farmers in the new polder survived the depression, however: Sicco Mansholt, the future Minister of Agriculture and European Commissioner, was born on a large farm in the Westpolder in which his great-grandfather had been one of the original investors. An illustration of their resilience is demonstrated by the fact that the deep channels in the dike breaches of 1877, which could not be reclaimed, were converted into duck decoys and in that way could provide revenue to the landowners on the polder.[17]

On the Wadden coast private reclamation projects continued until the interwar years. In Nordfriesland a consortium of farmers reclaimed the Sönke-Nissen-Koog near Bredstedt in 1924–6. The works were financed by Sönke Nissen, a railway engineer who had become rich from diamond mining in Namibia, a former German colony. In the 1930s the Third Reich took over land reclamation on the Schleswig-Holstein coast, resulting in polders with names like Hermann-Göring-Koog and Adolf-Hitler-Koog.[18] On the coast of the Wash in eastern England private reclamation schemes continued for even longer. During the twentieth century some 4000 hectares of land were reclaimed here, mostly by farmers living along the coast. The last intake was completed in 1979. Here too, reclamation of saltings was a risky enterprise, with new seawalls sometimes eroding or breaching shortly after completion of the works.[19]

---

[16] I. Noordhoff, *Schaduwkust: Vier generaties hertekenen de rand van de Waddenzee* (Amsterdam, 2018), pp. 25–31, 43, 45, 50–2.
[17] Ibid., pp. 65–6.
[18] Bantelmann et al., *Geschichte Nordfrieslands*, pp. 298, 335, 367–8.
[19] R. Sly, *Soil in their Souls: A History of Fenland Farming* (Stroud, 2010), pp. 56–62.

Reorganisation of Water Management

During the twentieth century the states in the North Sea area strengthened their grip on the water authorities and introduced far-reaching reorganisations. This began in England, where the organisation of water management was chaotic. By 1930, there were 49 Commissions of Sewers, 198 authorities established by private acts, and 114 elective boards created under enclosure awards or by the Board of Agriculture. During the inter-war years, it became clear that this chaotic institutional framework was itself an obstacle to the improvement of drainage the government pursued. Reorganisation of water management was required to ensure that upland areas and non-agricultural interests would contribute to the financing of water management. The Land Drainage Act of 1930 brought this about by creating 46 Catchment Boards, each responsible for the maintenance of drainage in the whole catchment area of a river. The boards were comprised of up to 31 members, two thirds of whom represented county councils or county borough councils. The Catchment Boards maintained the 'main rivers'; at the local level, Internal Drainage Boards (IDBs) remained responsible for the maintenance of smaller drains and ditches. IDBs remained organisations of local landowners.[20]

The Catchment Boards became River Boards in 1948, then River Authorities in 1965 and finally were amalgamated into ten Regional Water Authorities in 1974. This last change was the most far-reaching. The new Regional Water Authorities were not only responsible for drainage. Their major tasks were the development and control of water resources, distribution of water in bulk and its supply to consumers, control of water quality and the treatment of sewage. Land drainage was only a secondary task for these organisations. This did not mean, however, that the agricultural interest was not served by them. Decisions on drainage were not taken by the main board of the Authorities, but by a regional land drainage committee, the majority of whose members were farmers. Moreover, the IDBs responsible for the 'non-main' rivers were also controlled by farmers.[21]

England had over 360 water authorities of all kinds in 1930. That number sounds impressive, but it was modest compared to the Netherlands; by 1935 there were 2652 water authorities in that country.[22] Some were large, many were small. Some of them existed within the same territory but had

---

[20] Sheail, 'Arterial Drainage', pp. 254, 264; Bowers, 'Inter-war Land Drainage', p. 68; G. Bankoff, 'Of Time and Timing: Internal Drainage Boards and Water Level Management in the River Hull Valley', *Environmental History* 27 (2022), pp. 88–9.

[21] Summers, *The Great Level*, pp. 243–5; Purseglove, *Taming the Flood*, pp. 68, 216–21; Bankoff, 'Of Time and Timing', pp. 99–100.

[22] D. R. Mansholt, *Beschouwingen over een onderzoek naar de waterschapslasten in Nederland* ('s-Gravenhage, 1941), pp. 306, 325–81.

different tasks: one, for instance, was responsible for drainage, another for maintenance of a dike. Coordination was unsatisfactory or non-existent. The number of water authorities had increased from the Middle Ages onwards and slowly continued to increase in most of the country until the mid-twentieth century. Only in the province of Groningen had a thorough reorganisation taken place in the third quarter of the nineteenth century, as discussed in chapter seven. In the western Zeeland Flanders region, in the province of Zeeland, an area comprising some 29,000 hectares, there were no less than 76 water authorities, some of them smaller than ten hectares. An amalgamation into one water authority was achieved in 1941. It had taken the provincial administrator responsible for water management, Petrus Dieleman, many years to get his plans accepted. He was confronted with stubborn resistance from local farmers' elites, who did not want to give up offices that gave them some extra income and, more importantly, status and power.[23] This helps to explain why reform was more likely to be successful in nineteenth-century Groningen. There the plans were promoted by members of the liberal elite of the farming population and thus more easily accepted.

The disaster of 1953 served as a wake-up call. Like all major flooding catastrophes, it was partly man-made. Many, often small, water authorities had been liable for maintenance of stretches of dike. In the province of Zeeland there were still 305 water authorities in 1949. Maintenance was often suboptimal, and the smaller authorities lacked sufficient technical expertise. Often dikes were too low or the gradient of the dike on the landward side was too steep. Water that had streamed over the dike could then undermine it on the land side. This contributed to the bad state of repair of many sea walls and the large number of breaches that occurred in the night of 31 January/1 February 1953. Within weeks of the disaster the water authorities on some islands had already requested the province of Zeeland to found a single water authority for their island. Between 1959 and 1965 eight new water authorities were created that replaced most of the 300 that existed in 1949. In other provinces the same process took place, and by 1992 the number of water authorities in the Netherlands had been reduced to about 150.[24]

Improvement of flood protection was not the only reason for mergers of water authorities. The Pollution of Surface Waters Act of 1970 was the first legislation in the Netherlands that dealt with water pollution. Most provinces made water authorities responsible for sewage and prevention of water pollution. This required new expertise and large investments in

[23] M. de Vleesschauwer, *Het Vrije van Sluis: Polders en waterschappen in West-Zeeuws-Vlaanderen 1600–1699* (Utrecht, 2013), pp. 139–46.
[24] Slager, *De Ramp*, pp. 409–11; *Provinciale Almanak voor Zeeland voor het jaar 1949* (Middelburg, 1949), pp. 338–525; *Encyclopedie van Zeeland* (3 vols., Middelburg, 1982–4), vol. 3, pp. 282–92; H. J. M. Havekes, *Successful Decentralisation? A Critical Review of Dutch Water Governance* (Deventer, 2023), p. 10.

wastewater treatment plants. Larger authorities would be more suited to this task. This change also meant that a new group of rate payers was formed. No longer were landowners the only rate payers: the companies that were responsible for surface water pollution also had to pay rates. The Regional Water Authorities Act of 1992 extended the group of rate payers even further: from then, all households living in the territory of a water authority were obliged to pay rates. This also had consequences for the way board members were selected. Water authorities were no longer clubs of farmers and landowners who protected their own business. From now on elections were held in which all adult inhabitants could vote.[25]

The case of the Dutch province of Friesland deserves special attention here. Maintenance of flood defences was in the hands of water authorities here, as in other provinces, but the regional organisations liable for drainage were weak (see pp. 106–7, 159–60). From the seventeenth century onwards the province of Friesland reluctantly accepted liability for maintenance of individual sluices and canals and eventually for the reservoir system, the *boezem*. By the end of the nineteenth century this had become a competence that was jealously guarded by the provincial administrators, who rightly considered it as one of the few areas in which they could exert real political power. When pressure groups such as the Frisian Association of Agriculture argued in the 1890s for the founding of water authorities that were to take over the responsibility for drainage, the provincial administrators bluntly refused to consider these proposals.[26]

England and the Netherlands were the countries where the most far-reaching reorganisations were introduced. In Belgium, though, changes were very limited. The Napoleonic of 1811 remained the legal basis for water management. Attempts to introduce new legislation in the 1880s and 1930s came to nothing.[27] Eventually, the 1953 disaster was the motivation behind new legislation introduced in 1956 and 1957. Changes, however, continued to be limited. Water authorities remained associations of landowners with the right to levy rates as they had been from the late Middle Ages onwards. Neither was there much scaling-up. In the Belgian lowland zones there are still 58 polders and *wateringen* remaining today.[28] Germany fell in between the two extremes. Following the 1962 disaster, the government of the state of Lower Saxony decided that more substantial water management organisations were required to guarantee safety from flooding. Accordingly, several mergers of *Deichverbände* took place, but they were less far-reaching than in the Netherlands. By 1990 there were still 38 *Deichverbände* in Lower Saxony. In Schleswig-Holstein

---

[25] Havekes, *Successful Decentralisation*, pp. 13, 16; Van de Ven, *Man-Made Lowlands*, pp. 234–5, 238.
[26] Louman, *Fries waterstaatsbestuur*, pp. 351–8.
[27] Pauwels, *Polders en wateringen*, pp. 207–10.
[28] Vereniging van Vlaamse Polders en Wateringen, https://vvpw.be, retrieved 3 April 2023.

responsibility for maintenance of primary sea defences and for the sea defences of the *Halligen* was taken over by the state government in 1971.[29] Another consequence of the 1962 flood was that it finally brought an end to maintenance in kind.

The End of *Kabeldeichung*

After the mid-nineteenth century, maintenance in kind only survived in the German part of the North Sea Lowlands. Even there it was beating a slow retreat. In one area of Amt Ritzebüttel on the Lower Elbe, the city of Hamburg pushed through the abolition of maintenance in kind in 1840, and in another area this occurred in 1900. Land Hadeln introduced *Kommuniondeichung* in 1930. In some cases the landowners themselves began to discern the drawbacks of maintenance in kind and took the initiative to abolish it, as happened in a part of Osterstade on the eastern shore of the Lower Weser in 1882. The landowners here were convinced that monetisation was financially feasible and technically superior to maintenance in kind. They compared their dikes to those of the adjoining Land Würden, where maintenance had already become communal in 1722, and concluded that their own dikes presented *ein Bild eines jämmerlichen Stück- und Flickwercks* (an image of piteously improvised patchwork). Therefore they requested introduction of centrally organised and executed dike maintenance.[30]

Some areas, such as Kehdingen and the Alte Land on the Lower Elbe, stubbornly stuck to maintenance in kind until well after the Second World War. It only came to an end when the government of Lower Saxony decided to abolish it in the new regulation of water management of 1963, the contents of which were strongly influenced by the experiences of the flood disaster of the previous year. Moreover, the technical requirements for the new flood defences became so high and costly that maintenance in kind by individual farmers was simply no longer feasible.[31] Finally, the end had come for a maintenance system that had persisted for at least seven centuries.

Conclusion

The development of water management in the North Sea Lowlands in the 1880–1980 period was determined by increasingly efficient technology and by increasingly powerful central governments. Modern technology and huge state investments enabled implementation of huge schemes

---

[29] Peters, 'Entwicklung', pp. 196–7; Fischer, *Wassersnot und Marschengesellschaft*, p. 332.
[30] Ehrhardt, 'Kabel und Kommunion', pp. 148, 158–60.
[31] Fischer, *Wassersnot und Marschengesellschaft*, p. 333; Ehrhardt, *Ein Guldten Bandt*, pp. 561–4.

such as the Zuider Zee Works, the Delta Works and the Thames Barrier. At the local level, drainage could be improved through field drainage, construction of new drains and canals, and electric pumps. These investments were driven by two factors. Firstly, the flood disasters of 1916, 1953 and 1962 convinced policymakers that safety from flooding had to be enhanced by improving flood defences. Secondly, the Second World War in particular convinced them that food scarcity had to be prevented by increasing land productivity as much as possible. Therefore, drainage had to be promoted and subsidised.

Policymakers were also convinced that modern technology could only be applied efficiently by large, well-funded organisations. In England and the Netherlands, and to a lesser degree also in Germany, this led to far reaching reorganisations and mergers of water authorities. The search for more funding led to an extension of the group of rate payers. In the Netherlands after 1992, all inhabitants of the territory of a water authority became rate payers. These new rate payers were also represented on the board of the water authorities, which meant that farmers and landowners had to relinquish some power. The advantage for them was, of course, that they could share the burden of maintenance with a much larger group.

There was a strong belief during this period in the power of science and technology. Nature appeared to be tamed. Agricultural production had to grow. The environmental consequences of this were ignored, and even when they were acknowledged it was assumed that science would provide a solution. The voices of scientists and that section of the general public expressing concerns about the consequences of unlimited agricultural growth were not yet heard by those in power.

10

The Indian Summer of Lowland Farming, c. 1880–1980

Introduction

In 2005 the Dutch Agricultural History Society organised a symposium on biographies of leading figures of the Dutch agricultural sector. A lot of attention was paid to the Mansholt family, a wealthy family of Groningen farmers, and its most prominent member: Minister and European Commissioner Sicco Mansholt. One of the participants remarked ironically that the Society was somewhat behind the times. Nowadays, he said, in the agricultural sector the money is no longer being made by the arable farmers of lowland Groningen, but by the intensive animal farmers of the uplands. Therefore, he suggested, it would be more useful to focus on their leaders. Unsurprisingly, this remark was made by a professor at Maastricht University in Limburg, the most upland part of the Netherlands. He was right, however, and his remark illustrates the shifts that took place in the countryside. As early as 1910 the German economist Friedrich Swart had remarked that during the preceding decades the uplands had been catching up with the lowlands. The time of the prosperous lowland farmers, who lived like 'small kings' on their large farms and looked down upon the poor uplanders, was coming to an end.[1] Therefore, the 1880–1980 period is fundamentally different from the previous centuries. Although mechanisation and the use of artificial fertilisers and pesticides led to unprecedented levels of land and labour productivity, the lowland farmers lost their prominent position in rural society. Paradoxically, the agricultural sector itself prospered, but individual farmers lost wealth, power and status.

During the period discussed in this chapter, lowland farmers no longer benefitted from the much higher quality of their soils, because artificial fertilisers and pesticides made geographical differences less relevant. As a result, regional variation diminished, not only between lowlands and

---

[1] Swart, *Zur friesischen Agrargeschichte*, pp. 17, 33; the remark at the symposium was made by the late Hans Jansen. I thank Anton Schuurman for reminding me of it.

uplands, but also within the lowlands. This has also had its influence on historiography. Whereas regional studies of agrarian history abound for the period until about 1900, there are few for the twentieth century, particularly for the post-Second World War period. Regional variation is only attenuated; it has not completely disappeared, but it has disappeared from the view of most scholars studying the history of twentieth-century agriculture. The lack of regional studies forces the application of a somewhat different approach in this chapter. It is divided chronologically into two parts: 1880–1940 and 1940–1980. For the first period, sufficient regional studies are present. In addition, although there were significant changes in farming, basically the farming methods of 1940 were not that different from those in place 60 years earlier. A farmer who had died in 1880 would not have had much trouble finding his way around a farm if he were revived 60 years later. Had he come to life again in the 1980s, however, he would have suffered quite a shock, if only because of all the huge and noisy machinery he would find on the farm. It was in this period that agriculture changed unrecognisably into modern industrial farming, or, as Brassley et al. wrote, *The Real Agricultural Revolution* took place.[2] This will be discussed in the third section of the chapter, following a survey of the last modified landscapes.

The Last Decades of 'Traditional' Farming, 1880–1940

The North Sea Lowlands did not escape from the late nineteenth-century depression unscathed, in particular the arable regions such as Zeeland, Groningen, several German regions, and the English Fens. The lowlands showed remarkable resilience, however, and appear to have been less depressed than many other agricultural regions. Much of this resilience can be attributed to the mixed nature of farming in most of the lowlands, which permitted relatively easy adaptation to changing demand. Solutions to the problems of the 1880s were seldom clean breaks with the past. The way out of the depression was often found by extending the area under crops that were already cultivated, or by processing crops or waste products such as straw.

Farmers in Zeeland and the Flemish coastal plain, for instance, found a way out of the depression by strongly extending the area under sugar beet. As mentioned in chapter eight, sugar beet was already introduced in this area in the third quarter of the nineteenth century; in Zeeland this was as a substitute for the cultivation of madder, which had become unprofitable due to the invention of the chemical dyestuff alizarine in the 1860s. The presence of refineries, first in Flanders, later also in

[2] P. Brassley, D. Harvey, M. Lobley and M. Winter, *The Real Agricultural Revolution: The Transformation of English Farming 1939–1985* (Woodbridge, 2021).

the south-western Netherlands, was crucial. Beet could not be transported over large distances due to the high transport costs of this bulky product. Moreover, factory owners often provided farmers with seeds and advances. In 1866, when sugar beet are first mentioned in Zeeland agricultural statistics, 1400 hectares of sugar beet were cultivated in the province of Zeeland; by 1910 this had increased to 21,500 hectares, and beet had become the main cash crop on the Zeeland farms instead of wheat, the area of which was reduced from 20,000 to 12,000 hectares.[3] Conflicts between the growers and the sugar manufacturers over price and quality of the beet led to the founding of the first cooperative beet sugar factory in the Netherlands in 1899. The first years of this cooperative were difficult, but eventually it became a success, and in the period 1908–17 six more cooperative factories were founded.[4] Through these cooperatives, farmers could control the processing and marketing of their product and receive a good price.

Sugar beet also became a very popular crop in the most western part of the Flemish coastal plain in the department of Pas-de-Calais in northern France. In 1862 already, six per cent of arable land was under sugar beet here.[5] Afterwards, sugar beet cultivation increased strongly in this region. Between 1887 and 1900 the amount of beet processed in the three factories of the area increased from 53,000 to 288,000 tons.[6] In the Belgian part of the coastal plain sugar beet made headway more slowly. In 1895 only some 450 hectares of beet were grown in the polder municipalities of West Flanders.[7] By then, Zeeland farmers were already cultivating 12,000 hectares of sugar beet.[8] Zeeland farmers' more rapid adoption of sugar beet was almost certainly the result of their need to find a substitute for their suddenly lost cash crop, madder, but also of their being used to relatively intensive cultivation. Sugar beet demands careful soil preparation and intensive weeding, practices Zeeland farmers had been used to for centuries.

The farming sector in Zeeland weathered the economic storm of the 1880s and early 1890s quite well and remained viable, also because wheat prices recovered somewhat after 1895 and wheat yields increased after the successful introduction of the Wilhelmina wheat breed from 1899. This does not mean, however, that there was no suffering at the individual

[3] Priester, *Geschiedenis*, pp. 386–7, 401–2, 604–7.
[4] H. Zwarts, 'Knowledge, Networks, and Niches: Dutch Agricultural Innovation in an International Perspective, c. 1880–1970' (unpublished PhD dissertation, Wageningen University, 2021), pp. 131–3.
[5] R. H. Hubscher, *L'Agriculture et la société rurale dans le Pas-de-Calais du milieu du XIXe siècle à 1914* (2 vols., Arras, 1979–80), vol. 2, pp. 487–9.
[6] Blanchard, *La Flandre*, pp. 299–300.
[7] *Statistique de la Belgique. Agriculture. Recensement général de 1895* (Brussels, 1898), vol. 1, pp. 100–33.
[8] Priester, *Geschiedenis*, p. 606.

level. Evidence of this is the reduction in the share of owner-occupied land in Zeeland. During the late nineteenth century, some Zeeland farmers were forced to sell the land their ancestors had purchased during the eighteenth and early nineteenth centuries. In older literature a lot was made of this. Farmers were assumed to buy land during periods of prosperity at high prices and sell again during periods of depression, when land prices were low.[9] More recent research has demonstrated that, although such cases indeed occurred, this cannot be generalised. The most prosperous land-owning farmers' families in particular were remarkably resilient and managed to maintain themselves as landowners over several generations.[10]

The farmers in the Dutch River Area found it more difficult to adapt to the depression. Like all mixed and arable farmers, they were hit by the fall of cereal prices, but unlike the farmers in Flanders and Zeeland it was difficult for them to intensify. This was because, despite the early adoption of steam pumping, the drainage problems in the region were not resolved. Much of the land remained too wet for intensive farming and the application of artificial fertilisers. Sugar beet cultivation was increased in the west of the region, where it had been introduced in the 1860s, and in the east many orchards were planted. Fruit growers could profit from increasing export markets for apples, pears, cherries and plums. Market gardening also offered new opportunities. These activities were labour intensive, however, and the number of available labourers decreased because many of them opted for work in the growing brick industry. As a result, a divide occurred within the farming community. The large farmers, who had traditionally dominated the industry, concentrated on more extensive activities: stock fattening, horse breeding, and growing cereals as animal fodder. Smaller farmers concentrated on labour-intensive activities, such as fruit and vegetable cultivation. By 1910, when the rural economy was recovering, it was clear that these farmers were most successful. Much of the land remained difficult to cultivate, however, because the drainage problem was only solved after the Second World War.[11]

Although Zeeland farmers increased their cattle herds, they staunchly remained arable farmers. The same holds true for their colleagues in

[9] M. J. Wintle, 'Dearly Won and Cheaply Sold: The Purchase and Sale of Agricultural Land in Zeeland in the Nineteenth Century', *Economisch- en Sociaal-Historisch Jaarboek* 49 (1986), pp. 44–99.

[10] Priester, *Geschiedenis*, pp. 146–7, 162–4; Van Cruyningen, *Behoudend*, pp. 293–7.

[11] Brusse, 'De Gelderse landbouw', pp. 114–16; Brusse, *Leven en werken in de Lingestreek: De ontwikkeling van het platteland in een verstedelijkend land* (Utrecht, 2002), pp. 31–4; P. J. van Cruyningen, *Boeren aan de macht? Boerenemancipatie en machtsverhoudingen op het Gelderse platteland, 1880–1930* (Hilversum, 2010), p. 248.

Groningen. The income per hectare of the arable land in Groningen decreased until 1894.[12] Contrary to their colleagues on the Flemish coastal plain and the south-western archipelago, however, the Groningen farmers did not turn to the cultivation of alternative crops such as sugar beet. The area under cereals declined only slightly, and sugar beet cultivation only became of some importance after 1900.[13] This reluctance was due to the fact that there were no sugar processing plants in this region that could provide farmers here with incentives to grow sugar beet. Without these, transport costs of beet to the factories in the south-west of the country were prohibitive. Only after the founding of a sugar factory near the city of Groningen in 1896 did beet cultivation become a viable option in the northern provinces. By 1910, 2700 hectares of sugar beet were cultivated in the province, mostly on the friable young marine clay along the Wadden coast.[14] Groningen farmers were also able to compensate for some of their loss of revenue by selling their straw to strawboard factories. The first of those factories was founded in 1867 and by 1910 the Netherlands exported 160,000 tons of strawboard, almost all from Groningen. For farmers in the Groningen clay area, straw sales became an important source of income. In this case also, conflicts between farmers and manufacturers led to the founding of cooperative factories.[15]

Until the 1870s the development of farming in Groningen and the north-west German coastal zone showed many similarities, but then the two regions began to diverge. Whereas Groningen stuck to arable farming, north-west Germany shifted to animal husbandry. In Ostfriesland the number of cattle per 100 hectares of agricultural land increased from 47 in 1883 to 61 in 1912, while the share of land permanently under grass also increased.[16] Many East Frisian farmers, especially on the heavier clay soils, concentrated on stock breeding because there was a strong demand for the high-quality cattle of Ostfriesland in other parts of Germany.[17] In Hadeln, the large farms near the river which had in the past been mostly arable, shifted to fattening of slaughter cattle, and on the other shore of the Elbe river, in the Holsteinische Elbmarschen, dairy farming began to replace arable farming in the 1890s. Farmers in Eiderstedt focused on rearing of *shorthorn* cattle, which were exported in large numbers to Great Britain.[18] This change in orientation was, of course, in response

12 Priester, *De economische ontwikkeling*, pp. 387–8.
13 Ibid., p. 302.
14 Ibid., pp. 345–6.
15 Ibid., pp. 373–9.
16 Leemhuis, 'De landbouw', pp. 420, 427.
17 Ibid., p. 429.
18 Bierwirth, *Siedlung*, p. 75; K.-J. Lorenzen-Schmidt, 'Anschreibebuchforschungen aus den holsteinischen Elbmarschen', in K.-J. Lorenzen-Schmidt and B. Poulsen, eds., *Bäuerliche Anschreibebücher als Quellen zur Wirtschaftsgeschichte* (Neumünster, 1992) p. 151; Bantelmann et al., *Geschichte Nordfrieslands*, p. 267.

to the decreasing grain prices of the late nineteenth century, but it was also stimulated by the protectionist policy of the *Reich*, which introduced tariffs on cattle imports in 1879 and forbade cattle imports altogether in 1894.[19] The Dutch government did not introduce such protectionist measures, and therefore divergent government policies contributed to divergent developments in Groningen and north-west Germany in the late nineteenth century.

An impressive example of adaptability is offered by the English Fens. After the introduction of steam pumps, fenland farmers had concentrated on cereal cultivation. Therefore, they were initially heavily hit by the collapse of cereal prices. They managed to adapt, however, by responding to demand from the huge market of London and those of other urban centres, with which they had usually reliable railway connections. Farming in the Fens remained mixed: oriented towards cultivation of crops, combined with cattle which provided manure for the fields. The range of crops, however, was extended: potatoes, peas, vegetables, mustard, flowers and bulbs were added to cereals. Smallholders in particular turned to market gardening and fruit growing, and by the early twentieth century they were perceived as the group that had weathered the storm the best.[20]

Sugar beet was introduced here surprisingly late. As we have seen, the presence of a processing industry was crucial for the emergence of sugar beet cultivation due to the relatively high transport costs. Moreover, such an industry could induce farmers to grow beet by offering advances and seed, but beet sugar refineries only emerged very late in England. The United Kingdom mainly relied on imports of cane sugar from its colonies, and the countries on the continent dumped their excess sugar on the British market. Under British pressure, the exporting European nations agreed at the Brussels Convention of 1902 to limit the dumping of beet sugar. But as late as 1910 Britain still had no beet sugar factories, whereas Germany had 342 and France 213. In that year, Dutch manufacturers started to buy sugar beet from East Anglian farmers. Due to the high transport costs, this remained limited to farms close to the ports of Lowestoft, Yarmouth and King's Lynn, from which the beet were shipped to the Netherlands.[21] An attempt by a Dutch company to erect a factory in Maldon in Essex, also in 1910, failed. Two years later, the first beet sugar factory was built at Cantley in Norfolk, again with Dutch backing. Although the fen soils are

---

[19] Leemhuis, 'De landbouw', p. 430.
[20] Thirsk, *English Peasant Farming*, p. 315; E. J. T. Collins, 'Rural and Agricultural Change', in Collins, *The Agrarian History of England and Wales. Volume VII, 1850–1914* (Cambridge, 2000), p. 191.
[21] Bosma, *The World*, p. 228; H. D. Watts, 'The Location of the Beet-Sugar Industry in England and Wales, 1912–1936', *Transactions of the Institute of British Geographers* 53 (1971), pp. 98, 103.

very well-suited for beet cultivation, sugar beet only became significant here from the 1920s when several more factories were built.[22]

The regions discussed above were all characterised by mixed or arable farming. The fenlands of Holland, Utrecht and Friesland, however, were specialised dairying regions. As to be expected, initially the dairy farmers of these areas did not suffer from the depression caused by massive North American cereal imports. They could even profit from reduced prices of cattle feed. The Dutch dairy sector did experience problems, however, in the 1880s, due to the loss of the British butter market. Dutch butter exports to Britain increased until the Dutch had a 35 per cent share in British butter imports in 1870, but it was reduced to less than eight per cent in 1890, and never really recovered afterwards. During the same period Denmark increased its share in British butter imports from 11 per cent to 41 per cent.[23] This was due to the better and more uniform quality of Danish butter. Complaints about the quality of Frisian and Holland butter were already voiced by London butter importers in 1834, and in 1878 even a committee from the Frisian agricultural society had to admit that Danish butter was better than the product from Holland and Friesland. This dubious quality was the result of merchants adulterating the butter with other substances, and of selling inferior butter from the inland provinces as Holland or Frisian butter.[24] As a result, Dutch butter in general was considered as a low-quality product that had to compete with Irish butter and margarine, whereas the Danes could take over the market for high-quality butter. This resulted in a slump of the butter price in the Netherlands during the 1880s. By moving butter production to – often cooperative – creameries, from the 1890s Dutch dairy farmers also managed to sell better quality butter, supported by the government, which introduced quality control. By 1910 Germany had become the most important export market for Dutch butter, and dairy farming in Holland and Friesland was booming again. German farmers were also responding to the large demand for dairy products in the flourishing industrial cities. Farming in the Oldenburger Wesermarsch (Butjadingen and environs) had almost completely abandoned arable farming by the late nineteenth century and now concentrated on dairy production. Most of their milk was processed in cooperative creameries, the first of which was founded in 1885.[25]

---

[22] Collins, 'Rural and Agricultural Change', p. 198; Sly, *Soil in their Souls*, pp. 133–4.

[23] V. R. IJ. Croesen, *De geschiedenis van de ontwikkeling van de Nederlandsche zuivelbereiding in het laatst van de negentiende en het begin van de twintigste eeuw* (Wageningen, 1931), p. 192.

[24] Croesen, *De geschiedenis*, pp. 47–56.

[25] Ibid., pp. 57–60, 195; J. Gerriets-Purkswarf, *Die landwirtschaftlichen Verhältnisse der oldenburgischen Wesermarsch in den Ämtern Brake, Butjadingen und Varel* (Jena, 1913), pp. 54–5.

One reaction to the depression can be observed in all North Sea Lowlands: cutting of wage costs. This was done by shifting from labour-intensive arable farming to labour-extensive cattle farming or by simply spending less on activities such as ditching or weeding. Those farmers who could afford it also spent more on machinery which had already been introduced earlier, such as mechanical reapers and steam threshing machines. They could also let contractors thresh their harvests. Although this labour saving was partly offset by the introduction or expansion of the cultivation of labour-intensive crops such as sugar beet and potatoes, in general the demand for wage labour appears to have diminished. This led to a rural exodus, especially from regions such as Zeeland, Groningen, the Dutch River Area, northern Friesland, and the Elbe-Weser Triangle.[26] From the small region of western Zeeland Flanders, for instance, more than 6700 people emigrated to the United States alone in the 1870–1909 period. That was more than a quarter of its population at the beginning of that period. On the opposite shore of the North Sea another small region, the Isle of Axholme, lost 18 per cent of its population, in this case mostly to urban areas in England.[27] At first only surplus labourers emigrated, but later serious labour shortages began to occur. In Hadeln and Nordfriesland around the end of the nineteenth century, farmers had to employ labourers from as far away as the Russian part of Poland and Austro-Hungarian Galicia to replace emigrated labourers.[28]

The rural exodus led to concerns among politicians, and in England in 1892 legislation was introduced to keep labourers in the countryside by providing them with smallholdings. This legislation was not very successful, but here and there smallholdings were created. In the Fens this occurred around Spalding in south Lincolnshire in particular, where the Liberal politician Lord Carrington owned an estate. By 1914, 972 acres (394 hectares) of his 5100-acre (2065 hectares) estate were rented to small-holders through the South Lincolnshire Small Holders' Association.[29] In the Netherlands, providing smallholdings to labourers to keep them on

---

[26] H. de Vries, *Landbouw en bevolking tijdens de agrarische depressie in Friesland (1878–1895)* (Wageningen, 1971), pp. 184–7; Brusse, 'De Gelderse landbouw', p. 115; H. Rössler, 'Der Unterweserraum als Schauplatz von Wanderungsbewegungen (ca. 1750–1890)', in Bickelmann et al., eds., *Fluss, Land, Stadt*, p. 256.

[27] A. Vergouwe, *Emigranten naar Amerika uit West-Zeeuss-Vlaanderen* (n.p., 1993), p. 211; P. J. van Cruyningen, 'Bevolking en landbouw', in A. Bauwens and H. Krabbendam, eds., *Scharnierend gewest: 200 jaar Zeeuws-Vlaanderen 1814–2014*, Bijdragen tot de Geschiedenis van West-Zeeuws-Vlaanderen 42 (2014), pp. 56–60; Haresign, 'Agricultural Change', p. 441.

[28] Bierwirth, *Siedlung*, pp. 84–5; Bantelmann et al., *Geschichte Nordfrieslands*, p. 266.

[29] A. Adonis, 'Aristocracy, Agriculture and Liberalism: The Politics, Finances and Estates of the Third Lord Carrington', *Historical Journal* 31 (1988), pp. 888–9.

the land was also discussed, but without result. In a country without an estate system or entail, however, purchasing or leasing land was much easier than in England, also for poorer people. Therefore, the government wisely decided to leave this to the free land and lease market.[30]

The period of agricultural prosperity that began shortly before 1900 reached its zenith during the First World War and shortly after. From 1921, however, prices of agricultural products plummeted. War damage had been repaired and productivity was increasing, which led to overproduction. This time, depression was not limited to the arable sector. Fast refrigerated ships carried meat and dairy from the Americas, Australia and New Zealand to Europe, and high-quality margarine became a serious competitor for home-produced butter.[31] And worse was still to come: the Great Depression of the early 1930s drew agriculture even deeper into the morass. Prices of agricultural products reached their nadir around 1932, and under these circumstances even governments that had staunchly supported free market principles, such as those of the United Kingdom and the Netherlands, were forced to abandon them. Guaranteed prices, subsidies, marketing boards, and regulation of production were introduced to support a sector that seemed about to collapse.[32]

The fens and marshes in England appear to have suffered more from the interwar depression than the lowlands on the opposite shore of the North Sea. This may have to do with the extremely depressed condition of English farming in general. English agriculture during the interwar years has been described as low-input, low-output livestock farming with little investment in maintenance of buildings, hedges and drainage; 'for many, doing one's duty by the soil was no longer possible'.[33] In the fens and marshes, no longer 'doing one's duty to the soil' in the first place meant neglecting maintenance of the drainage system. This indeed occurred in many parts of the English lowlands. Ditches were no longer sufficiently dredged and that meant the start of a vicious cycle: 'Each year, because of neglected ditches, the winter flood water took longer to drain away. Cleaning had to be scamped to get seed in before it was too late. Harvests, always behind in the fens, were later yet and spoiled by autumn gales and

---

[30] *Rapporten en voorstellen betreffende den oeconomischen toestand der landarbeiders in Nederland* ('s-Gravenhage, 1906), pp. 133–66.

[31] Bieleman, *Five Centuries*, pp. 156–7; E. H. Whetham, *The Agrarian History of England and Wales. Volume VIII, 1914–1939* (Cambridge, 1978), p. 139.

[32] Bieleman, *Five Centuries*, pp. 157, 167–8; Whetham, *The Agrarian History*, pp. 243–58; G. Mahlerwein, *Grundzüge der Agrargeschichte: Die Moderne (1880–2010)* (Cologne, Weimar and Vienna, 2016), p. 172.

[33] J. Martin, 'The Structural Transformation of British Agriculture: The Resurgence of Progressive High-Input Arable Farming', in B. Short, C. Watkins and J. Martin, eds., *The Front Line of Freedom: British Farming in the Second World War* (Exeter, 2006), p. 17.

rain …'.[34] Maintenance of flood defences was also neglected. In 1932 the sea dikes in Essex were reported to be in 'an abandoned state'.[35] This was to have dire consequences in January 1953.

The picture painted above is too bleak, for the fens in particular. In eastern England intensive arable farming remained more important than elsewhere. In the fens of Cambridgeshire and southern Lincolnshire, by 1939, more than 70 per cent of agricultural land was arable. On the Isle of Ely most of this land was used to grow wheat, potatoes and sugar beet.[36] The latter crop had been introduced and diffused rapidly during the 1920s. This was the result of the introduction of a government subsidy for the production of sugar beet. This subsidy was to be paid for ten years, and during the last five years of that term it was to be gradually reduced to zero. The subsidy was to be paid to beet sugar factories for each hundredweight of sugar they produced, and the extent of the subsidy was reflected in the price the factories paid to the growers. It was extremely successful: from 1924/5 to 1927/8 the area under sugar beet increased more than tenfold from 9100 hectares to 94,300 hectares. By 1928, 17 factories were processing the beet, five of them run by an Anglo-Dutch company.[37] Cultivation was strongly concentrated in East Anglia and the fens. The area under sugar beet in the Cambridgeshire and Lincolnshire fens has been estimated at c. 30,000 hectares in the 1930s. The factory at Ely possessed a fleet of 50 barges to transport the voluminous product from farm to factory through the fens where road transport was difficult.[38]

Although sugar beet demands more fertilisation and labour than most other crops, its cultivation in East Anglia turned out to be lucrative, as on the continent. It also fitted well into the mixed farming system prevalent in the fens. The leaf could be used as cattle fodder and the waste product beet pulp could be returned to the farm, also to be used to feed animals. It was dubious, however, whether beet cultivation would be profitable without the subsidy.[39] This became apparent when the subsidy was gradually diminished in the early 1930s while at the same time the price of sugar on the world market was falling due to overproduction. In 1935 the government decided to instill the subsidy on a permanent basis and

---

[34] Whetham, *The Agrarian History*, p. 186.
[35] Summers, *The East Coast Floods*, p. 35.
[36] Martin, 'The Structural Transformation', p. 29; J. R. Borchert, 'The Agriculture of England and Wales, 1939–1946', *Agricultural History* 22 (1948), p. 60.
[37] Whetham, *The Agrarian History*, pp. 166–7; Watts, 'The Location', pp. 98–9.
[38] Watts, 'The Location', p. 107; Sly, *Soil in their Souls*, p. 134.
[39] R. McG. Carslaw, *Four Years of Farming in East Anglia, 1923–1927: Being a Detailed Investigation into the Costs and Returns on Twenty-Six Farms* (Cambridge, 1929), pp. 55–8.

all factories were combined into the British Sugar Corporation to reduce manufacturing costs.[40]

A factor that may have contributed to the economic problems of farmers in the English Lowlands was the shift in landownership that occurred between c. 1910 and the early 1920s. During those years, many aristocratic landlords sold large parts of their possessions, mostly to their former tenants. As a result, the share of land in England and Wales farmed by owner-occupiers increased from 12 per cent in 1912 to 36 per cent in 1927.[41] The new owner-occupiers had purchased their farms during a period of prosperity, when land prices were high. Many of them may have had to borrow money and mortgage their farm. Then, from the early 1920s onwards, they were confronted with steeply declining prices for their products. For farmers in the marshes and fens this may have necessitated cutting costs of maintenance of the drainage system in order to be able to pay interest on mortgages.

On the fens and marshes large estates appear not to have been as dominant as in England as a whole, although reliable data is scarce. Some estate sales did occur here. In 1920 Lord Carrington sold his 3000-acre (c. 1200 hectares) marshland estate near Grimsby – most of it to the sitting tenants – but he held on to his estate near Spalding.[42] The most spectacular estate sale in the Fens was that of the c. 7500-hectare Thorney estate by the Duke of Bedford in 1909. The farms on the estate were sold to the tenants. Some of them did 'back-to-back' deals, buying the farm and then immediately selling to a new landlord; others became owner-occupiers.[43] These new owner-occupiers may have got into financial trouble during the depression of the interwar years.

Of course, farmers in the Dutch lowlands also suffered during the depressed 1920s and 1930s. Farming reached its nadir here in the harvest year 1931/2. Farmers in the river area incurred the highest losses in that year: 100 guilders per hectare. Cattle farmers in the peat and clay-on-peat zones lost 66 guilders per hectare, and arable and mixed farms in the marine clay zone lost 48 guilders per hectare. Farmers in the uplands incurred losses of 82–86 guilders per hectare. The situation was improved with support measures that were introduced by the government from 1932 onwards, and by 1936/7 farm incomes were again positive and even higher than in the late 1920s.[44] In spite of economic adversity, Dutch lowland farmers were still able to increase production. Between 1923/7 and 1933/7 yields of marketable crops increased by 13 per cent

---

[40] Whetham, *The Agrarian History*, pp. 244–5.
[41] S. G. Sturmey, 'Owner-Farming in England and Wales, 1900–1950', in W. Minchinton, ed., *Essays in Agrarian History* (Exeter, 1968), p. 287.
[42] Adonis, 'Aristocracy, Agriculture and Liberalism', p. 887.
[43] Sly, *Soil in their Souls*, pp. 49–50.
[44] Bieleman, *Five Centuries*, p. 157.

in the marine clay areas. In Zeeland this increase was partly the result of farmers investing in underdrainage of their fields.[45] In the river clay zones, production per hectare increased by even more than 25 per cent over this decade. In the Tielerwaard area in western Gelderland, for instance, arable yields per hectare increased by more than 30 per cent and wheat yields even by 50 per cent. At the same time, arable acreage was reduced, but the increasing yields more than compensated for that. It appears the smaller acreage was cultivated more intensively, while at the same time fruit and vegetable cultivation was also increased. These increases in productivity coincided with investments in underdrainage.[46] In the historiography, the river area has usually been described as a rather backward region until the land consolidation schemes of the 1950s and later. It appears, however, that farmers were already investing in improvement during the interwar years.

The End of Farming from *Terpen*

Until the late nineteenth and early twentieth centuries in some corners of the North Sea Lowlands the wetland landscape was not transformed into intensive farmland, but only modified. The most spectacular example of this were the *Halligen*, the islands in Nordfriesland that were not protected by dikes. As already noted, the population lived on artificial dwelling mounds and derived its income from a combination of extensive animal husbandry and working for the (mostly Dutch) merchant navy. As late as 1794, on *Hallig* Hooge, 96 men from a population of 480 were sailors.[47] However, by the early nineteenth century, Dutch shipping was in decline and employment opportunities there were reduced.

During the eighteenth and nineteenth centuries the *Halligen* were suffering very much from erosion. Nordstrandischmoor, for instance, was reduced from c. 600 to only 157 hectares. The church on this tiny island had to be rebuilt several times after the mound on which it was built was wiped away by storms. The last church, built in 1815, was destroyed in the 1825 flood.[48] From that year on, the community no longer had its own church. Nordstrandischmoor was not the only *Hallig* that was eroded: by the early nineteenth century the size of all *Halligen* was reduced. This, combined with the consequences of the 1825 flood, was cause for the Danish government to ask advice from experts on the future of the islands. The government seriously considered moving the

---

[45] M. B. Smits, 'Productie van marktbare gewassen', *Landbouwkundig Tijdschrift* 53 (1941), pp. 48–9; Van Cruyningen, 'Bevolking en landbouw', p. 68.
[46] Smits, 'Productie', p. 49; Brusse, *Leven en werken*, pp. 91, 98; Bijl, *Het Gelderse water*, p. 197.
[47] Meier, *Die Halligen*, p. 96.
[48] Meier, Kühn and Borger, *Der Küstenatlas*, pp. 156–7, 168.

population to nearby islands with flood defences and using the *Halligen* only for grazing. Experts, however, considered this impractical and instead advised raising the height of the *Warften*.[49]

The erosion of the *Halligen* was a natural process, and if left to the sea and the currents, the unprotected islands would eventually have disappeared. In 1867 Schleswig-Holstein became a Prussian province, and the Prussian government, realising the value of the *Halligen* as wave-breakers for the mainland, began to invest in flood protection. The shores of some *Halligen* were protected with stones, and several were connected to the mainland by dams built in the 1890s, with Hooge even being protected by a low dike, constructed in 1911–13, which protects the island from all but the highest floods.[50] Life on the *Halligen* became safer, but society and economy lost much of their unique characteristics. Hooge was the first *Hallig* where in 1941 communal ownership of the land by the inhabitants of the *Warften* was abolished. It was also the first of the islands where in 1933 four farmers successfully cultivated oats and barley.[51]

In 1883 most farmers on the *Halligen* held less than ten head of cattle.[52] The scale of farming on Kampereiland was very different, although the circumstances were similar. Kampereiland was protected by low dikes, but these could not prevent regular flooding, which could last for weeks. Therefore, all farms were built on dwelling mounds. Contrary to the *Halligen*, however, the farms of Kampereiland were not small marginal holdings, but large (30–50 hectares) commercial enterprises.[53] Their main product was hay, which was of high quality thanks to the regular deposition of sediment by the floods. In 1862 the height of the dikes was increased, but new farms were still being built on mounds as late as 1879, because the raised dikes did not provide full protection from floods.[54] The greater height of the dikes meant that the land flooded much less often and for shorter periods. It also meant, however, that much less sediment was deposited, and pasture thus profited less from the fertilising effect of flooding. Therefore, the floods that occurred after 1862 did not contribute much to increasing soil fertility, but the power of the waves did cause damage to the land.[55] This was not enough reason for the owner of Kampereiland, the city of Kampen, to raise the dikes to a safer level. Maybe the loss of sediment fertilisation was compensated for by the application of artificial fertilisers, and the damage was something the

---

[49] Meier, *Die Halligen*, p. 110.
[50] Ibid., pp. 97, 111–12.
[51] H. Koehn, *Die nordfriesischen Inseln: Die Entwicklung ihrer Landschaft und die Geschichte ihres Volkstums* (Hamburg, 1961), p. 177.
[52] Postma, *De Friesche kleihoeve*, p. 141.
[53] G. Hendriks, *Een stad en haar boeren* (Kampen, 1953), p. 94.
[54] Dirkx, Hommel and Vervloet, *Kampereiland*, p. 61.
[55] De Kraker, 'Sustainable Coastal Management', p. 157; Directie van den Landbouw, 'Waterafvoer', p. 88.

inhabitants had lived with for many generations. Farming appears to have been rather intensive in the late nineteenth century. Around 1870 a portion of the land was even used to grow crops. Later this higher land was turned into pasture, while the lower parts were still used for hay production.[56]

On the opposite shore of the Zuider Zee, for the inhabitants of the island of Marken, agriculture was of secondary importance. The islanders fished for herring, flounder, anchovy, shrimp, eel, and smelt, on the Zuider Zee. They hardly held any cattle; in 1870 the c. 1000 islanders had no more than 23 cows altogether. Still, the 215 hectares of grassland on Marken were not unimportant for the island economy. They were used as meadows, and the hay was sold in villages and towns around the Zuider Zee. In good years in the early 1870s it could yield some 40,000 guilders. The reason why the Markers did not farm the land themselves was probably the distribution of landownership. A plot of one hectare could easily have 40 to 60 owners, which made it impossible to form real agricultural holdings. As a consequence, the islanders decided to farm the meadows communally and sell the hay.[57] The income from the hay sales was used to purchase winter provisions, but sometimes, as in 1895, a summer flood destroyed the hay harvest and caused poverty on the island. Such events led to calls to construct higher dikes.

For both Kampereiland and Marken the end of farming in a modified landscape came with the completion of the *Afsluitdijk* in 1932, which changed the Zuider Zee into the freshwater lake IJsselmeer. For the Kampereiland farmers, the main consequence was that flood risk was reduced. Apart from that, not much changed; they could continue farming as they had done in the past. They also continued leasing their farms from the city of Kampen, which was in the early 1950s one of the largest landowners in the Netherlands, with an estate of c. 5600 hectares.[58] For the fishermen of Marken, however, the closing off of the Zuider Zee meant the end of a way of life. Although many complaints were voiced about this in the 1930s, it had already been obvious to most fishermen from the beginning of the century that their industry was in decline and that it was hardly possible to make a living from fishing on the Zuider Zee. Tourism, commuting, and agriculture provided alternative employment. Following a land consolidation project in the early 1930s, 15 dairy farms were formed and in 1934 the island got its own creamery.[59]

On the Ooijpolder near Nijmegen the practice of opening sluices during the winter season to fertilise the pasture with sediment was continued well into the nineteenth century. During the second half of the nineteenth century doubts arose, however, about the benefits of winter

[56] Directie van den Landbouw, 'Het grondgebruik in Nederland', *Verslagen en Mededeelingen van de Directie van den Landbouw* 1912 no. 3, p. 132.
[57] Schutte and Weitkamp, *Marken*, pp. 58–9.
[58] Hendriks, *Een stad*, p. 59.
[59] Schutte and Weitkamp, *Marken*, pp. 241, 249–51.

flooding. Often the land stayed inundated until well into spring; in 1867 even until July, and then the losses caused by flooding were not compensated by sediment fertilisation. A lecture of c. 1885 even claimed that in dry years the hay harvest was better than in years with more flooding. In 1893 the board of the polder ordered an engineering bureau to create a drainage plan for the polder, including a steam pumping station that would keep the land dry during spring and summer.[60] The land was still permitted to flood during winter. The end of this centuries-old system of fertilisation came shortly after the First World War. In 1919 the local farmers' union requested the water authority to also keep the polder dry during winter. The union's argumentation is interesting. Although the farming population was living safely on mounds, they complained about being isolated in winter. Doctors and midwives had trouble reaching them, and their children could not always go to school when the water was very high.[61] So, their reasons for wanting to end winter flooding were more social than agronomic. Society was changing and the farmers of the Ooijpolder wanted to participate in new opportunities. As a result of all these developments, on the eve of the Second World War most modified landscapes were transformed into intensively farmed land. Only the landscape of the *Halligen* partly retained its modified character.

The 'Real Agricultural Revolution' in the Lowlands

There has been a lot of discussion amongst historians of the periodisation of an 'agricultural revolution' in western European and English agriculture in particular. Usually this is situated somewhere in the 1500–1850 period. Recently, however, Brassley et al. have claimed that the 'Real Agricultural Revolution' occurred in the 1939–85 period.[62] And indeed, compared to what happened in that period, all earlier 'agricultural revolutions' were mere ripples in the development of agriculture. Artificial fertilisers, pesticides, and new seed and cattle varieties caused unprecedented increases in production and productivity. Land consolidation projects enabled farmers to use these new inputs optimally. In this way the role of the natural environment in farming was reduced and standardised to enable farmers to become 'entrepreneurs'.[63] This also wiped out much – though not all – regional variety in agriculture. It is also reflected in historiography: regional studies, on which most of this book is based, are almost absent for this period. There is something to be said for this, because

---

[60] J. van Eck, *Historische atlas*, p. 11; Regionaal Archief Nijmegen, Polderdistrict Circul van de Ooij 2295.
[61] Regionaal Archief Nijmegen, Polderdistrict Circul van de Ooij 2298.
[62] Brassley et al., *The Real Agricultural Revolution*, pp. 1–3.
[63] J. D. van der Ploeg, *The New Peasantries: Struggles for Autonomy and Sustainability in an Era of Empire and Globalization* (London and Sterling, 2008), pp. 114–15.

agriculture, the landscape, and even biodiversity became more uniform and were standardised after the Second World War.[64] Therefore, the post-1940 period demands a different approach.

The far-reaching changes of this period were to a high degree the result of government policy. In all states surrounding the North Sea, government interfered deeply in agricultural production and marketing from the early 1930s onwards. The extreme circumstances of the Second World War only served to strengthen this tendency.[65] After the War, there was no return to a free market economy for agriculture. Food shortages and lack of foreign currency propelled governments onto the road to continued state intervention in order to increase agricultural output. During the first three decades after the War, all states on the southern North Sea shores promoted increased production and efficiency in agriculture, leading to an impressive increase in productivity.[66] Within that framework, however, different choices were made. The British government chose to increase production of basic food products such as cereals, potatoes, sugar beet, milk and eggs with the aim of becoming more self-sufficient and having to spend less foreign currency on food imports. The Dutch government, however, aimed at developing Dutch agriculture into an export sector of mainly animal products in order attract foreign currency.[67]

The divergent preferences of governments were mirrored by developments in the Lowlands. In the English Lowlands vegetable production was increased. On the Isle of Ely, for instance, the part of the land under grass was even further reduced. The area under wheat and sugar beet was reduced here, and cultivation of vegetables and fruit expanded. On Romney Marsh, where for centuries large flocks of the famous marsh sheep had roamed, more and more land was drained and changed into arable. Between 1939 and 1950 alone the percentage of arable land on Romney Marsh increased from 9 to 48.[68] On the opposite shore of the North Sea, in the central fenlands of Friesland, for instance, drainage was

---

[64] A. Schuurman, 'Agricultural Policy and the Dutch Agricultural Institutional Matrix during the Transition from Organized to Disorganized Capitalism', in P. Moser and T. Varley, eds., *Integration through Subordination: The Politics of Industrial Modernisation in Europe* (Turnhout, 2013), p. 72.
[65] Brassley et al., *The Real Agricultural Revolution*, pp. 92–3; U. Kluge, *Agrarwirtschaft und Ländliche Gesellschaft im 20. Jahrhundert* (Munich, 2005), pp. 32–4.
[66] Kluge, *Agrarwirtschaft*, pp. 38–9.
[67] Brassley et al., *The Real Agricultural Revolution*, pp. 94–101; J. van Merriënboer, *Mansholt: Een biografie* (Amsterdam, 2006), p. 116; J. S. C. Wiskerke, *Zeeuwse akkerbouw tussen verandering en continuïteit: Een sociologische studie naar de diversiteit in landbouwbeoefening, technologieontwikkeling en plattelandsvernieuwing* (Wageningen, 1997), pp. 40–1.
[68] Borchert, 'The Agriculture', p. 60; Sheail and Mountford, 'Changes', pp. 48–9.

improved to expand dairy farming. As a result, the number of cattle per hectare almost doubled from 1.7 in 1950 to 3.1 in 1980.[69]

The post-War development of agriculture was also promoted by land consolidation, especially in the Netherlands and Germany. Originally, land consolidation (Dutch: *ruilverkaveling*, German: *Flurbereiniging*) meant no more than an exchange of plots between landowners to create larger plots that were easier to cultivate and situated closer to the farmyard. After the Second World War, however, land consolidation schemes were used to completely re-arrange the countryside by building new roads and farms and creating new drainage systems. In the Dutch river area this resulted in the disappearance of the *komgronden*, the back swamps with impermeable heavy clay soils that had for centuries only been used for production of low-quality hay. They were changed into fertile pasture and even arable fields.[70] Both the Dutch and German governments spent huge sums on subsidies for these schemes. In the Dutch province of Zeeland many land consolidation schemes were implemented. Much land had been flooded there at the end of the War and during the 1953 flood disaster. The reconstruction after those disasters was used to change the structure of agriculture in the province. Land consolidation strongly improved production and productivity, but also resulted in uniform, standardised landscapes.[71]

The increase in agricultural production went hand in hand with plummeting numbers of workers in agriculture, resulting in an enormous increase in labour productivity. First the farm labourers disappeared. Large arable farms especially had employed dozens of labourers until the War. By 1970 most labourers had left agriculture and were replaced by machinery – as were horses, once the pride of any arable farmer. This development was caused by strongly increased wages.[72] The fact that they no longer were the major employers in lowland rural society reduced the status and power of those farmers. They were toiling in the fields themselves now instead of ordering their labourers about.

After the farm labourers, it was the turn of the smallholders to disappear from the scene of lowland farming. Although farms in the lowlands in general were larger than those in the uplands, in some regions, such as the English Fens, there were also many small farms. As

---

69 Vellema, 'De landbouw', pp. 25, 32.
70 Brusse, *Leven en werken*, pp. 138–45.
71 Kluge, *Agrarwirtschaft*, p. 38; P. Brusse and W. van den Broeke, *Provincie in de periferie: De economische geschiedenis van* Zeeland (Utrecht, 2005), pp. 307–8; K. Ditt, 'Zwischen Markt, Agrarpolitik und Umweltschutz: Die deutsche Landwirtschaft und ihre einflüsse auf Natur und Landschaft im 20. Jahrhundert', in K. Ditt, R. Gudermann and N. Rüße, eds., *Agrarmodernisierung und ökologische Folgen: Westfalen vom 18. bis zum 20. Jahrhundert* (Paderborn, 2001), pp. 114–15.
72 Bieleman, *Five Centuries*, p. 240.

The Indian Summer of Lowland Farming, c. 1880–1980

FIGURE 9. Revolutionised farming: wheat harvest with a Massey-Harris combine harvester in Kats on the island of Noord-Beveland, Zeeland, 1956. Photo F. Coolman. Special Collections, Wageningen University & Research – Library.

late as 1963, of the 2800 holdings over 2 hectares on the Isle of Ely, 2000 (71 per cent) were smaller than 20 hectares. These were intensive arable farms cultivating wheat, potatoes and vegetables such as onions, celery, red beet and lettuce. On these smallholdings all labour was provided by the farmer and his wife. On such small farms, investing in modern machinery such as tractors caused excessive capital expenditure, leading to heavy depreciation charges and financing problems. Moreover, these smallholdings suffered from fragmentation, often consisting of five or six widely scattered plots. The authors of the report on the small fen farm on which this paragraph is based concluded that the prospects for the future of the smallholdings were bleak. According to them, the optimal size of a full-time holding would soon be 75–100 acres (c. 30–40 hectares).[73] Investing in expensive new technology was also problematic for small dairy farms in the Netherlands. The compulsory introduction of the milk cooling tank from the 1960s onwards forced many small dairy farmers to give up farming.[74]

[73] C. M. Williams and J. B. Hardaker, *The Small Fen Farm: Problems and Prospects of Intensive Arable Samllholdings on Skirt Soils in the Fens* (Cambridge, 1966), pp. 7–9.
[74] Bieleman, *Five Centuries*, p. 283.

The disappearance of the smallholder was also promoted by the Dutch government, particularly from the late 1950s onwards. A fund was created in 1963 to stimulate small farmers into leaving agriculture.[75] This policy was mainly aimed at the smallholders of the sandy uplands, but small farmers on the marshes and fens were also considered redundant. Land consolidation schemes were used to gently convince smallholders to give up farming. Therefore, land consolidation not only led to larger plots, but also to increased holdings. On the marshland island of Noord-Beveland, for instance, there were 233 farmers at the start of land consolidation in 1961. When the project was finished in 1975, 147 holdings were left. Average holding size had increased from 24.8 to 41.4 hectares.[76]

The standardisation of agriculture resulted in the disappearance of much of the difference in crop yields between lowlands and uplands. This becomes obvious when we compare crop yields in the south-western marine clay area of the Netherlands with the southern sandy regions around 1990. In the south-western region wheat yielded on average c. 8000 kilograms per hectare, in the south 6500 kilograms. Wheat yielded some 23 per cent more in the fertile marine clay zone, but two centuries earlier that difference would have been in the order of 100 per cent. For sugar beet the difference was even smaller: 63 tons per hectare in the south-west, 61 tons in the south.[77] New crop varieties, pesticides and fertilisers had wiped out much of the difference between uplands and lowlands.

The arable farmers of the south-western Netherlands still did somewhat better than their counterparts in the sandy uplands, but in animal husbandry the difference had completely disappeared. Milk yields were no longer higher in the lowlands. By 1980 dairy farmers all over the country had the same highly productive Holstein-Frisian cows.[78] Moreover, the huge pig and poultry farms, with thousands or even tens of thousands of animals, were almost entirely concentrated in the uplands. The 'agricultural entrepreneurs' who owned these rural industrial plants now looked down upon the ordinary farmers of the lowlands. The tables had been turned.

The farmers of the English and Dutch lowlands were losing their lead on the upland farmers during the 1950s and 1960s, but they still managed to substantially increase production and productivity. A 1954 report on agriculture in the German lowlands, however, sounded quite pessimistic. According to the author, mixed farming with mostly arable, such as in parts of Dithmarschen and Ostfriesland, was doing well. Where stock farming was predominant, like in Eiderstedt or Hadeln,

---

[75] Ibid., p. 244.
[76] Wiskerke, *Zeeuwse akkerbouw*, pp. 50–2.
[77] P. C. M. Hoppenbrouwers, 'Crop yields in Dutch agriculture, 1850–1990', in Van Bavel and Thoen, eds., *Land Productivity and Agro-Systems*, p. 126.
[78] Bieleman, *Five Centuries*, pp. 282–92.

however, prospects were bleak. For generations farmers in these regions had made a comfortable living from fattening of oxen and cows. This was very labour extensive, so wage costs were low. These farmers also invested very little in land improvement. They hardly purchased fertilisers, and underdrainage was seldom practised. Most excess water was still drained through open ditches. This was inefficient as well as being an obstacle for modern machinery. Moreover, demand for the fatty meat of oxen and adult cows declined. Post-War consumers preferred lean meat. Therefore, the both labour- and capital-extensive farming that had been predominant in many German marshland areas became unprofitable. The author advised farmers to intensify cattle farming on better drained and fertilised land or to shift to arable farming, depending on the quality of the soil.[79]

Conclusion

The Austrian environmental historian Verena Winiwarter relates in one of her publications a meeting with a farmer who told her that the Middle Ages had lasted until the introduction of the tractor.[80] Like Winiwarter, I am convinced that this remark is much less silly than it may seem at first glance. Until Brassley et al.'s *Real Agricultural Revolution* – of which the tractor was the most powerful symbol – changes and innovations occurred in agriculture, sometimes important ones, but basically farming methods remained the same and the natural environment remained a major constraint on farmers' decision-making. The revolutionary developments after c. 1940 changed that completely: new crop and animal varieties, fertilisers, pesticides, and machinery wiped out most environmental constraints. As a result, regional variation within the lowlands was reduced, and the uplands were able to catch up with the lowlands and in some cases even surpass them.

The new technology that was diffused widely after 1940 strongly reduced the effects of the natural environment on farming. Given application of the right technological fixes, farming in the uplands could be as lucrative – or even more so – than farming in the lowlands. Of the two constraints identified by Ogilvie – environment and institutions – one, the natural environment, had now become almost irrelevant.[81] The second one, the institutional environment, had become even more important.

---

[79] G. Blohm, 'Die Marsch und ihre betriebswirtschaftlichen Probleme', *Schriftenreihe der landwirtschaftlichen Fakultät der Universität Kiel* 12 (1954), pp. 43–64.

[80] V. Winiwarter, 'Landwirtschaft, Natur und Ländliche Gesellschaft im Umbruch: Eine umwelthistorische Perspektive zur Agrarmodernisierung', in Ditt, Gudermann and Rüße, eds., *Agrarmodenisierung*, p. 733.

[81] Ogilvie, 'Choices and Constraints', p. 277.

However, the state was now the most important creator of institutions. Chapter nine showed that the state had got a grip on water management. Most water authorities had become cogs in the wheels of the state machinery. This chapter shows that the developments in agriculture from 1940 onwards were to a high degree determined by state policy, which aimed at improvement of production and productivity. Institutions were more important than ever, but they were no longer designed at the local or regional level of the water authorities, but at the level of the national state. Therefore, the post-1940 period is fundamentally different from the preceding millennium of farming.

Conclusion

The major constraints on farmers' decision-making have been identified as the natural environment and the human environment of institutions determining behaviour in rural societies. This book has attempted to gauge the effects of these two constraints for the reclaimed wetlands along the southern North Sea coast over the period from c. 900 to the end of the twentieth century. Humans attempted to control the waterlogged landscape by creating physical infrastructure, which in turn demanded the creation of institutional infrastructure – formal and informal rules and organisations – with which to organise maintenance of the waterworks. This institutional infrastructure was required because wetland reclamation and maintenance of infrastructure were collective efforts of farmers, lords and communities that needed to be coordinated. Over the millennium under study, humans were able to transform the landscape into productive farmland through creating physical and institutional infrastructure, but they also caused unintended changes in the landscape that forced them to adapt their infrastructure and institutions. Institutional arrangements were embedded in the continuously changing natural environment and as such were also required to be adjusted continuously.

The above certainly does not mean that this book aims to reintroduce environmental determinism into historiography. Throughout the areas that were transformed, physical infrastructure was uniform: dikes to protect the land from flooding; ditches, canals, culverts and sluices to drain excess water from the fields. The institutional arrangements for the organisation of maintenance showed a considerable variation, however. Divergent institutions could be applied successfully to the same environmental issue. This makes the North Sea Lowlands a fascinating laboratory within which to study the effects of different institutions on the development of agriculture. Therefore, this book not only investigates the interaction between man and nature in the form of wetland landscapes, but also the effects variegated institutions had on agricultural development in those landscapes. It can do so over a very long period, because codification of regulations for water management began as early as around 1100 AD.

At the end of the first millennium AD most of the wetlands along the southern North Sea coast were already in use by humans. This use took the form of either exploitation – simply using the available resources – or modification – modestly changing the landscape by digging ditches or constructing low banks. Under these circumstances, the ever-present threat of flooding limited the opportunities for land use to more extensive activities such as fishing, fowling, grazing cattle and sheep, and growing spring-sown crops on some higher places. This changed drastically between c. 900 and 1300 AD, during the so-called Great Reclamation. Most of the wetland landscapes were transformed – complex drainage systems were created and seawalls built that were intended to prevent flooding by sea and rivers. The goal of this transformation, or reclamation, was to make the land safe for more intensive settlement and agriculture. In the short run, this goal was achieved. Hundreds of thousands of hectares of wetlands were reclaimed, mostly used as pasture, but partly also sown with crops, and settled.

This impressive achievement was enabled by local and territorial lords who managed to convince the peasantry to leave their beloved high, light and well-draining soils to settle in the fertile, but also soggy and heavy, marshes, fens and floodplains of the Lowlands. Population pressure may partly explain why people were now willing to settle there, but more important was the fact that lords made settlement in the Lowlands attractive by granting the pioneers advantageous conditions: free personal status and free ownership of land. Although there was regional variation, in general the peasantry in the Lowlands enjoyed greater access to wider freedoms than their counterparts in the uplands. At a time when most of western Europe saw the peasantry merged into a mass of dependent serfs and villeins, the Lowlands offered another path. A class of free farmers emerged here, and that class – or the more prosperous layers of it – was to play an important part in the development of agriculture in the Lowlands.

Maintenance of physical infrastructure was generally performed in kind by owner-occupiers, tenants or landlords who provided labour and materials such as earth or wood. On the continent this was organised by the general administration at the local level. Each landowner had to maintain a stretch of dike proportional to the amount of land he or she owned. The village court inspected the work performed by those liable for maintenance and could fine those whose work was deemed insufficient. Less is known about the organisation of water management in England. It appears this was mostly in the hands of the village communities, and in some areas abbeys and monasteries also played an important part. Apparently, in England water management was not integrated into the local administration as in the territorial principalities on the eastern shore of the North Sea. From the thirteenth century onwards, specialised water management organisation began to emerge on the Flemish coastal plain and on Romney Marsh, called *wateringen* in Flanders and waterings on Romney Marsh. In these areas water management was no longer the task of the local authorities.

Elsewhere, local authorities began to cooperate for the maintenance of infrastructure of supra-local importance, such as sluices or canals. In Holland and Zeeland, these regional water authorities were integrated into the administrative structure of the principality at an early stage and were granted charters that gave them authority over the cooperating local authorities. In the Frisian lands (with the exception of Groningen) and in England they remained informal cooperative units without coercive authority over the local organisations. In the course of time, this could lead to conflicts and free riding behaviour. In England, the ensuing problems were solved by ad hoc royal commissions, the Commissions of Sewers. Therefore, in England and the Frisian lands no strong water authorities developed: in England probably because the Crown did not want to devolve power to local and regional authorities; in the Frisian lands because there was no central authority that could grant charters to water management organisations. In Holland, Zeeland and Flanders, however, the territorial lords could use the chartered water authorities to strengthen their own position by integrating them into existing power structures. Obviously, the development of water authorities was strongly influenced by the process of state formation.

Not all wetlands along the southern North Sea were transformed into intensively farmed land during the Great Reclamation. The most important exception were the English peat fens, which were only modified by improving drainage to make better use of the existing resources of the fens, such as reed, sedge, fish, fowl, fuel, and grazing during the summer season. The fenland village communities, which had rights of common to the peat fens, jealously guarded these resources. In the coastal plain of the Low Countries and north-west Germany, where territorial lords had laid their hands on the rights to the 'wilderness', rights of common had all but disappeared. Therefore, they could grant concessions to pioneers willing to reclaim and settle the peat fens. To attract those pioneers, they offered advantageous conditions: free personal status and free landownership with the right to sell, mortgage and bequeath the land. As a result, these lowlands were not feudalised, and feudalism was not very strongly developed in other parts of the North Sea Lowlands either. Where it did exist, as in Zeeland or the Guelders river area, it disappeared early on.

It was fortuitous of the English fenlanders not to transform their peat fens, because in the fourteenth century the unintended consequences of drainage and reclamation came to light. In a period characterised by a cooling climate, disease and warfare, the inhabitants of the Lowlands were confronted with a lowering surface level of the soil due to shrinkage and oxidation of the peat caused by drainage. Moreover, they were hit by an increasing number of devastating floods, which were at least partly caused by a reduction of storage capacity of sea inlets due to the construction of embankments to protect new land. The 900–1300 period of expansion was followed by a contraction period which lasted until c. 1550 and was accompanied by the abandoning of land and villages. In order to survive, the lowlanders had to adapt to the new circumstances. After humans remade nature, nature now remade human society and economy.

During the fifteenth and sixteenth centuries, technological means were found to respond to the late medieval problems: drainage mills to pump water from low peatlands to open water, dikes with less steep slopes and strengthened by groynes to better withstand the onslaught of the waves, osier mats to protect the underwater part of dikes from erosion. These were mostly rather simple adaptations of existing technological knowledge. Their implementation, however, was expensive. The impoverished rural population was not able to finance these investments. In the Holland fenlands, capital from wealthy urbanites was required to erect drainage mills. From the sixteenth century onwards the inhabitants of the booming cities were able and willing to provide this money, and hundreds of mills were built. Urban capital was also used to build new and stronger dikes along the sea inlets in the south-west of the Netherlands. In fact, Dutch urban capital ushered in a new period in the history of the North Sea Lowlands, the Second Reclamation, which reached its zenith between 1590 and 1660. It was characterised by large-scale, capital-intensive drainage projects, often of land that had been lost during the Middle Ages. The Second Reclamation was not limited to the Low Countries: it also reached north-west Germany and England.

The availability of capital was a necessary but not entirely sufficient condition for the success of reclamation projects in the early modern period. Projects could only succeed under the right institutional arrangements. For the correct functioning of the drainage mills, for instance, in Holland a two-tiered organisational structure was designed. At the local level, mill polders were created, in which landlords and farmers erected and maintained a drainage mill to drain their land. To prevent those mill polders from applying the simplest solution to their problems – to pump excess water to the land of the neighbours – the regional water authorities that had been created in the Middle Ages supervised the mill polders, instructing them where to build a mill and where to pump excess water. The regional authorities also created a reservoir and transport system to get the excess water to open water and maintained it by levying rates. Thus, a combination of old – regional water authorities – and new – mill polders – institutions contributed to the resurrection of farming in the Holland fenlands.

In the Frisian fens both capital and beneficent institutions were lacking. There, farmers and landlords built small mills to drain the plot of land surrounding the farmhouse. In the low central part of Friesland, the farms were small islands in the midst of vast stretches of undrained land. The Frisian mills drained onto a partly natural reservoir system in the centre of the province. Farming was less intensive here than in Holland, where intensive dairy farming became predominant. The Frisian farmers concentrated on cattle breeding and hay production. Farming in central Friesland may have been less productive than in Holland, but it can be perceived as a successful adaptation to the natural and institutional environment. The weak spot was that there was a lack of clarity regarding the liability for maintenance of the canals and sluices of the reservoir system. This led to increasing flooding from the late eighteenth century, and in the nineteenth century to the province of Friesland accepting responsibility for maintenance.

Efficient institutions were also required for the many costly land reclamation projects that were implemented during the sixteenth and seventeenth centuries, particularly because promoters of drainage were often confronted with conflicting rights and interests. The marshes, fens and lakes that were drained during this period were not wastelands, but were rather used in different grades of intensity: marshes and fens for grazing, fishing, fowling and fuel gathering; lakes for shipping and fishing. Moreover, many of the reclamations of this period were re-reclamations, meaning that the rights of former landowners had to be dealt with. If not compensated sufficiently, resistance from those parties in the form of litigation or rioting was to be expected. The 'core principalities' of the Low Countries – Holland, Friesland, Flanders and Zeeland – probably had the most efficient institutions in this respect. Commons had disappeared from these places already in the Middle Ages, and therefore there were no commoners that could contest drainage schemes. Permission for drainage was granted by a so-called *octrooi* (patent), which was based on research by the Audit Office or the Council of State of the legal and hydrological circumstances in the area concerned, and which provided compensation for damage to third parties and tax exemptions for the drainers. All this guaranteed a relatively smooth implementation of reclamation projects. In north-west Germany, commons had also all but disappeared and the *octrooi* was introduced here as well in the seventeenth century.

In England the situation was very different. Very large-scale drainage projects in the peat fens were implemented during the reign of Charles I. Commons still existed and the commoners were prepared to go to court or to riot to defend their rights. There was also no procedure that was aimed at finding a balance between the rights of the commoners and the drainage projectors. The commoners lost much of their grazing rights, which were crucial to their pastoral economy, and received no compensation. They revolted, not because they were opposed to drainage as such, but because of the loss of their rights. As a result, most of the Carolean drainage projects failed during the civil war of the 1640s, when the commoners managed to regain possession of the peat fens. Continental institutions were obviously more efficient in the sense that they enabled the smooth implementation of reclamation schemes.

Most reclamation projects of the early modern period required a large amount of capital. Therefore investors formed drainage companies. By purchasing a share in different companies, investors could spread risks. Such companies originated in the Low Countries, but were also formed in north-west Germany and England, often with Dutch capital. After completion of the project, the investors each received a part of the drained land in proportion to their share in the total investment. Some investors sold their land, but most decided to keep it. Since most investors were urbanites, this resulted in urban shares in landownership of up to 90 per cent. On these new polders a type of farming emerged that could be characterised as 'agrarian capitalism': large specialised farms (c. 40–100 hectares), often run by tenants with wage labourers, producing for the market. In Holland, Zeeland, coastal Flanders, much of Friesland and German regions such as Ostfriesland, Hadeln and Dithmarschen, this type of farming became predominant from the seventeenth century onwards. For Holland, Bas van Bavel has interpreted the emergence of 'agrarian capitalism' as a consequence of increasing urban landownership, not only on the new, but also on the old land. He imputes the increase in large, specialised, commercial holdings, which indeed also occurred on the old land, to the actions of urban landlords. Nonetheless, the data on landownership of the old land available at this moment does not confirm this. There are no indications that urban landownership strongly increased there after c. 1560. In my view, the emergence of these large commercial holdings was the result of the actions of the rural population itself, or at least the most prosperous part of that population. Most large holdings were created by wealthy farmers accumulating tenancies, sometimes combined with some owner-occupied land. These prosperous country dwellers could make use of the thriving land markets of the Lowlands.

In the core principalities of the Low Countries, much of what happened during the Second Reclamation can be perceived from an economic perspective as positive developments, leading to the emergence of large, specialised farms that produced more efficiently. There were also some negative developments, one of which was increasing oligarchisation of the water authorities. They were increasingly dominated by absentee urban landlords, and participation in decision-making by rural people was reduced. In the regional and medium-sized water authorities only the most prosperous farmers could sometimes play a part. Much of the rural population could only participate in small local organisations. Rule by oligarchy was not necessarily detrimental to good water management, so long as the interests of the landowning elite coincided with the common good. It went wrong in areas where offices in water authorities were sold to capitalists who often had no land in the territory of the organisation. They perceived their office as an investment on which they had to make a profit by squeezing as much as they could from the rural population by imposing unjustified fines and neglecting maintenance. The Ommelanden of Groningen were infamous in this respect. The oppressed rural population could not even appeal to a higher court. The resulting inadequate dike maintenance contributed to the large number of victims of the storm surges of 1686 and 1717 in Groningen.

Another negative aspect of the period after 1600 was that the shift from dike maintenance in kind to monetised maintenance – in which the landowners paid rates to the water authority which then contracted maintenance out or performed it itself – came to a halt. Performance in kind is considered as technically inferior to monetised maintenance because of the lack of uniformity of the work of the individual landowners. It also led to glaring inequality between landowners who had to spend large amounts of labour and materials on endangered stretches of dike and those who were liable for stretches that were easy to maintain. Monetisation was introduced in Flanders and in England in the Middle Ages. From the sixteenth century it slowly spread northwards from Flanders to Friesland, but around 1700 it stopped. In some German regions maintenance in kind survived until 1963. An explanation that has been put forward for this is that so long as the market economy was underdeveloped farmers often had insufficient cash to pay rates. The main problem with that explanation is that whereas by 1800 at the latest the whole of the North Sea Lowlands had a well-developed market economy, in several areas, in north-west Germany in particular, maintenance in kind persisted. This persistence can be partly explained by the fact that measures were taken to mitigate the consequences of the system: by limiting it to simple activities and contracting out more difficult tasks, by re-allotting stretches every seven years over the owners, etcetera. There also may have been less trust in the eastern regions where maintenance in kind persisted longest. Elites there perceived offices as an investment – as in Groningen – or managed to get exemptions from maintenance – as in Butjadingen. In the areas to the west of Groningen, elites were probably not less greedy, but simply in less of a position to enrich themselves at the cost of the common people.

The difference between maintenance in kind and monetised maintenance did not influence long-term agricultural development. The areas with maintenance in kind appear not to have suffered very much from their inferior way of keeping dikes in repair. They probably suffered more from storm surges, but although those events often had tragic consequences – some people died, more animals died, some farmers went bankrupt – their long-term consequences were limited. Rural society recovered, even after the disastrous Christmas Flood of 1717, maybe with the exception of Butjadingen. Therefore the persistence of maintenance in kind may also have been the result of a combination of fatalism and pragmatism: floods occurred every 30 years or so, that was a fact of life, and afterwards life would somehow return to normal. That recovery cost a lot of money, but the people to the west, who tried to prevent floods, had to spend a lot of money on flood protection. Both ways of dealing with flood risks had their own rationality, and development in the long run was the same. More important for long-term development was drainage. Insufficient drainage was an impediment to agricultural development, as evidenced by the peat fens in England, central Friesland and central Groningen. Good drainage was a necessary condition for intensive farming, arable husbandry in particular. Parts of the drainage system, in particular sluices, also demanded expertise to build and maintain, which farmers did not possess. It is not surprising then that in most places maintenance of the drainage system was monetised earlier than dike maintenance, and that it was also introduced in areas that persisted in maintenance in kind of flood defences.

Until 1870 the trend towards larger holdings in the Lowlands continued. Leasing or purchasing more land required that farmers be able compete on the land market, and therefore it has been assumed that large farmers tried to cut down on labour costs by farming more extensively and so increase profit. In the Lowlands this clearly was not the case. Wherever soil and hydrology permitted, a shift took place from animal to arable husbandry, which required more labour input. Moreover, there was also a shift towards crops such as wheat, rape, and madder, which demanded more care and therefore more labour. It appears farmers were more concerned with raising revenue – wheat, rape and madder were crops which provided high financial yields per hectare – than with cutting expenditure. They were able to intensify farming because, due to a growing population in the 1750–1850 period and the influx of seasonal labourers from Flanders and Westphalia, wages were relatively low and the modest rise in wage costs was more than compensated for by the increase in income. Increasing labour productivity only became interesting for the lowland farmers in the second half of the nineteenth century, when wages began to increase and prices dropped in the 1880s.

The two major regions that had long lagged behind in the development of large-scale, commercial farming, the English Fens and Groningen, caught up from the late eighteenth century onwards. In the case of Groningen, the driver of this was institutional change. In the first place, new regulations in 1755 put an end to the worst forms of corruption in water management. Secondly, in the second half of the eighteenth century, most Groningen tenants became *de facto* owners of their farms, which – in combination with improved water management – made it possible and worthwhile for them to invest in improvement of their land and to plough up grassland. Thirdly, after the constitution of 1848, the tenants-turned-landowners seized power in the provincial administration and effected a complete overhaul of the organisation of water management in the province. This reorganisation led to the founding of water authorities that were able to invest in steam pumping stations which strongly improved drainage.

Steam pumps were also crucial for the catching up of the English Fens. They made it finally possible to realise the potential that seventeenth-century drainers had seen in this region. Drainage mills had insufficient capacity to keep the Fens dry, but from the 1810s steam pumps made the Fens dry enough even for arable farming, and crop yields were high. The fenland farmers, who had stubbornly long defended their pastoral economy, now shifted to arable farming and stuck to it even after the prices of arable products declined. It was this technological breakthrough, and not institutional change, that enabled the Fens to catch up. The responsibility for maintenance of infrastructure remained divided over many small organisations, which were mostly busy defending their own parochial interests. Enormous sums were spent by drainage boards on fighting each other in court, money that might have been better spent on maintaining and improving drains and sluices. The case of the English Fens shows that development without institutional change was possible, but that a considerable price had to be paid for it.

Efficient institutions – especially in combination with a good organisational structure – were indispensable for the development of farming in the North Sea Lowlands towards large-scale, productive, specialised and commercial farming. In combination with technological change, they made the area into one of the most productive agricultural regions in the world. Until c. 1940 these institutions were embedded in the natural environment and had to be adapted to changes in that environment. Then, as Winiwarter's farmer said, the tractor heralded the end of the Middle Ages. Science and technology made it possible to overcome the constraints the natural environment had imposed for centuries. Fertilisers, pesticides, machinery and land consolidation enabled lucrative farming on any soil, however nutrient-poor or wet. Institutions became more important than ever for farming, but they were now imposed by the state and the European Community in the form of agricultural and environmental policy. The role of local and regional water authorities in the design of institutional arrangements was much reduced. As far as they still designed regulations, they did it within the framework of the legislation of the national state.

Epilogue: The North Sea Lowlands after 1980

From 1967 onwards, various plans were developed for the future of the small island of Tiengemeten in the south-western Dutch delta: it was to become a recreation park, or an airport, or a depot for polluted soil, or factories were to be built on the island, or residential neighbourhoods. Eventually, in 1997 the island was sold to the Dutch Society for the Preservation of Natural Monuments and changed into a nature reserve.[1] One thing was never considered by the planners: to continue to use the land on Tiengemeten for agriculture, the reason why it had been reclaimed during the eighteenth and nineteenth centuries. This oversight had nothing to do with the quality of the land. The young friable marine clay soils of Tiengemeten are among the most fertile soils in the world. It was instead caused by the changed position of agriculture in the Dutch economy, and by the changed position of lowland agriculture within the Dutch agricultural sector.

According to Otto Knottnerus, the Lowlands on the Wadden coast have become peripheral areas. Firstly, transport over sea is no longer that important, and new traffic infrastructure is leading traffic inland and to the south, where the economic and cultural centre is now situated.[2] What Knottnerus writes about the Wadden area can be applied to most of the North Sea Lowlands. Only Holland is situated in the economic, cultural and political heart of the Netherlands, but agriculture does not profit very much from that location. The farming sector in Holland is squeezed between growing cities. Haarlemmermeer, for instance, once a prosperous arable district, is now mostly covered by residential neighbourhoods, business parks and Schiphol Airport. Secondly, due to continuing mechanisation, employment in agriculture was reduced enormously. Apart from Holland and Flanders, the Lowlands have hardly industrialised. Where possible, tourism has become the most prominent branch of the economy in the Lowlands. As an employer, agriculture has

---

[1] D. Bosscher, 'Tiengemeten: Zandplaat, landbouwgebied, speculatieobject en natuurreservaat', in J. Bank and D. Bosscher, *Omringd door water: De geschiedenis van de 25 Nederlandse eilanden* (Amsterdam, 2021), pp. 578–9.

[2] O. S. Knottnerus, 'Räume und Raumbeziehungen im Ems Dollart Gebiet', in Knottnerus et al., eds., *Rondom Eems en Dollard*, p. 29.

become irrelevant. Market gardening, prominent in the Dutch Westland region and in the English Fens, does still require a lot of labour, but the work is unskilled and not well paid. Market gardeners need to attract labourers from eastern Europe. Since most work in the tourism sector is also low-skilled and low-paid, the Lowland areas now have little higher skilled and better paid work to offer, which contributes to a brain drain from the Lowlands. Many rural areas suffer from such problems, but in the North Sea Lowlands they are felt more strongly because these regions were once very prosperous. The farmers in particular, the 'kings' of the polder land, used to look down upon the inhabitants of the uplands. Farming in the Lowlands is still extremely productive, but that does not count for much anymore in a globalised economy in which cheap food can be imported from all over the world.

As noted in chapter ten, during the first three decades after the Second World War these changes did not have serious consequences for the lowland farmers, because everything in the western European states was geared towards increasing agricultural production. Around 1985 the expansionist policy was ended by the EEC and its member states.[3] During the 1970s the drawbacks of the single-minded policy of increasing agricultural productivity had become more and more obvious. Increasing numbers of people became aware of air and water pollution, of the standardisation of the landscape and of the reduced biodiversity that resulted from modern agriculture. Moreover, landscapes – tropical rain forests in particular – in other parts of the world are being destroyed to provide intensive animal husbandry in Europe with fodder. Dutch farmers alone need three million hectares of 'ghost acres' to feed their cattle, pigs and chickens.[4] Although the departments of agriculture in the western European countries tried to continue to defend the interests of the sector, during the 1980s they were forced to introduce new environmental policies.[5]

High-productivity farming was not only environmentally, but also economically, costly. The EEC had to spend huge sums on subsidising surplus production, and by the 1980s it had become obvious that this could not be continued indefinitely. In 1992 the EEC substituted income support for price support and tried to cut the costs of the CAP.[6] These developments

---

[3] Bieleman, *Five Centuries*, p. 317; Brassley et al., *The Real Agricultural Revolution*, pp. 109–10.
[4] Van der Ploeg, *Gesloten vanwege stikstof*, p. 181.
[5] E. Karel, *Boeren tussen markt en maatschappij: Essays over effecten van de modernisering van het boerenbestaan (1945–2012)* (Groningen and Wageningen, 2013), p. 157; Ditt, 'Zwischen Markt', pp. 117–22; T. C. Smout, *Nature Contested: Environmental History in Scotland and Northern England since 1600* (Edinburgh, 2000), p. 164.
[6] Bieleman, *Five Centuries*, p. 251.

FIGURE 10. Dark clouds gathering above *Hallig* Nordstrandischmoor, Nordfriesland. Photo Dirk Meier. The image shows the vulnerability of this landscape, but also the ingenuity and resilience of the people who managed to survive here for centuries.

also had consequences for the water authorities, who now no longer simply supported agricultural expansion by improving drainage. The grip of the farming sector on the water authorities was reduced and other groups were represented in their boards, especially in the Netherlands. Water authorities paid more and more attention to the environment and tried to balance agricultural and environmental interests.

The biggest challenge the Lowlands is now facing is sea level rise in combination with land subsidence. The latter is worst in the peat regions of Holland, Friesland and eastern England. Year-round pumping has enabled intensive farming in these areas but also sped up shrinkage and oxidation of the soil. As a result, drainage must be continually adjusted to the lowering level of the land. Eventually, the peat layers will disappear completely, and the soil layers below the peat are often of mediocre quality.[7] Moreover, this process damages buildings and infrastructure in the fenlands. For the Netherlands alone, that damage is estimated at c. 19–22 billion euros for the period until 2050.[8] This process is forcing the water authorities to consider new options for draining the land, and most of those options will mean less drainage and therefore less intensive and

[7] M. Woestenburg, *Waarheen met het veen: Kennis voor keuzes in het westelijk veenweidegebied* (Wageningen, 2009), p. 9; Purseglove, *Taming the Flood*, pp. 83–4; Sly, *Soil in their Souls*, pp. 21, 23.

[8] G. J. van den Born, F. Kragt, D. Henkens, B. Rijken, B. van Bemmel and S. van der Sluis, *Dalende bodems, stijgende kosten: Mogelijke maatregelen tegen veenbodemdaling in het landelijk en stedelijk gebied* (Den Haag, 2016), p. 8.

less lucrative agriculture. In the clay areas there is less shrinkage, but there too the level of the land becomes lower.

The consequences of shrinkage and wastage are exacerbated by climate change, which is causing increased sea level rise. Over the 1000 years discussed in the previous chapters, sea level rise remained limited, but from the twentieth century onwards it has increased. A rise of one metre over the next century is considered to be a realistic estimate, and sea level rise is expected to continue after that.[9] It is possible to protect the lowering land from the rising sea by constructing higher sea defences, at a very high cost. But there are additional problems resulting from sea level rise: it will become increasingly difficult to drain excess water to the sea, and salinisation of the groundwater will become worse. Considering these problems, the Dutch hydraulic engineer Johan van Veen had already concluded 80 years ago that at some time in the far future we would 'met een zucht van verlichting het land weer aan de zee prijsgeven' (with a sigh of relief abandon the land to the sea).[10] Had Van Veen known about climate change he might have put that moment in the not-so-far future. It is difficult to imagine technological fixes for these daunting challenges. On the other hand, in the seventeenth century no one could have imagined that the Lowlands would now be pumped dry by powerful pumping stations or that a brilliant twentieth-century engineer – Johan van Veen – would design a successful plan to close the sea inlets in the south-western Netherlands with huge dams. Pandora's box appears to be opened, but hope remains.

---

[9] J. Beersma, J. Bessembinder, R. Bintanja, J. Burgers, R. van Dorland, B. Overbeek, P. Siegmund and A. Sterl, eds., *Klimaatsignaal'21: Hoe het klimaat in Nederland snel verandert* (De Bilt, 2021), pp. 26–31; M. Schirmer, 'Küstenschutz im Klimawandel – ein Blick in die Zukunft', in Fischer, ed., *Zwischen Wattenmeer und Marschenland*, p. 199.

[10] Quoted in Van der Ham, *Meester van de zee*, p. 114.

Bibliography

Archives

Nationaal Archief, The Hague
Raad van State, 2145 I

Regionaal Archief Nijmegen, Nijmegen
Polderdistrict Circul van de Ooij, 2295, 2298

Bedfordshire Archives and Records, Bedford
Russell Collection, R4/4029, R4/4033, R4/4034
Transcribed Primary Sources retrieved from www.hogenda.nl

Penningkohieren
Bergambacht 1556
Berkenwoude 1561
Gouderak 1553
Haastrecht 1553
Krimpen aan den IJssel 1561
Krimpen aan de Lek 1561
Ouderkerk aan den IJssel 1561
Stolwijk 1556
Vlist en Bonrepas 1556

Morgenboeken
Oegstgeest 1588
Rijnsburg 1588

Other
Cattle census Delfland 1672

Printed Sources

Directie van den Landbouw, 'Het grondgebruik in Nederland', *Verslagen en Mededeelingen van de Directie van den Landbouw* 1912 no. 3.

Directie van den Landbouw, 'De invloed van den waterafvoer op het Nederlandsche landbouwbedrijf', *Verslagen en Mededeelingen van de Directie van den Landbouw* 1917 no. 1.

Grootte der gronden tijdens de invoering van het kadaster ('s-Gravenhage, 1875).

Provinciale Almanak voor Zeeland voor het jaar 1949 (Middelburg, 1949).

Rapporten en voorstellen betreffende den oeconomischen toestand der landarbeiders in Nederland ('s-Gravenhage, 1906).

Statistique de la Belgique: Agriculture. Recensement general de 1846 (Brussels, 1850).

Statistique de la Belgique: Agriculture. Recensement général de 1895 (Brussels, 1898).

Secondary Sources

Note: Dutch names beginning with 'de' or 'van' are listed under the last part of the name. Thus 'Van Dam' is listed under 'Dam, van', etcetera.

Abel, W., *Geschichte der deutschen Landwirtschaft vom frühen Mittelalter bis zum 19. Jahrhundert* (3rd. edn., Stuttgart, 1978).

Adonis, A., 'Aristocracy, Agriculture and Liberalism: The Politics, Finances and Estates of the Third Lord Carrington', *Historical Journal* 31 (1988), pp. 871–97.

Aerts, R., *Thorbecke wil het: Biografie van een staatsman* (Amsterdam, 2018).

Albright, M., 'The Entrepreneurs of Fen Drainage in England under James I and Charles I: An Illustration of the Uses of Influence', *Explorations in Entrepreneurial History* 8 (1955), pp. 51–65.

Albright Knittl, M., 'The Design for the Initial Drainage of the Great Level of the Fens: An Historical Whodunit in Three Parts', *Agricultural History Review* 55 (2007), pp. 23–50.

Allemeyer, M. L., *"Kein Land ohne Deich ...": Lebenswelten einer Küstengesellschaft in der Frühen Neuzeit* (Göttingen, 2006).

Andreae, B., *Betriebsformen in der Landwirtschaft: Entstehung und Wandlung von Bodennutzungs-, Viehhaltungs- und Betriebssystemen in Europa und Übersee sowie neue Methoden ihrer Abgrenzung: Systematischer Teil einer Agrarbetriebslehre* (Stuttgart, 1964).

Ash, E., *The Draining of the Fens: Projectors, Popular Politics, and State Building in Early Modern England* (Baltimore, 2017).

Asselen, S. van, *Peat Compaction in Deltas: Implications for Holocene Delta Evolution* (Utrecht, 2010).

Aston, T. H., and C. H. E. Philpin, eds., *The Brenner Debate: Agrarian Class Structure and Economic Development in Pre-Industrial Europe* (Cambridge, 1985).

Aten, D., *'Als het gewelt comt ...': Politiek en economie in Holland benoorden het IJ 1500–1800* (Hilversum, 1995).

Bibliography

——'Een afgerond geheel: Waterstaat en waterschappen ten noorden van het IJ tot 1800', in E. Beukers, ed., *Hollanders en het water: Twintig eeuwen strijd en profijt* (Hilversum, 2007), pp. 23–60.

Aubin, H., 'Von den Ursachen der Freiheit der Seelande an der Nordsee', *Nachrichten der Akademie der Wissenschaften in Göttingen I. philologisch-historische Klasse* (1953), pp. 29–45.

Augustyn, B., *Zeespiegelrijzing, transgressiefasen en stormvloeden in maritiem Vlaanderen tot het einde van de XVIe eeuw: Een landschappelijke, ecologische en klimatologische studie in historisch perspektief* (Brussels, 1992).

Ayers, B., *The German Ocean: Medieval Europe around the North Sea* (Sheffield, 2016).

Baars, C., *De geschiedenis van de landbouw in de Beijerlanden* (Wageningen, 1973).

Bailey, M., 'Per Impetum Maris: Natural Disaster and Economic Decline in Eastern England, 1275–1350', in B. M. S. Campbell, ed., *Before the Black Death: Studies on the 'Crisis' of the Early Fourteenth Century* (Manchester and New York, 1991), pp. 184–208.

——*The Decline of Serfdom in Late Medieval England: From Bondage to Freedom* (Woodbridge, 2014).

Bakels, C., R. Kok, L. I. Kooistra and C. Vermeeren, 'The Plant Remains from Gouda-Oostpolder, a Twelfth Century Farm in the Peatlands of Holland', *Vegetation History and Archaeobotany* 9 (2000), pp. 147–60.

Bakker, G., 'Het ontstaan van het Sneekermeer in relatie tot de ontginning van een laagveengebied 950–1300', *Tijdschrift voor Waterstaatsgeschiedenis* 10 (2001), pp. 54–66.

Bankoff, G., 'In the Eye of the Storm: The Social Construction of the Forces of Nature and the Climatic and Seismic Construction of God in the Philippines', *Journal of Southeast Asian Studies* 35 (2004), pp. 91–111.

——'Cultures of Disaster, Cultures of Coping: Hazard as a Frequent Life Experience in the Philippines', in Mauch and Pfister, eds., *Natural Disasters*, pp. 265–84.

——'The "English Lowlands" and the North Sea Basin System: A History of Shared Risk', *Environment and History* 19 (2013), pp. 1–37.

——'Of Time and Timing: Internal Drainage Boards and Water Level Management in the River Hull Valley', *Environmental History* 27 (2022), pp. 86–112.

Bantelmann, A., A. Panten, R. Kuschert and T. Steensen, *Geschichte Nordfrieslands* (Bredstedt, 1996).

Bavel, B. J. P. van, *Transitie en continuïteit: De bezitsverhoudingen en de plattelandseconomie in het westelijke gedeelte van het Gelderse rivierengebied, ca. 1300 – ca. 1570* (Hilversum, 1999).

——'Land, Lease and Agriculture: The Transition of the Rural Economy in the Dutch River Area from the Fourteenth to the Sixteenth Century', *Past and Present* 172 (2001), pp. 3–43.

——'People and Land. Rural Population Developments and Property Structures in the Low Countries, c. 1300–c. 1600', *Continuity and Change* 17 (2002), pp. 9–37.

——'Structures of Landownership, Mobility of Land and Farm Sizes. Diverging Developments in the Northern Part of the Low Countries, c. 1300–c. 1650', in B. J. P. van Bavel and P. Hoppenbrouwers, eds., *Landholding and Land Transfer in the North Sea Area (Late Middle Ages – 19th Century)*, CORN Publication Series 5 (Turnhout, 2004), pp. 131–48.

——'The Organization and Rise of Land and Lease Markets in Northwestern Europe and Italy, c. 1000–1800', *Continuity and Change* 23 (2008), pp. 13–53.

——'The Emergence and Growth of Short-Term Leasing in the Netherlands and Other Parts of Northwestern Europe (Eleventh–Seventeenth centuries): A Chronology and Tentative Investigation into its Causes', in B. van Bavel and P. R. Schofield, eds., *The Development of Leasehold in Northwestern Europe, ca. 1200–1600*, CORN Publication Series 10 (Turnhout, 2008), pp. 179–213.

——'Rural Development and Landownership in Holland, c. 1400–1650' in O. Gelderblom, ed., *The Political Economy of the Dutch Republic* (Farnham, 2009), pp. 167–96.

——*Manors and Markets: Economy and Society in the Low Countries, 500–1600* (Oxford, 2010).

——'History as a Laboratory to Better Understand the Formation of Institutions', *Journal of Institutional Economics* 11 (2015), pp. 69–91.

——*The Invisible Hand? How Market Economies Have Emerged and Declined since AD 500* (Oxford, 2016).

Bavel, B. J. P. van, P. J. van Cruyningen and E. Thoen, 'The Low Countries, 1000–1750', in B. J. P. van Bavel and R. W. Hoyle, eds., *Social Relations: Property and Power. Rural History in Northwestern Europe, 500–2000* (Turnhout, 2010), pp. 169–97.

Bavel, B. J. P. van, and R. W. Hoyle, 'Introduction: Social Relations, Property and Power around the North Sea, 500–2000', in B. J. P. van Bavel and R. W. Hoyle, eds., *Social Relations: Property and Power. Rural History in Northwestern Europe, 500–2000* (Turnhout, 2010), pp. 1–23.

Bavel, B. J. P. van, and J. L. van Zanden, 'The Jump-Start of the Holland Economy during the Late-Medieval Crisis, c. 1350–c. 1500', *Economic History Review* 57 (2004), pp. 503–32.

Beekman, A. A., *Het dijk- en waterschapsrecht in Nederland vóór 1795* (2 vols., 's-Gravenhage, 1905–7).

Beekman, F., *De Kop van Schouwen onder het zand: Duizend jaar duinvorming en duingebruik op een Zeeuws eiland* (Utrecht, 2007).

Beenakker, J. J. J. M., *Van Rentersluze tot strijkmolen: De waterstaatsgeschiedenis en landschapsontwikkeling van de Schager- en Niedorperkoggen tot 1653* (Alphen aan den Rijn, 1988).

Bibliography

Beersma, J., J. Bessembinder, R. Bintanja, J. Burgers, R. van Dorland, B. Overbeek, P. Siegmund and A. Sterl, eds., *Klimaatsignaal'21: Hoe het klimaat in Nederland snel verandert* (De Bilt, 2021).

Behre, K.-E., *Landschaftsgeschichte Norddeutschlands: Umwelt und Siedlung von der Steinzeit bis zur Gegenwart* (Neumünster, 2008).

Behringer, W., *Kulturgeschichte des Klimas: Von der Eiszeit bis zur globalen Erwärmung* (Munich, 2007).

——*Tambora und das Jahr ohne Sommer: Wie ein Vulkan die Welt in die Krise stürzte* (Munich, 2015).

Benders, J. F., *Een economische geschiedenis van Groningen: Stad en lande, 1200–1575* (Assen, 2011).

Berg, B. K. van den, *Het laagveengebied van Friesland* (Utrecht, 1933).

Berghmans, S., 'War, Taxation and the Enlargement of Farms in Coastal Flanders (Seventeenth–Eighteenth Centuries)', *Agricultural History Review* 70 (2022), pp. 193–218.

Besteman, J. C., 'Van Assendelft naar Amsterdam: Occupatie en ontginning van de Noordhollandse veengebieden in de Middeleeuwen', in D. E. H. de Boer, E. H. P. Cordfunke and H. Sarfatij, eds., *Holland en het water in de Middeleeuwen: Strijd tegen het water en beheersing en gebruik van het water* (Hilversum, 1997), pp. 21–39.

Bickelmann, H., H.-E. Dannenberg, N. Fischer, F. Kopitzsch and D. J. Peters, eds., *Fluss, Land, Stadt: Beiträge zur Regionalgeschichte der Unterweser* (Stade, 2011).

Bicker Caarten, A., *Middeleeuwse watermolens in Hollands polderland 1407/'08 – rondom 1500* (Wormerveer, 1990).

Bieleman, J., 'Farming System Research as a Guideline in Agricultural History', in B. J. P. van Bavel and E. Thoen, eds., *Land Productivity and Agro-Systems in the North Sea Area Middle Ages – 20th Century*, CORN Publication Series 2 (Turnhout, 1999), pp. 235–50.

——'"De ossen zijn hier seer schoon en groot". De landbouw in Holland tijdens de Republiek', in T. de Nijs and E. Beukers, eds., *Geschiedenis van Holland 1572–1795* (Hilversum, 2002), pp. 79–107.

——*Boeren in Nederland: Geschiedenis van de landbouw 1500–2000* (Amsterdam, 2008).

——*Five Centuries of Farming: A Short History of Dutch Agriculture 1500–2000* (Wageningen, 2010).

Bierwirth, L., *Siedlung und Wirtschaft in Lande Hadeln: Eine kulturgeographische Untersuchung* (Bad Godesberg, 1966).

Bijl, A., *Het Gelderse water: Waterstaatkundige en sociaal-economische ontwikkelingen in de polders van de westelijke Tielerwaard (1809–1940)* (Leiden, 1997).

Blanchard, R., *La Flandre: Etude géographique de la plaine Flamande en France, Belgique et Hollande* (Dunkirk, 1906).

Blauw, M. J. E., *Van Friese grond: Agrarische eigendoms- en gebruiksverhoudingen*

Bibliography

en de ontwikkelingen in de Friese landbouw in de negentiende eeuw (Leeuwarden, 1995).

Blickle, P., 'German Agrarian History during the Second Half of the Twentieth Century', in E. Thoen and L. Van Molle, eds., *Rural History in the North Sea Area: An Overview of Recent Research (Middle Ages – Twentieth Century)*, CORN Publication Series 1 (Turnhout, 2006), pp. 147–75.

Bloch, M., *Les caractères originaux de l'histoire rurale française* (2nd edn., Paris 1952, reprint 1999).

——*Feudal Society*, translation L. A. Manyon (1939, reprint London and New York, 2014).

Blockmans, W. P., 'The Social and Economic Effects of Plague in the Low Countries, 1349–1550', *Belgisch Tijdschrift voor Filologie en Geschiedenis* 58 (1980), pp. 833–63.

——*Metropolen aan de Noordzee. De geschiedenis van Nederland, 1100–1560* (Amsterdam, 2010).

Blockmans, W. P., and P. Hoppenbrouwers, *Introduction to Medieval Europe, 300–1550* (London and New York, 2007).

Blockmans, W. P., G. Pieters, W. Prevenier and R. W. M. van Schaïk, 'Tussen crisis en welvaart: sociale veranderingen 1300–1500', in *Algemene Geschiedenis der Nederlanden* 4 (Haarlem, 1980), pp. 42–86.

Blohm, G., 'Die Marsch und ihre betriebswirtschaftlichen Probleme', *Schriftenreihe der landwirtschaftlichen Fakultät der Universität Kiel* 12 (1954), pp. 43–64.

Boer, D. E. H. de, *Graaf en grafiek: Sociale en economische ontwikkelingen in het middeleeuwse 'Noordholland' tussen ca. 1345 en ca. 1415* (Leiden, 1978).

Boer, T. J. de, 'De Friesche grond in 1511 (Leeuwarderadeel – Ferwerderadeel volgens het register van den Aanbreng)', in *Historische Avonden* (Groningen, 1907), pp. 95–114.

Bölts, J., 'Die Rindviehhaltung im oldenburgisch-ostfriesischen Raum vom Ausgang des 16. bis zum Beginn des 19. Jahrhunderts', in H. Wiese and J. Bölts, *Rinderhandel und Rinderhaltung im nordwesteuropäischen Küstengebiet vom 15. bis zum 19. Jahrhundert* (Stuttgart, 1966), pp. 133–271.

Bont, C. de, *Delft's Water: Two Thousand Years of Habitation and Water Management in and around Delft* (Delft and Zutphen, 2000).

——*Amsterdamse boeren: Een historische geografie van het gebied tussen de duinen en het Gooi in de Middeleeuwen* (Hilversum, 2014).

Borchert, J. R., 'The Agriculture of England and Wales, 1939–1946', *Agricultural History* 22 (1948), pp. 56–62.

Borger, G. J., *De Veenhoop: Een historisch-geografisch onderzoek naar het verdwijnen van het veendek in een deel van West-Friesland* (Amsterdam, 1975).

——'Draining – Digging – Dredging: The Creation of a New Landscape in the Peat Areas of the Low Countries', in J. T. A. Verhoeven, ed., *Fens and Bogs in the Netherlands: Vegetation History, Nutrient Dynamics and Conservation* (Dordrecht, 1992), pp. 131–71.

——'Henk van der Linden en de historische geografie: De betekenis van het werk van prof. mr. H. van der Linden (Alphen a/d Rijn 14 februari 1922–Velp 27 maart 2012)', *Historisch-Geografisch Tijdschrift* 33 (2015), pp. 67–76.

Borger, G. J., F. H. Horsten and J. F. Roest, *De dam bij Hoppenesse: Gevolgen voor de afwatering van het gebied tussen Oude Rijn en Hollandsche IJssel, 1250–1600* (Hilversum, 2016).

Borger, G. J., and W. A. Ligtendag, 'The Role of Water in the Development of the Netherlands – A Historical Perspective', *Journal of Coastal Conservation* 4 (1998), pp. 109–14.

Born, G. J. van den, F. Kragt, D. Henkens, B. Rijken, B. van Bemmel and S. van der Sluis, *Dalende bodems, stijgende kosten: Mogelijke maatregelen tegen veenbodemdaling in het landelijk en stedelijk gebied* (Den Haag, 2016).

Bos, J. M., B. van Geel and J. P. Pals, 'Waterland – Environmental and Economic Changes in a Dutch Bog Area, 1000 A.D. – 2000 A.D.', in H. H. Birks, H. J. B. Birks, P. E. Kaland and D. Moe, eds., *The Cultural Landscape – Past, Present and Future* (Cambridge, 1988), pp. 321–31.

Bosch, T., *Om de macht over het water: De nationale waterstaatsdienst tussen staat en samenleving 1798–1849* (Zaltbommel, 2000).

Bosch van Rosenthal, E. J., *De ontwikkeling der waterschappen in Gelderland* ('s-Gravenhage, 1930).

Boschma-Aarnoudse, C., *Tot verbeteringe van de neeringe deser Stede: Edam en de Zeevang in de late Middeleeuwen en de 16de eeuw* (Hilversum, 2003).

——*Boer en poorter in het veen: Waterland en de Zeevang in de middeleeuwen* (Heerhugowaard and Wormer, 2022).

Bosma, U., *The World of Sugar: How the Sweet Stuff Transformed our Politics, Health, and Environment over 2,000 Years* (Cambridge, MA and London, 2023).

Bosscher, D., 'Tiengemeten: Zandplaat, landbouwgebied, speculatie-object en natuurreservaat', in J. Bank and D. Bosscher, *Omringd door water: De geschiedenis van de 25 Nederlandse eilanden* (Amsterdam, 2021) pp. 571–83.

Botke, IJ., *Boer en heer: 'De Groninger boer' 1760–1960* (Assen, 2002).

Bowers, J., 'Inter-War Land Drainage and Policy in England and Wales', *Agricultural History Review* 46 (1998), pp. 64–80.

Bowring, J., 'Between the Corporation and Captain Flood: the Fens and Drainage after 1663', in R. W. Hoyle, ed., *Custom, Improvement and the Landscape in Early Modern Britain* (Farnham, 2011), pp. 235–61.

Brand, K. J. J., 'Van een poging tot opheffen van polders en waterschappen in 1797, tot handhaving van deze publiekrechtelijke instellingen door de keizerlijke decreten van 1811', in A. M. J. de Kraker, H. Van Royen

Bibliography

and M. E. E. De Smet, eds., *Over den Vier Ambachten. 750 jaar Keure 500 jaar Graaf Jansdijk* (Kloosterzande, 1993), pp. 585–96.

Brassley, P., D. Harvey, M. Lobley and M. Winter, *The Real Agricultural Revolution: The Transformation of English Farming 1939–1985* (Woodbridge, 2021).

Brenner, R., 'Agrarian Class Structure and Economic Development in Pre-Industrial Europe', in Aston and Philpin, eds., *The Brenner Debate*, pp. 10–63.

——'The Agrarian Roots of European Capitalism', in Aston and Philpin, eds., *The Brenner Debate*, pp. 213–327.

——'The Low Countries in the Transition to Capitalism', in Hoppenbrouwers and Van Zanden, eds., *Peasants into Farmers?*, pp. 275–338.

Breuker, Ph., *Het landschap van de Friese klei 800–1800* (Leeuwarden, 2017).

Britnell, R. H., *The Commercialisation of English Society 1000–1500* (Cambridge, 1993).

Bruijn, E. de, *De hoeve en het hart: Een boerenfamilie in de Gouden Eeuw* (Amsterdam, 2019).

Bruin, H. P. de, 'Wording van het land', in H. P. de Bruin, ed., *Het Gelders rivierengebied uit zijn isolement: Een halve eeuw plattelandsontwikkeling* (Tiel, 1988), pp. 11–15.

Brümmel, P., *Die Dienste und Abgaben bäuerlicher Betriebe im ehemaligen Herzogtum Bremen-Verden während des 18. Jahrhunderts* (Göttingen, 1975).

Brusse, P., 'De Gelderse landbouw in de periode 1850–1920', in J. Bieleman, W. R. Foorthuis, F. Keverling Buisman and P. Thissen, eds., *Anderhalve eeuw Gelderse landbouw. De geschiedenis van de Geldersche Maatschappij van Landbouw en het Gelderse platteland* (Groningen, 1995), pp. 92–121.

——*Overleven door ondernemen: De agrarische geschiedenis van de Over-Betuwe*, A. A. G. Bijdragen 38 (Wageningen, 1999).

——*Leven en werken in de Lingestreek: De ontwikkeling van het platteland in een verstedelijkt land* (Utrecht, 2002).

——'Property, Power and Participation in Local Administration in the Dutch Delta in the Early Modern Period', *Continuity and Change* 33 (2018), pp. 59–86.

——'Economie: De effecten van de Deltawerken en regionaal overheidsbeleid op de economische ontwikkeling van Zuidwest-Nederland', in Van der Ham et al., *Modern wereldwonder*, pp. 260–97.

Brusse, P., and W. van den Broeke, *Provincie in de periferie: De economische geschiedenis van Zeeland* (Utrecht, 2005).

Brusse, P., and W. W. Mijnhardt, *Towards a New Template for Dutch History* (Zwolle, 2011).

Buisman, J., *Duizend jaar weer, wind en water in de Lage Landen* (7 vols., Franeker, 1995–2019).

Buitelaar, A. L. P., *De Stichtse ministerialiteit en de ontginningen in de Utrechtse Vechtstreek* (Hilversum, 1993).
Buma, W. J., and W. Ebel, eds., *Westerlauwerssches Recht: Jus municipale frisonum* (Göttingen, 1977).
Byford, D., 'Agricultural Change in the Lowlands of South Yorkshire with Special Reference to the Manor of Hatfield 1600–c. 1875' (unpublished PhD thesis, University of Sheffield, 2005).
Byres, T. J., 'Differentiation of the Peasantry under Feudalism and the Transition to Capitalism: In Defence of Rodney Hilton', *Journal of Agrarian Change* 6 (2006), pp. 17–68.
Camenisch, C., 'Endless Cold: Seasonal Reconstruction of Temperature and Precipitation in the Burgundian Low Countries during the 15th Century based on Documentary Evidence', *Climate of the Past* 11 (2015), pp. 1049–66.
Campbell, B. M. S., 'Nature as an Historical Protagonist: Environment and Society in Pre-Industrial England', *Economic History Review* 62 (2010), pp. 281–314.
——*The Great Transition: Climate, Disease and Society in the Late-Medieval World* (Cambridge, 2016).
Carey, H. C., *Principles of Social Science* (3 vols., Philapdelphia, 1858).
Carslaw, R. McG., *Four Years of Farming in East Anglia, 1923–1927: Being a Detailed Investigation into the Costs and Returns on Twenty-Six Farms* (Cambridge, 1929).
Ciriacono, S., *Building on Water: Venice, Holland and the Construction of the European Landscape in Early Modern Times*, translation J. Scott (New York and Oxford, 2006).
Collins, E. J. T., 'Rural and Agricultural Change', in E. J. T. Collins, ed., *The Agrarian History of England and Wales. Volume VII, 1850–1914* (Cambridge, 2000), pp. 72–223.
Cook, H., 'Soil and Water Management: Principles and Purposes', in Cook and Williamson, eds., *Water Management*, pp. 15–27.
Cook, H., and T. Williamson, *Water Management in the English Landscape: Field, Marsh and Meadow* (Edinburgh, 1999).
Cracknell, B. E., *Canvey Island: The History of a Marshland Community* (Leicester, 1959).
Croesen, V. R. IJ., *De geschiedenis van de ontwikkeling van de Nederlandsche zuivelbereiding in het laatst van de negentiende en het begin van de twintigste eeuw* (Wageningen, 1931).
Cronshagen, J., *Einfach vornehm: Die Hausleute der nordwestdeutschen Küstenmarsch in der frühen Neuzeit* (Göttingen, 2014).
Croot, P., *The World of the Small Farmer: Tenure, Profit and Politics in the Early Modern Somerset Levels* (Hatfield, 2017).
Cruyningen, P. J. van, *Behoudend maar buigzaam: Boeren in West-Zeeuws-Vlaanderen 1650–1850*, A. A. G. Bijdragen 40 (Wageningen, 2000).

Bibliography

——'Profits and Risks in Drainage Projects in Staats-Vlaanderen c. 1590–1665', *Jaarboek voor Ecologische Geschiedenis* (2005/6), pp. 123–42.

——*Boeren aan de macht? Boerenemancipatie en machtsverhoudingen op het Gelderse platteland, 1880–1930* (Hilversum, 2010).

——'Land en water', in P. Brusse and W. Mijnhardt, eds., *Geschiedenis van Zeeland II 1550–1770* (Zwolle, 2012), pp. 15–49.

——'State, Property Rights and Sustainability of Drained Areas along the North Sea Coast, Sixteenth – Eighteenth Centuries', in B. J. P. van Bavel and E. Thoen, eds., *Rural Societies and Environments at Risk. Ecology, Property Rights and Social Organisation in Fragile Areas (Middle Ages – Twentieth Century)*, Rural History in Europe 9 (Turnhout, 2013), pp. 181–207.

——'From Disaster to Sustainability: Floods, Changing Property Relations and Water Management in the South-Western Netherlands, c. 1500–1800', *Continuity and Change* 29 (2014), pp. 241–65.

——'A Cat with Nine Lives: Rural History in the Netherlands after 1900', *TSEG/The Low Countries Journal of Social and Economic History* 11/2 (2014), pp. 131–52.

——'Bevolking en landbouw', in A. Bauwens and H. Krabbendam, eds., *Scharnierend gewest: 200 jaar Zeeuws-Vlaanderen 1814–2014*, Bijdragen tot de Geschiedenis van West-Zeeuws-Vlaanderen 42 (2014), pp. 55–70.

——'Dealing with Drainage: State Regulation of Drainage Projects in the Dutch Republic, France, and England during the Sixteenth and Seventeenth Centuries', *Economic History Review* 68 (2015), pp. 420–40.

——'Dutch Investors and the Drainage of Hatfield Chase, 1626 to 1656', *Agricultural History Review* 64 (2016), pp. 17–37.

——'Sharing the Cost of Dike Maintenance in the South-Western Netherlands: Comparing "Calamitous Polders" in Three "States" 1715–1795', *Environment and History* 23 (2017), pp. 363–83.

——'Water Management and Agricultural Development in the Coastal Areas of the Low Countries, c. 1200 – c. 1800', in *Gestione dell'acqua in Europa (XII–XVIII secc.)/Water Management in Europe (12th–18th Centuries)* (Florence, 2018), pp. 63–80.

——'Régulation des eaux, investissement urbain et croissance agricole: l'agriculture dans les provinces littorales des Pays-Bas (1400–1900)', in L. Herment, ed., *Histoire rurale de l'Europe, XVIe–XXe siècle* (Paris, 2019), pp. 47–68.

Curtis, D. R., *Coping with Crisis: The Resilience and Vulnerability of Pre-Industrial Settlements* (Farnham, 2014).

——'Danger and Displacement in the Dollard: The 1509 Flooding of the Dollard Sea (Groningen) and its Impact on Long-Term Inequality in the Distribution of Property', *Environment and History* 22 (2016), pp. 103–35.

Dam, P. J. E. M. van, *Vissen in veenmeren: De sluisvisserij op aal tussen Haarlem en Amsterdam en de ecologische transformatie in Rijnland 1440–1530* (Hilversum, 1998).

——'Sinking Peat Bogs: Environmental Change in Holland, 1350–1550', *Environmental History* 6 (2001), pp. 32–45.

——'Ecological Challenges, Technological Innovations: The Modernization of Sluice Building in Holland, 1300–1600', *Technology and Culture* 43 (2002), pp. 500–20.

——'Harnessing the Wind: The History of Windmills in Holland, 1300–1600', in P. Galetti and P. Racine, eds., *I mulini nell' Europa medievale* (Siena, 2003), pp. 34–53.

——'Schijven en beuken balken: Een sociaal-ecologische transformatie in de Riederwaard', in Wouda, ed., *Ingelanden als uitbaters*, pp. 11–43.

——'Fuzzy Boundaries and Three-Legged Tables: A Comment on Ecological and Spatial Dynamics in Bas van Bavel's *Manors and Markets*', *Tijdschrift voor Sociale en Economische Geschiedenis* 8/2 (2011), pp. 103–13.

——'An Amphibious Culture: Coping with Floods in the Netherlands', in P. Coates, D. Moon and P. Warde, eds., *Local Places, Global Processes* (Oxford, 2016), pp. 78–93.

Dam, P. J. E. M. van, P. J. van Cruyningen and M. van Tielhof, 'A Global Comparison of Pre-Modern Institutions for Water Management', *Environment and History* 23 (2017), pp. 335–40.

Danner, H. S., B. van Rijswijk, C. Streefkerk and F. D. Zeiler, *Polderlands: Glossarium van waterstaatstermen* (Wormerveer, 2009).

Darby, H. C., *The Medieval Fenland* (Cambridge, 1940; reprint 2011).

——*The Draining of the Fens* (2nd edn., Cambridge, 1956; reprint 2011).

Darby, H. C., R. E. Glasscock, J. Sheail and G. R. Versey, 'The Changing Geographical Distribution of Wealth in England: 1086–1334–1525', *Journal of Historical Geography* 5 (1979), pp. 247–62.

David, P. A., 'Clio and the Economics of QWERTY', *American Economic Review* 75 (1985), pp. 332–7.

Degroot, D., *The Frigid Golden Age: Climate Change, the Little Ice Age, and the Dutch Republic, 1560–1720* (Cambridge, 2018).

——'Climate Change and Society in the 15th to 18th Centuries', *WIREs Climate Change* 9 (2018).

Deike, L., *Die Entstehung der Grundherrschaft in den Hollerkolonien an der Niederweser* (Bremen, 1959).

Dekker, C., *Zuid-Beveland: De historische geografie en de instellingen van een Zeeuws eiland in de Middeleeuwen* (Assen, 1971).

——*Het Kromme Rijngebied in de Middeleeuwen: Een institutioneel-geografische studie* (Zutphen, 1983).

Dekker, C., and R. Baetens, *Geld in het water: Antwerps en Mechels kapitaal in Zuid-Beveland na de stormvloeden van de 16e eeuw* (Hilversum, 2011).

Bibliography

Derville, A., *L'agriculture du Nord au Moyen Age (Artois, Cambrésis, Flandre Wallonne)* (Villeneuve d'Ascq, 1999).

Devroey, J.-P., and A. Nissen, 'Early Middle Ages, 500–1000', in Thoen and Soens, eds., *Struggling with the Environment*, pp. 11–68.

Diepeveen, W. J., *De vervening in Delfland en Schieland tot het einde der zestiende eeuw* (Leiden, 1950).

Dierßen, K., 'Ökosysteme der Nordseemarschen', in L. Fischer, ed., *Kulturlandschaft Nordseemarschen* (Bredstedt and Westerhever, 1997), pp. 15–26.

Dimmock, S., *The Origin of Capitalism in England, 1400–1600* (Leiden and Boston, MA, 2014).

Dirkx, G. H. P., and J. A. J. Vervloet, *'Oude Leede', een historisch-geografische beschrijving, inventarisatie en waardering van het cultuurlandschap* (Wageningen, 1989).

Dirkx, J., P. Hommel and J. A. J. Vervloet, *Kampereiland: Een wereld op de grens van zout en zoet* (Utrecht, 1996).

Dissel, E. F. van, 'Grond in eigendom en in huur in de ambachten van Rijnland omstreeks 1545', *Handelingen en Mededeelingen van de Maatschappij der Nederlandsche Letterkunde te Leiden, over het jaar 1896/1897* (Leiden, 1897), pp. 152–4.

Ditt, K., 'Zwischen Markt, Agrarpolitik und Umweltschutz: Die deutsche Landwirtschaft und ihre Einflüsse auf Natur und Landschaft im 20. Jahrhundert', in Ditt, Gudermannn and Rüße, eds., *Agrarmodernisierung*, pp. 85–125.

Ditt, K., R. Gudermann and N. Rüße, eds., *Agrarmodernisierung und ökologische Folgen: Westfalen vom 18. bis zum 20. Jahrhundert* (Paderborn, 2001).

Draper, G., 'The Farmers of Canterbury Cathedral Priory and All Souls College Oxford on Romney Marsh, c. 1443–1545', in J. Eddison, M. Gardiner and A. Long, eds., *Romney Marsh: Environmental Change and Human Occupation in a Coastal Lowland* (Oxford, 1998), pp. 109–28.

Dugdale, W., *The History of Imbanking and Draining of Divers Fens and Marshes in Foreign Parts and this Kingdom and of the Improvements thereby* (2nd edn., London, 1772).

Duyverman, J. J., *De landbouwscheikundige basis van het streekplan: Het centrale veengebied Utrecht en Zuid-Holland* (Wageningen, 1948).

Dyer, C., *Making a Living in the Middle Ages: The People of Britain 850–1520* (New Haven and London, 2002).

——*An Age of Transition? Economy and Society in England in the Later Middle Ages* (Oxford, 2005).

Eck, J. van, *Historische atlas van Ooijpolder & Duffelt: Een rivierengebied in woord en beeld* (Amsterdam, 2005).

Edelman, T., *Bijdrage tot de historische geografie van de Nederlandse kuststrook* ('s-Gravenhage, 1974).

Ehbrecht, W., 'Gemeinschaft, Land und Bund im Friesland des 12. bis 14. Jahrhunderts', in H. van Lengen, ed., *Die Friesische Freiheit des Mittelalters – Leben und Legende* (Aurich, 2003), pp. 134–93.

Ehrhardt, M., *Ein Guldten Band des Landes: Zur Geschichte der Deiche im Alten Land* (Stade, 2003).

——'Eine kleine Territorialgeschichte der Region Unterweser', in Bickelmann et al., eds., *Fluss, Land, Stadt*, pp. 147–96.

——'Zur Geschichte der Deiche an der Unterweser', in Bickelmann et al., eds., *Fluss, Land, Stadt*, pp. 305–28.

——'Kabel und Kommunion – Formen des individuellen und genossenschaftlichen Deichens im Elbe-Weser-Raum', in Fischer, ed., *Zwischen Wattenmeer und Marschenland*, pp. 125–64.

Elvin, M., *The Retreat of the Elephants: An Environmental History of China* (New Haven and London, 2004).

Encyclopedie van Zeeland (3 vols., Middelburg, 1982–4).

Epstein, S. R., *Freedom and Growth: The Rise of States and Markets in Europe, 1300–1750* (London and New York, 2000).

——'Craft Guilds in the Pre-Modern Economy: A Discussion', *Economic History Review* 61 (2008), pp. 155–74.

Ettema, W., 'Boeren op het veen (1000–1500): Een ecologisch-historische benadering', *Holland* 37 (2005), pp. 239–58.

Faber, J. A., *Drie eeuwen Friesland: Economische en sociale ontwikkelingen van 1500 tot 1800* (Wageningen, 1972).

——'Bildtboer met ploeg en pen', in P. Gerbenzon, ed., *Het aantekeningenboek van Dirck Jansz (1604–1636)* (Hilversum, 1993), pp. 7–51.

Feenstra, H., and H. H. Oudman, *Een vergeten plattelandselite: Eigenerfden in het Groninger Westerkwartier van de vijftiende tot de zeventiende eeuw* (Leeuwarden, 2004).

Feikes, E., *Die geschichtliche Entwicklung der Deichlast in Nordfriesland* (Stuttgart and Berlin, 1937).

Fischer, N., *Wassersnot und Marschengesellschaft: Zur Geschichte der Deiche in Kehdingen* (Stade, 2003).

——*Im Antlitz der Nordsee: Zur Geschichte der Deiche in Hadeln* (Stade, 2007).

——'Deich und Marschengesellschaft: Fallstudien zur Mentalitäts-, Technik- und Landschaftsgeschichte des Wasserbaus', in Fischer, ed., *Zwischen Wattenmeer und Marschenland*, pp. 93–123.

——ed., *Zwischen Wattenmeer und Marschenland: Deiche und Deichforschung an der Nordseeküste* (Stade, 2021).

Fleet, P. F., 'The Isle of Axholme, 1540–1640: Economy and Society' (unpublished PhD thesis, University of Nottingham, 2002).

Fliedner, D., *Die Kulturlandschaft der Hamme-Wümme Niederung. Gestalt und Entwicklung des Siedlungsraumes nördlich von Bremen* (Göttingen,1970).

Fockema Andreae, S. J., 'Embanking and Drainage Authorities in the Netherlands during the Middle Ages', *Speculum* 27 (1952), pp. 158–67.

Bibliography

——'L'eau et les hommes de la Flandre maritime', *Tijschrift voor Rechtsgeschiedenis* 28 (1960), pp. 181–96.

——'Dijken of wijken', in J. J. Kalma, J. J. Spahr van der Hoek and K. de Vries, eds., *Geschiedenis van Friesland* (Drachten, 1968), pp. 182–200.

——*Warmond, Valkenburg en Oegstgeest* (Dordrecht, 1976).

Fowler, G., 'Old River-Beds in the Fenlands', *Geographical Journal* 79 (1932), pp. 210–12.

——'Shrinkage of the Peat-Covered Fenlands', *Geographical Journal* 81 (1933), pp. 149–50.

Freist, D., and F. Schmekel, eds., *Hinter dem Horizont. Band 2: Projektion und Distinktion ländlicher Oberschichten im europäischen Vergleich, 17.–19. Jahrhundert* (Münster, 2013).

Gallé, P. H., *Beveiligd bestaan: Grondtrekken van het middeleeuwse waterstaatsrecht in Z.W. Nederland en hoofdlijnen van de geschiedenis van het dijksbeheer in dit gebied (1200–1963)* (Delft, 1963).

Galloway, J. A., 'Storm Flooding, Coastal Defence and Land Use around the Thames Estuary and Tidal River c. 1250–1450', *Journal of Medieval History* 35 (2009), pp. 171–88.

——'Storms, Economics and Environmental Change in an English Coastal Wetland: the Thames Estuary c. 1250–1550', in Thoen et al., eds., *Landscapes or Seascapes?*, pp. 379–96.

Gardiner, M., 'Settlement Change on Denge and Walland Marshes, 1400–1550', in J. Eddison, M. Gardiner and A. Long, eds., *Romney Marsh: Environmental Change and Human Occupation in a Coastal Lowland* (Oxford, 1998), pp. 129–45.

Gasser, A., *Geschichte der Volksfreiheit und der Demokratie* (Aarau, 1939).

Gerriets-Purkswarf, J., *Die landwirtschaftlichen Verhältnisse der oldenburgischen Wesermarsch in den Ämtern Brake, Butjadingen und Varel* (Jena, 1913).

Ghosh, S., 'Rural Economies and Transitions to Capitalism: Germany and England Compared (c. 1200–c. 1800)', *Journal of Agrarian Change* 16 (2016), pp. 255–90.

Gijsbers, W., *Kapitale ossen: De internationale handel in slachtvee in Noordwest-Europa (1330–1750)* (Hilversum, 1999).

Godwin, H., *Fenland: Its Ancient Past and Uncertain Future* (Cambridge, 1978).

Gottschalk, M. K. E., *Stormvloeden en rivieroverstromingen in Nederland/Storm Surges and River Floods in the Netherlands* (3 vols., Assen, 1971–7).

——*Historische geografie van westelijk Zeeuws-Vlaanderen* (2 vols., 2nd. edn., Dieren 1983).

——*De Vier Ambachten en het Land van Saaftinge in de Middeleeuwen* (Assen, 1984).

Gouw, J. L. van der, *De Ring van Putten: Onderzoekingen over een hoogheemraadschap in het Deltagebied* ('s-Gravenhage, 1967).

Bibliography

Griffiths, E., *Managing for Posterity: The Norfork Gentry and their Estates c. 1450–1700* (Hatfield, 2022).

Haar, G. ter, and P. L. Polhuis, *De loop van het Friese water: Geschiedenis van het waterbeheer en de waterschappen in Friesland* (Franeker, 2004).

Hagoort, W., *Het hoofd boven water: De geschiedenis van de Gelderse zeepolder Arkemheen 1356 (806) – 1916. Gemeenten Nijkerk en Putten* (Nijkerk, 2018).

Hallam, H. E., 'The Fen Bylaws of Spalding and Pinchbeck', *Lincolnshire Architectural and Archaeological Society Reports and Papers* 10 (1963), pp. 40–56.

——*Settlement and Society: A Study of the Early Agrarian History of South Lincolnshire* (Cambridge, 1965).

——ed., *The Agrarian History of England and Wales: Volume II 1042–1350* (Cambridge, 1988).

Ham, W. van der, *De Grote Waard: Geschiedenis van een Hollands landschap* (Rotterdam, 2003).

——*Verover mij dat land: Lely en de Zuiderzeewerken* (Amsterdam, 2007).

——*Hollandse polders* (Amsterdam, 2009).

——*Johan van Veen, meester van de zee: Grondlegger van het Deltaplan* (Amsterdam, 2020).

——'Veiligheid. Het Deltaplan: nieuwe bescherming voor een kwetsbaar land', in Van der Ham et al., *Modern wereldwonder*, pp. 22–119.

Ham, W. van der, E. Berkers, P. Brusse, H. Buiter, A. van Heezik and B. Toussaint, *Modern wereldwonder: Geschiedenis van de Deltawerken* (Amsterdam, 2018).

Hardin, G., 'The Tragedy of the Commons', *Science* 162 (1968), pp. 1243–8.

Haresign, S. R., 'Agricultural Change and Rural Society in the Lincolnshire Fenlands and the Isle of Axholme, 1870–1914' (unpublished PhD thesis, University of East Anglia, 1980).

Harris, L. E., *Vermuyden and the Fens: A Study of Sir Cornelius Vermuyden and the Great Level* (London, 1953).

Hart, M. 't, *The Dutch Wars of Independence: Warfare and Commerce in the Netherlands, 1570–1680* (London and New York, 2014).

Harten, J. D. H., 'De invloed van de mens op het landschap', in H. P. de Bruin, ed., *Het Gelders rivierengebied uit zijn isolement: Een halve eeuw plattelandsontwikkeling* (Tiel, 1988), pp. 16–37.

Hatcher, J., 'Peasant Productivity and Welfare in the Middle Ages and Beyond', *Past and Present* 262 (2024), pp. 282–314.

Hauschildt, H., *Zur Geschichte der Landwirtschaft im Alten Land: Studien zur bäuerlichen Wirtschaft in einem eigenständigen Marschgebiet des Erzstifts Bremen am Beginn der Neuzeit (1500–1618)* (Hamburg, 1988).

Havekes, H. J. M., *Successful Decentralisation? A Critical Review of Dutch Water Governance* (Deventer, 2023).

Heidinga, H. A., 'Indications of Severe Drought during the Tenth Century

Bibliography

AD from an Inland Dune Area in the Central Netherlands', *Geologie en Mijnbouw* 63 (1984), pp. 241–8.

Henderikx, P. A., 'The Lower Delta of the Rhine and the Maas: Landscape and Habitation from the Roman Period until c.1000', *Berichten van de Rijksdienst voor het Oudheidkundig Bodemonderzoek* 36 (1986) pp. 445–599.

——'De zorg voor de afwatering en dijken op Walcheren voor circa 1400', in P. A. Henderikx, J. A. Lantsheer, A. C. Meijer, J. A. van Werkum and A. Wiggers, eds., *Duizend jaar Walcheren: Over gelanden, heren en geschot, over binnen- en buitenbeheer* (Middelburg, 1996), pp. 23–36.

——*Land, water en bewoning: Waterstaats- en nederzettingsgeschiedenis in de Zeeuwse en Hollandse delta in de Middeleeuwen* (Hilversum, 2001).

——'Land, bewoning, sociale structuren', in P. Brusse and P. A. Henderikx, eds., *Geschiedenis van Zeeland. Prehistorie–1550* (Zwolle, 2012), pp. 91–106.

——'Economische geschiedenis', in P. Brusse and P. A. Henderikx, eds., *Geschiedenis van Zeeland. Prehistorie–1550* (Zwolle, 2012), pp. 125–45.

Hendriks, G., *Een stad en haar boeren* (Kampen, 1953).

Hilkens, B., 'Land Inequality, the Land Market, and Urban Capital in Early Modern Holland: A Case Study, Hazerswoude 1555–1684', paper presented at the Conference 'Feeding the Citizens?', Ghent, 12 April 2024.

Hills, R. L., *Machines, Mills and Uncountable Costly Necessities: A Short History of the Drainage of the Fens* (Norwich, 1967).

Hinrichs, E., R. Krämer and C. Reinders, *Die Wirtschaft des Landes Oldenburg in vorindustrieller Zeit: Eine Regionalgeschichtliche Dokumentation für die Zeit von 1700 bis 1850* (Oldenburg, 1988).

Hipkin, S., 'Tenant Farming and Short-Term Leasing on Romney Marsh, 1587–1705', *Economic History Review* 53 (2000), pp. 646–76.

——'The Structure of Landownership in the Romney Marsh Region, 1646–1834', *Agricultural History Review* 51 (2003), pp. 69–94.

Hodgson, G. M., 'Introduction to the Douglass C. North Memorial Issue', *Journal of Institutional Economics* 13 (2017), pp. 1–23.

Hoffmann, R. C., *An Environmental History of Medieval Europe* (Cambridge, 2014).

Hofmeister, A. E., *Besiedlung und Verfassung der Stader Elbmarschen im Mittelalter* (2 vols., Hildesheim, 1979–81).

——*Seehausen und Hasenbüren im Mittelalter: Bauer und Herrschaft im Bremer Vieland* (Bremen, 1987).

Hollestelle, L., 'Polder- en waterschapsarchieven', in A. C. Meijer, L. R. Priester and H. Uil, eds., *Gids voor historisch onderzoek in Zeeland* (Amsterdam, 1991), pp. 137–47.

Hoon, A. De, *Mémoire sur les polders de la rive gauche de l'Escaut et du littoral Belge* (Brussels, 1853).

Hopcroft, R. L., *Regions, Institutions, and Agrarian Change in European History* (Ann Arbor, 1999).

Hoppenbrouwers, P. C. M., 'Grondgebruik en agrarische bedrijfsstructuur in het Oldambt na de vroegste inpolderingen (1630–1720)', in J. N. H. Elerie and P. C. M. Hoppenbrouwers, eds., *Het Oldambt, deel 2: Nieuwe visies op geschiedenis en actuele problemen*, Historia Agriculturae 22 (Groningen, 1991), pp. 73–94.

——*Een middeleeuwse samenleving: Het Land van Heusden, ca. 1300 – ca. 1515*, Historia Agriculturae 25 (Groningen, 1992).

——'Crop yields in Dutch agriculture, 1850–1990', in Van Bavel and Thoen, eds., *Land Productivity and Agro-Systems*, pp. 113–35.

——'Mapping an Unexplored Field: The Brenner Debate and the Case of Holland', in Hoppenbrouwers and van Zanden, eds., *Peasants into Farmers?*, pp. 41–66.

——'Van waterland tot stedenland: De Hollandse economie ca. 975–ca. 1570', in T. de Nijs and E. Beukers, eds., *Geschiedenis van Holland tot 1572* (Hilversum, 2002), pp. 103–48.

——'Dutch Rural Economy and Society in the Later Medieval Period (c. 1000–1500): An Historiographical Survey', in E. Thoen and L. Van Molle, eds., *Rural History in the North Sea Area: An Overview of Recent Research (Middle Ages – Twentieth Century)*, CORN Publication Series 1 (Turnhout, 2006), pp. 249–82.

Hoppenbrouwers, P. C. M., and J. L. van Zanden, eds., *Peasants into Farmers? The Transformation of Rural Economy and Society in the Low Countries (Middle Ages – Nineteenth Century) in Light of the Brenner Debate*, CORN Publication Series 4 (Turnhout, 2001).

Hoyle, R. W., 'Introduction: Custom, Improvement and Anti-Improvement', in R. W. Hoyle, ed., *Custom, Improvement and the Landscape in Early Modern Britain* (Farnham, 2011), pp. 1–38.

Hubscher, R. H., *L'Agriculture et la société rurale dans le Pas-de-Calais du milieu du XIXe siècle à 1914* (2 vols., Arras, 1979–80).

Hudig, J., and J. J. Duyverman, 'De cultuur der zogenaamde laagveengronden en hun moeilijkheden', *Mededelingen van de Nederlandsche Heidemaatschappij* 7 (1949), pp. 1–29.

Huiting, J. H., 'Domeinen in beweging: Samenleving, bezit en exploitatie in het West-Utrechtse landschap tot in de Nieuwe Tijd' (unpublished PhD thesis, University of Groningen, 2020).

Ibelings, B., 'Turfwinning en waterstaat in het Groene Hart van Holland vóór 1530', *Tijdschrift voor Waterstaatsgeschiedenis* 5 (1996), pp. 74–80.

——'Aspects of an Uneasy Relationship: Gouda and its Countryside (15th–16th centuries)', in Hoppenbrouwers and Van Zanden, eds., *Peasants into Farmers?*, pp. 256–74.

Jakubowski-Tiessen, M., *Sturmflut 1717: Die Bewältigung einer Naturkatastrophe in der frühen Neuzeit* (Munich, 1992).

Bibliography

Janse, A., *Ridderschap in Holland: Portret van een adellijke elite in de late Middeleeuwen* (Hilversum, 2001).

Jones, E. L., and S. J. Woolf, 'Introduction: The Historical Role of Agrarian Change in Economic Development', in E. L. Jones and S. J. Woolf, eds., *Agrarian Change and Economic Development: The Historical Problems* (London, 1969), pp. 1–21.

Jongmans, A. G., M. W. van den Berg, M. P. W. Sonneveld, G. J. W. C. Peek and R. M. van den Berg van Saparoea, *Landschappen van Nederland: Geologie, bodem en landgebruik* (Wageningen, 2013).

Kaijser, A., 'System Building from below: Institutional Change in Dutch Water Control Systems', *Technology and Culture* 43 (2002), pp. 521–48.

Kain, R. J. P., and E. Baigent, *The Cadastral Map in the Service of the State: A History of Property Mapping* (Chicago, 1992).

Kan, F. J. W. van, *Sleutels tot de macht: De ontwikkeling van het Leidse patriciaat tot 1420* (Hilversum, 1988).

Karel, E., and R. Paping, 'In the Shadow of the Nobility: Local Farmer Elites in the Northern Netherlands from the 17th to the 19th Century', in Freist and Schmekel, eds., *Hinter dem Horizont*, pp. 43–54.

Klerk, A. P. de, 'Water naar de zee dragen II: Last en bestrijding van het binnenwater op Walcheren na 1795', in A. P. de Klerk, *Het Nederlandse landschap, de dorpen in Zeeland en het water op Walcheren: Historisch-geografische en waterstaatshistorische bijdragen* (Utrecht, 2003), pp. 245–54.

Kluge, U., *Agrarwirtschaft und Ländliche Gesellschaft im 20. Jahrhundert* (Munich, 2005).

Knibbe, M., *Lokkich Fryslân: Landpacht, arbeidslonen en landbouwproductiviteit in het Friese kleigebied 1505–1830* (Groningen and Wageningen, 2006).

——'De staat, de kerk en het vredesdividend: De pachtopbrengsten van kerkelijke goederen in Friesland, 1511–1543', *De Vrije Fries* 94 (2014), pp. 251–78.

Knollmann, W., and H. Bauer, *Die oldenburger Seekante im 17. Jahrhundert: Zur Geschichte des II. oldenburgischen Deichbandes* (Oldenburg, 1995).

Knottnerus, O. S., 'Räume und Raumbeziehungen im Ems Dollart Gebiet', in Knottnerus et al., eds., *Rondom Eems en Dollard*, pp. 11–42.

——'Deicharbeit und Unternehmertätigkeit in den Nordseemarschen um 1600', in Steensen, ed., *Deichbau und Sturmfluten*, pp. 60–72.

——'Agrarverfassung und Landschaftsgestaltung in den Nordseemarschen', in L. Fischer, ed., *Kulturlandschaft Nordseemarschen* (Bredstedt and Westerhever, 1997), pp. 87–105.

——'De Waddenzeeregio, een uniek cultuurlandschap', in D. van Marrewijk and A. Haartsen, eds., *Waddenland. Het landschap en cultureel erfgoed in de Waddenzeeregio* (Groningen and Leeuwarden, 2001), pp. 15–87.

——'Yeomen and Farmers in the Wadden Sea Coastal Marshes, c. 1500–c.

1900', in B. J. P. van Bavel and P. Hoppenbrouwers, eds., *Landholding and Land Transfer in the North Sea Area (Late Middle Ages – 19th Century)*, CORN Publication Series 5 (Turnhout, 2004), pp. 149–86.

——'Culture and Society in the Frisian and German North Sea Coastal Marshes (1500–1800)', in S. Ciriacono, ed., *Eau et développement dans l'Europe moderne* (Paris, 2004), pp. 139–54.

——'Die Verbreitung neuer Deich- und Sielbautechniken entlang der südlichen Nordseeküste im 16. und 17. Jahrhundert', in C. Endlich, ed., *Kulturlandschaft Marsch: Natur Geschichte Gegenwart* (Oldenburg, 2005), pp. 161–7.

Knottnerus, O. S., P. Brood, W. Deeters and H. van Lengen, eds., *Rondom Eems en Dollard/Rund um Ems und Dollart: Historische verkenningen in het grensgebied van Noordoost-Nederland en Noordwest-Duitsland* (Groningen and Leer, 1992).

Koehn, H., *Die nordfriesischen Inseln: Die entwicklung ihrer Landschaft und die Geschichte ihres Volkstums* (Hamburg, 1961).

Kole, H., *Polderen of niet? Participatie in het bestuur van de waterschappen Bunschoten en Mastenbroek vóór 1800* (Hilversum, 2017).

Konersmann, F., and W. Troßbach, 'Die Bevölkerungsverluste des Spätmittelalters', in R. Kießling, F. Konersmann and W. Troßbach, eds., *Grundzüge der Agrargeschichte. Band 1: Vom Spätmittelalter bis zum Dreißigjährigen Krieg (1350–1650)* (Cologne, Weimar and Vienna, 2016), pp. 17–33.

——'Agrarverfassung im Übergang', in R. Kießling, F. Konersmann and W. Troßbach, eds., *Grundzüge der Agrargeschichte. Band 1: Vom Spätmittelalter bis zum Dreißigjährigen Krieg (1350–1650)* (Cologne, Weimar and Vienna, 2016), pp. 182–227.

Kooistra, L. I., *Delfgauw vindplaats PZPD2: Een middeleeuwse boerderij met een stads sausje?*, BIAXiaal 151 (Zaandam, 2002).

Korf, J., 'Het tijdvak van 1838 tot 1954', in O. Moorman van Kappen, J. Korf and O. W. A. van Verschuer, *Tieler- en Bommelerwaarden 1327–1977: Grepen uit de geschiedenis van 650 jaar waterstaatszorg in Tieler- en Bommelerwaard* (Tiel and Zaltbommel, 1977), pp. 235–424.

Korthals Altes, J., *Sir Cornelius Vermuyden: The Lifework of a Great Anglo-Dutchman in Land-Reclamation and Drainage* (London and The Hague, 1925).

Krabath, S., 'Mittelalterlicher Deich- und Landesausbau im Nordwestdeutschen', in Fischer, ed., *Zwischen Wattenmeer und Marschenland*, pp. 75–92.

Kraker, A. M. J. de, *Landschap uit balans: De invloed van de natuur, de economie en de politiek op de ontwikkeling van het landschap in de Vier Ambachten en het Land van Saeftinghe tussen 1488 en 1609* (Utrecht, 1997).

——'Flood Events in the Southwestern Netherlands and Coastal Belgium, 1400–1953', *Hydrological Sciences Journal* 21 (2006), pp. 913–29.

Bibliography

——'Sustainable Coastal Management: Past, Present and Future or How to Deal with the Tides', *Water History* 3 (2011), pp. 145–62.

——'Storminess in the Low Countries, 1390–1725', *Environment and History* 19 (2013), pp. 149–71.

——'Two Floods Compared. Perception of and Response to the 1682 and 1715 Flooding Disasters in the Low Countries', in K. Pfeifer and N. Pfeifer, eds., *Forces of Nature and Cultural Responses* (Dordrecht, 2013), pp. 185–202.

Kramer, J., 'Entwicklung der Deichbautechnik an der Nordseeküste', in Kramer and Rohde, eds., *Historischer Küstenschutz*, pp. 63–109.

——'Binnenentwässerung und Sielbau im Küstengebiet der Nordsee', in Kramer and Rohde, eds., *Historischer Küstenschutz*, pp. 111–38.

——'Küstenschutz und Binnenentwässerung zwischen Ems und Weser', in Kramer and Rohde, eds., *Historischer Küstenschutz*, pp. 207–53.

Kramer, J., and H. Rohde, eds., *Historischer Küstenschutz: Deichbau, Inselschutz und Binnenentwässerung an Nord- und Ostsee* (Stuttgart, 1992).

Kühn, H. J., 'Sieben Thesen zur Frühgeschichte des Deichbaus in Nordfriesland', in Steensen, ed., *Deichbau und Sturmfluten*, pp. 9–12.

Kuiken, K., 'Van "copers" tot compagnons: De aannemers en aandeelhouders van het Bildt 1505–1555', *Jaarboek van het Centraal Bureau voor Genealogie* 57 (2003), pp. 78–112.

Langdon, J., *Horses, Oxen and Technological Innovation: The Use of Draught Animals in English Farming from 1066 to 1500* (Cambridge, 1986).

Langen, G. J. de, *Middeleeuws Friesland: De economische ontwikkeling van het gewest Oostergo in de vroege en volle Middeleeuwen* (Groningen, 1992).

Langen, G. de, and J. A. Mol, *Friese edelen, hun kapitaal en boerderijen in de vijftiende en zestiende eeuw: De casus Rienck Hemmema* (Amsterdam, 2022).

Langen, G. de, and J. A. Mol, 'The Distribution of Farmland on the Medieval and Prehistoric Salt Marshes of the Northern Netherlands: A Retrogressive Model of the (Pre-)Frisian Farm, Based on Historical Sources from the Early Modern Period', in Nicolay and Schepers, eds., *Embracing the Salt Marsh*, pp. 27–56.

——'Vroege benedictijner kloosterboerderijen in Zuidwest-Friesland', in Nieuwhof and Buursma, eds., *Van Drenthe*, pp. 161–74.

Laveleye, E. De, *L'agriculture Belge: rapport présenté au nom des sociétés agricoles de Belgique et sous les auspices du gouvernement* (Brussels, 1878).

Leemhuis, L., 'De landbouw op de klei in de 19ᵉ eeuw: een vergelijking tussen ontwikkelingen in Groningen en Oost-Friesland', in Knottnerus et al., eds., *Rondom Eems en Dollard*, pp. 416–32.

Leenders, K. A. H. W., *Verdwenen venen: Een onderzoek naar de exploitatie van thans verdwenen venen in het gebied tussen Antwerpen, Turnhout, Geertruidenberg en Willemstad (1250–1750)* (Wageningen, 1989).

——'Zoe lanck ende breedt alst oijt hadde geweest bij staende lande: Het landschap van de Grote Waard vóór 1421', in *'Nijet dan water ende wolcken'. De onderzoekscommissie naar de aanwassen in de Verdronken Waard (1521–1523)* (Tilburg, 2009), pp. 33–43.

Lengen, H. van, 'Bauernfreiheit und Häuptlingsherrlichkeit', in K.-E. Behre and H. van Lengen, eds., *Ostfriesland: Geschichte und Gestalt einer Kulturlandschaft* (Aurich, 1996), pp. 113–34.

L'Epie, Z., *Onderzoek over de oude en tegenwoordige natuurlyke gesteltheyt van Holland en West-Vriesland* (Amsterdam, 1734).

Le Roy Ladurie, E., *Histoire des paysans français: De la Peste noire à la Révolution* (Paris, 2002).

——*Histoire humaine et comparée du climat: canicules et glaciers XIIIe–XVIIIe siècles* (Paris, 2004).

Lesger, C. M., *Hoorn als stedelijk knooppunt: Stedensystemen tijdens de late Middeleeuwen en vroegmoderne tijd* (Hilversum, 1990).

——*Handel in Amsterdam: Kooplieden, commerciële expansie en verandering in de ruimtelijke economie van de Nederlanden ca. 1550–ca. 1630* (Hilversum, 2001).

Levine, D., *At the Dawn of Modernity: Biology, Culture, and Material Life in Europe after the Year 1000* (Berkeley, 2001).

Lewis, A. R., 'The Closing of the Mediaeval Frontier 1250–1359', *Speculum* 33 (1958), pp. 475–83.

Ligtendag, W., *De Wolden en het water: De landschaps- en waterstaatsontwikkeling in het lage land ten oosten van de stad Groningen vanaf de volle Middeleeuwen tot ca. 1870* (Groningen, 1995).

Linden, H. van der, *De cope: Bijdrage tot de geschiedenis van de openlegging der Hollands-Utrechtse laagvlakte* (Assen, 1955).

——'Het platteland in het Noordwesten met nadruk op de occupatie circa 1000–1300', in *Algemene Geschiedenis der Nederlanden*, vol. 2 (Haarlem, 1982), pp. 46–82.

——'Geschiedenis van het waterschap als instituut van waterstaatsbestuur', in B. de Goede, J. H. M. Kienhuis, H. van der Linden and J. G. Steenbeek, eds., *Het waterschap: Recht en werking* (Deventer, 1982), pp. 9–34.

——'History of the Reclamation of the Western Fenlands and of the Organizations to Keep them Drained', in H. de Bakker and M. W. van den Berg, eds., *Peat Lands Lying below Sea Level in the Western Part of the Netherlands, their Geology, Reclamation, Soils, Management and Land Use* (Wageningen, 1982), pp. 42–73.

Lindley, K., *Fenland Riots and the English Revolution* (London, 1982).

Lorenzen-Schmidt, K.-J., 'Anschreibebuchforschungen aus den holsteinischen Elbmarschen', in K.-J. Lorenzen-Schmidt and B. Poulsen, eds., *Bäuerliche Anschreibebücher als Quellen zur Wirtschaftsgeschichte* (Neumünster, 1992), pp. 147–63.

——'Reiche Bauern: Agrarproduzenten der holsteinischen Elbmarschen

Bibliography

als Betriebsführer und Konsumenten', in Freist and Schmekel, eds., *Hinter dem Horizont*, pp. 29–41.

——'Northwest Germany, 1000–1750', in Thoen and Soens, eds., *Struggling with the Environment*, pp. 309–38.

Louman, J. P. A., *Fries waterstaatsbestuur: Een geschiedenis van de waterbeheersing in Friesland vanaf het midden van de achttiende eeuw tot omstreeks 1970* (Amsterdam, 2007).

Lyon, B., 'Medieval Real Estate Development and Freedom', *American Historical Review* 63 (1957), pp. 47–61.

Maenhout, J., *Het polderbeleid in het Scheldedepartement (1794–1814): Juridische – technische – financiële problematiek. De publieke opinie tegenover dit beleid* (Eeklo, 2021).

Mahlerwein, G., *Grundzüge der Agrargeschichte: Die Moderne* (Cologne, Weimar and Vienna, 2016).

Mansholt, D. R., *Beschouwingen over een onderzoek naar de waterschapslasten in Nederland* ('s-Gravenhage, 1941).

Martin, J., 'The Structural Transformation of British Agriculture: The Resurgence of Progressive, High-Input Arable Farming', in B. Short, C. Watkins and J. Martin, eds., *The Front Line of Freedom: British Farming in the Second World War* (Exeter, 2006), pp. 16–35.

Martin, R., and P. Sunley, 'Path Dependence and Regional Economic Evolution', *Journal of Economic Geography* 6 (2006), pp. 395–437.

Mauch, C., and C. Pfister, eds., *Natural Disasters, Cultural Responses: Case Studies toward a Global Environmental History* (Lanham, 2009).

Mauelshagen, F., 'Disaster and Political Culture in Germany since 1500', in Mauch and Pfister, eds., *Natural Disasters*, pp. 41–75.

Mayhew, A., *Rural Settlement and Farming in Germany* (London, 1973).

McCants, A. E. C., *Civic Charity in a Golden Age: Orphan Care in Early Modern Amsterdam* (Urbana and Chicago, 1997).

McCloskey, D. N., *Bourgeois Dignity: Why Economics Can't Explain the Modern World* (Chicago and London, 2010).

Meer, P. L. G. van der, *Opkomst en ûndergong fan in boerebedriuw ûnder Achlum. De famylje Hibma, 1697–1824* (Ljouwert/Leeuwarden, 2001).

Meer, Z. Y. van der, *Het opkomen van den waterstaat als taak van het landsbestuur in de Republiek der Vereenigde Provinciën* (Delft, 1939).

Meier, D., 'Man and Environment in the Marsh Area of Schleswig-Holstein from Roman until Late Medieval Times', *Quaternary International* 112 (2004), pp. 55–69.

——'From Nature to Culture: Landscape and Settlement History of the North-Sea Coast of Schleswig-Holstein, Germany', in Thoen et al., eds., *Landscapes or Seascapes?*, pp. 85–110.

——*Die Halligen in Vergangenheit und Gegenwart* (Heide, 2020).

Meier, D., H. J. Kühn and G. J. Borger, *Der Küstenatlas: Das schleswig-holsteinische Wattenmeer in Vergangenheit und Gegenwart* (Heide, 2013).

Merriënboer, J. van, *Mansholt: Een biografie* (Amsterdam, 2006).
Mertens, J., *De laat-middeleeuwse landbouweconomie in enkele gemeenten van het Brugse platteland* (Brussels, 1970).
Miller, E., ed., *The Agrarian History of England and Wales: Volume III 1348–1500* (Cambridge, 1991).
Miller, E., and J. Hatcher, *Medieval England: Rural Society and Economic Change 1086–1348* (London, 1978).
Mol, J. A., *De Friese huizen van de Duitse Orde: Nes, Steenkerk en Schoten en hun plaats in het Middeleeuwse Friese kloosterlandschap* (Ljouwert, 1991).
——'Mittelalterliche Klöster und Deichbau im westerlauwersschen Friesland', in Steensen, ed., *Deichbau und Sturmfluten*, pp. 46–59.
——'De middeleeuwse veenontginningen in Noordwest Overijssel en Zuid-Friesland', *Jaarboek voor Middeleeuwse Geschiedenis* 14 (2011) pp. 49–90.
Moor, T. De, 'Avoiding Tragedies: A Flemish Common and its Commoners under the Pressure of Social and Economic Change during the Eighteenth Century', *Economic History Review* 62 (2009), pp. 1–22.
Moorman van Kappen, O., 'De historische ontwikkeling van het waterschapswezen, dijk- en waterschapsrecht in de Tieler- en Bommelerwaarden tot het begin der negentiende eeuw', in O. Moorman van Kappen, J. Korf and O. W. A. van Verschuer, *Tieler- en Bommelerwaarden 1327–1977: Grepen uit de geschiedenis van 650 jaar waterstaatszorg in Tieler- en Bommelerwaard* (Tiel and Zaltbommel, 1977), pp. 1–233.
Moree, J. M., 'Barendrecht bouwt Carnisselande, een dijk van een wijk: Archeologisch onderzoek in een nieuwbouwwijk naar het middeleeuwse dijkdorp Carnisse in de verdwenen Riederwaard', in Wouda, ed., *Ingelanden als uitbaters*, pp. 45–69.
Morera, R., *L'Assèchement des marais en France au XVIIe siècle* (Rennes, 2011).
——'Conquest and Incorporation: Merging French-Style Central Government Practices with Local Water Management in Seventeenth-Century Maritime Flanders', *Environment and History* 23 (2017), pp. 341–62.
Morgan, J. E., 'The Micro-Politics of Water Management in Early Modern England: Regulation and Representation in Commissions of Sewers', *Environment and History* 23 (2017), pp. 409–30.
——'Funding and Organising Flood Defence in Eastern England, c. 1570–1700', in *Gestione dell'acqua in Europa (XII–XVIII secc.)/Water Management in Europe (12th–18th centuries)* (Florence, 2018), pp. 413–31.
Müller, F., and O. Fischer, *Das Wasserwesen an der schleswig-holsteinischen Nordseeküste* (3 vols., Berlin, 1917–57).
Newfield, T. P., 'A Cattle Panzootic in Early Fourteenth Century Europe', *Agricultural History Review* 57 (2009), pp. 155–90.
Nicholas, D., 'Of Poverty and Primacy: Demand, Liquidity, and the

Bibliography

Flemish Economic Miracle', *American Historical Review* 96 (1991), pp. 17–41.

Nicolay, J., and H. Huisman, 'Ploughing the Salt Marsh: Cultivated Horizons and their Relation to the Chronology and Techniques of Ploughing', in Nicolay and Schepers, eds., *Embracing the Salt Marsh*, pp. 57–75.

Nicolay, J., and G. de Langen, eds., *Friese terpen in doorsnede: Landschap, bewoning en exploitatie* (Groningen, 2023).

——'Synthese: de steilkanten onderling vergeleken', in Nicolay and De Langen, eds., *Friese terpen in doorsnede*, pp. 423–38.

Nicolay, J., and M. Schepers, eds., *Embracing the Salt Marsh: Foraging, Farming and Food Preparation in the Dutch–German Coastal Area up to AD 1600* (Eelde, 2022).

Nierop, H. F. K. van, *Van ridders tot regenten: De Hollandse adel in de zestiende en de eerste helft van de zeventiende eeuw* (Amsterdam, 1990).

Nieuwhoff, A., and A. Buursma, eds., *Van Drenthe tot aan 't Wad: Over landschap, archeologie en geschiedenis van Noord-Nederland. Essays ter ere van Egge Knol* (Groningen, 2023).

Nijdam, H., 'Preface to the Edition and Translation of the Old Frisian Main Text', in H. Nijdam, J. Hallebeek and H. de Jong, eds., *Frisian Land Law: A Critical Edition of the Freeska Landriucht* (Leiden and Boston, MA, 2023), pp. 3–61.

Noordam, D. J., *Leven in Maasland: Een hoogontwikkelde plattelandssamenleving in de achttiende en het begin van de negentiende eeuw* (Hilversum, 1986).

——*Geringde buffels en heren van stand: Het patriciaat van Leiden, 1574–1700* (Hilversum, 1994).

Noordegraaf, L., *Hollands welvaren? Levensstandaard in Holland 1450–1650* (Bergen, 1985).

Noordhoff, I., *Schaduwkust: Vier generaties hertekenen de rand van de Waddenzee* (Amsterdam, 2018).

Noort, R. Van de, *North Sea Archaeologies: A Maritime Biography, 10,000 BC – AD 1500* (Oxford, 2011).

Norden, W., *Eine Bevölkerung in der Krise: Historisch-demographische Untersuchungen zur Biographie einer norddeutschen Küstenregion (Butjadingen 1600–1850)* (Hildesheim, 1984).

North, D. C., *Structure and Change in Economic History* (New York, 1981).

——*Institutions, Institutional Change and Economic Performance* (Cambridge, 1990).

——*Understanding the Process of Economic Change* (Princeton and Oxford, 2005).

North, D. C., J. J. Wallis and B. R. Weingast, *Violence and Social Orders: A Conceptual Framework for Interpreting Recorded Human History* (Cambridge, 2009).

Numan, K. C., *Burghorn van kwelder tot polder: De eerste bedijking in Hollands Noorderkwartier* (Heerhugowaard and Wormer, 2020).
Ogilvie, S., 'Whatever Is, Is Right? Economic Institutions in Pre-Industrial Europe', *Economic History Review* 60 (2007), pp. 649–84.
——'Choices and Constraints in the Pre-Industrial Countryside', in C. Briggs, P. Kitson and S. J. Thompson, eds., *Population, Welfare and Economic Change in Britain, 1290–1834* (Woodbridge, 2014), pp. 269–305.
Olson, M., *The Logic of Collective Action: Public Goods and the Theory of Groups* (2nd edn., Cambridge, MA and London, 1971).
Ostrom, E., *Governing the Commons: The Evolution of Institutions for Collective Action* (Cambridge, 1990).
Owen, A. E. B., '"The Levy Book of the Sea": The Organization of the Lindsey Sea Defences in 1500', *Lincolnshire Architectural and Archaeological Society Reports and Papers* 9 (1961), pp. 35–48.
Panten, A., 'Deiche und Sturmfluten in der geschichtichen Darstellung Nordfrieslands', in Steensen, ed., *Deichbau und Sturmfluten*, pp. 13–19.
Paping, R. F. J., *Voor een handvol stuivers: Werken, verdienen en besteden: de levensstandaard van boeren, arbeiders en middenstanders op de Groninger klei, 1770–1860* (Groningen, 1995).
Pauwels, A., *Polders en wateringen* (Brussels, 1935).
Peters, K.-H., 'Entwicklung des Deich- und Wasserrechts im Nordseeküstengebiet', in Kramer and Rohde, eds., *Historischer Küstenschutz*, pp. 183–206.
Pfister, C., and H. Wanner, *Klima und Gesellschaft in Europa: Die letzten tausend Jahre* (Bern, 2021).
Pieken, H. A., *Die Osterstader Marsch: Werden und Wandel einer Kulturlandschaft* (Bremen, 1991).
Pierson, P., *Politics in Time: History, Institutions, and Social Analysis* (Princeton and Oxford, 2004).
Pleijter, G., and J. A. J. Vervloet, *Kromakkers en bol liggende percelen in de ruilverkaveling Schalkwijk: In het bijzonder bij Tull en 't Waal en bij Honswijk*, Rapport Stichting voor Bodemkartering 1703 (Wageningen, 1983).
Ploeg, J. D. van der, *The New Peasantries: Struggles for Autonomy and Sustainability in an Era of Empire and Globalization* (London and Sterling, 2008).
——*Gesloten vanwege stikstof: Achtergronden, uitwegen en lessen* (Gorredijk, 2023).
Pons, L. J., 'Passen en meten: De landinrichting bij de herdijking van de polders Oud- en Nieuw Reijerwaard in respectievelijk 1404/05 en 1442/3', in Wouda, ed., *Ingelanden als uitbaters*, pp. 71–111.
Popta, Y. T. van, *When the Shore becomes the Sea: New Maritime Archaeological Insights on the Dynamic Development of the Northeastern Zuyder Zee Region (AD 1100–1400), the Netherlands* (Eelde, 2021).

Bibliography

Postan, M. M., *Fact and Relevance: Essays on Historical Method* (Cambridge, 1971).
Postma, O., *De Friesche kleihoeve: Bijdrage tot de geschiedenis van den cultuurgrond vooral in Friesland en Groningen* (Leeuwarden, 1934).
——*De Fryske boerkerij en it boerelibben yn de 16e en 17e iuw* (Snits, 1937).
Poulsen, B., 'Rural Credit and Land Market in the Duchy of Schleswig c. 1450–1660', in B. J. P. van Bavel and P. Hoppenbrouwers, eds., *Landholding and Land Transfer in the North Sea Area (Late Middle Ages – 19th Century)*, CORN Publication Series 5 (Turnhout, 2004), pp. 203–17.
Prak, M., and J. L. van Zanden, *Pioneers of Capitalism: The Netherlands 1000–1800* (Princeton and Oxford, 2023).
Priester, P. R., *De economische ontwikkeling van de landbouw in Groningen 1800–1910. Een kwalitatieve en kwantitatieve analyse*, A. A. G. Bijdragen 31 (Wageningen, 1991).
——*Geschiedenis van de Zeeuwse landbouw circa 1600–1910*, A. A. G. Bijdragen 37 (Wageningen, 1998).
Prummel, W., and H. C. Küchelmann, 'The Use of Animals on the Dutch and German Wadden Sea Coast, 600 BC–AD 1500', in Nicolay and Schepers, eds., *Embracing the Salt Marsh*, pp. 109–27.
Purseglove, J., *Taming the Flood: A History and Natural History of Rivers and Wetlands* (Oxford, 1988).
Putnam, R. D., and K. A. Goss, 'Introduction', in R. D. Putnam, ed., *Democracies in Flux: The Evolution of Social Capital in Contemporary Society* (Oxford, 2002), pp. 3–19.
Putnam, R. D., R. Leonardi and R. Nanetti, *Making Democracy Work: Civic Traditions in Modern Italy* (Princeton, 1993).
Rasmussen, C. P., 'Yeoman Capitalism and Smallholder Liberalism: Property Rights and Social Realities of Early Modern Schleswig Marshland Societies', in L. Egberts and M. Schroor, eds., *Waddenland Outstanding: History, Landscape and Cultural Heritage of the Wadden Sea Region* (Amsterdam, 2018), pp. 227–37.
Ravensdale, J. R., *Liable to Floods: Village Landscape on the Edge of the Fens A.D. 450–1850* (Cambridge, 1974).
Reinders-Düselder, C., 'Zur Landwirtschaft in Ostfriesland um 1800', in Knottnerus et al., eds., *Rondom Eems en Dollard*, pp. 396–415.
Renes, J., 'The Fenlands of England and the Netherlands: Some Thoughts on their Different Histories' in T. Unwin and T. Spek, eds., *European Landscapes: From Mountain to Sea* (Tallinn, 2003), pp. 101–7.
Ricardo, D., *On the Principles of Political Economy and Taxation* (Georgetown, 1819).
Richardson, H. G., 'The Early History of Commissions of Sewers', *English Historical Review* 34 (1919), pp. 385–93.
Rienks, K. A., and G. L. Walther, *Binnendiken en slieperdiken in Fryslân* (Bolswert, 1954).

Rij, H. van, ed., *Alpertus van Metz: Gebeurtenissen van deze tijd. Een fragment over bisschop Diederik I van Metz. De mirakelen van de heilige Walburg in Tiel* (Hilversum, 1999).

Rippon, S., *The Transformation of Coastal Wetlands: Exploitation and Management of Marshland Landscapes in North West Europe during Roman and Medieval Periods* (Oxford, 2000).

——'Human Impact on the Coastal Wetlands of Britain in the Medieval Period', in Thoen et al., eds., *Landscapes or Seascapes*, pp. 333–49.

Ritson, K., *The Shifting Sands of the North Sea Lowlands: Literary and Historical Imaginaries* (London and New York, 2019).

Robson, E. D., 'Improvement and Environmental Conflict in the Northern Fens, 1560–1665' (unpublished PhD thesis, University of Cambridge, 2018).

Rodengate Marissen, J. Z., *Bijzondere plantenteelt*, vol. 1 (Groningen, 1907).

Roessingh, H. K., and A. H. G. Schaars, *De Gelderse landbouw beschreven omstreeks 1825. Een heruitgave van het landbouwkundig deel van de Statistieke beschrijving van Gelderland* (Wageningen, 1996).

Roosbroeck, F. Van, and A. Sundberg, 'Culling the Herds? Regional Divergences in Rinderpest Mortality in Flanders and South Holland, 1769–1785', *Tijdschrift voor Sociale en Economische Geschiedenis* 14/3 (2017), pp. 31–55.

Roosen, J., and D. R. Curtis, 'The "Light Touch" of the Black Death in the Southern Netherlands: An Urban Trick?', *Economic History Review* 72 (2019), pp. 32–56.

Rössler, H., 'Der Unterweserraum als Schauplatz von Wanderungsbewegungen (ca. 1750–1890)', in Bickelmann et al., *Fluss, Land, Stadt*, pp. 249–70.

Rotherham, I. D., *Peatlands: Ecology, Conservation and Heritage* (London and New York, 2020).

Sayers, D. L., *The Nine Tailors* (London, 1982).

Schaïk, R. W. M. van, 'Een samenleving in verandering: de periode van de elfde en twaalfde eeuw', in M. G. J. Duijvendak, H. Feenstra, M. Hillenga and C. G. Santing, *Geschiedenis van Groningen. Prehistorie – Middeleeuwen* (Zwolle, 2008), pp. 125–67.

Schepers, M., 'Variatie in cultuurplanten in Friese terpen', in Nicolay and De Langen, eds., *Friese terpen in doorsnede*, pp. 305–79.

Schepers, M., and K. E. Behre, 'A Meta-Analysis of the Presence of Crop Plants in the Dutch and German Terp Area between 700 BC and AD 1600', *Vegetation History and Archaebotany* 32 (2023), pp. 305–19.

Schepers, M., C. Smit and J. F. Scheepens, 'Akkeren op de kwelder', in Nieuwhof and Buursma, eds., *Van Drenthe*, pp. 41–52.

Schepper, H. De, 'De burgerlijke overheden en hun permanente kaders, 1480–1579', in *Algemene Geschiedenis der Nederlanden*, vol. 5 (Haarlem, 1980), pp. 311–405.

Bibliography

Schirmer, M., 'Küstenschutz im Klimawandel – ein Blick in die Zukunft', in Fischer, ed., *Zwischen Wattenmeer und Marschenland*, pp. 193–208.

Schoorl, H., *Isaäc Le Maire: Koopman en bedijker* (Haarlem, 1969).

——*Zeshonderd jaar water en land: Bijdrage tot de historische geo- en hydrografie van de Kop van Noord-Holland 1150–1750* (Groningen, 1973).

Schorer, J. A., *De geschiedenis der calamiteuse polders in Zeeland tot het reglement van 20 januari 1791* (Leiden, 1897).

Schothorst, C. J., 'Drainage and Behaviour of Peat Soils', in H. de Bakker and M. W. van den Berg, eds., *Peat Lands Lying below Sea Level in the Western Part of the Netherlands, their Geology, Reclamation, Soils, Management and Land Use* (Wageningen, 1982), pp. 130–63.

Schroor, M., *Wotter: Waterstaat en waterschappen in de provincie Groningen, 1850–1995* (Groningen, 1995).

Schuijtemaker, L., 'Stedelijke weides: Landbouwgrondbezit van stedelingen en de conjunctuur van de Westfriese agrarische sector' (unpublished BA thesis, University of Amsterdam, 2016).

——'Koeien, kaas, kentering. De agrarische geschiedenis van oostelijk Westfriesland tot 1811' (unpublished MA thesis, University of Amsterdam, 2018).

Schürmann, T., *Die Inventare des Landes Hadeln: Wirtschaft und Haushalte einer Marschenlandschaft im Spiegel überlieferter Nachlassverzeichnisse* (Stade and Otterndorf, 2005).

Schutte, G. J., and J. B. Weitkamp, *Marken: De geschiedenis van een eiland* (Amsterdam, 1998).

Schuur, J. R. G., 'De plaatsimg van de Schoutenrechten in hun historische context', *It Beaken* 76 (2014), pp. 1–33.

Schuurman, A., 'Agricultural Policy and the Dutch Agricultural Institutional Matrix during the Transition from Organized to Disorganized Capitalism', in P. Moser and T. Varley, eds., *Integration through Subordination: The Politics of Industrial Modernisation in Europe* (Turnhout, 2013), pp. 65–84.

Scott, J., *How the Old World Ended: The Anglo-Dutch-American Revolution, 1500–1800* (New Haven and London, 2019).

Searle, J. R., 'What Is an Institution?', *Journal of Institutional Economics* 1 (2005), pp. 1–22.

Segers, Y., and L. Van Molle, *Leven van het land: Boeren in België 1750–2000* (Leuven, 2004).

Serruys, M.-W., 'The Societal Effects of the Eighteenth-Century Shipworm Epidemic in the Austrian Netherlands (c. 1730–1760)', *Journal for the History of Environment and Society* 6 (2021), pp. 95–128.

Sheail, J., 'Arterial Drainage in Inter-War England: The Legislative Perspective', *Agricultural History Review* 50 (2003), pp. 253–70.

Bibliography

———'Water Management Systems: Drainage and Conservation', in Cook and Williamson, *Water Management*, pp. 227–43.
Sheail, J., and J. O. Mountford, 'Changes in the Perception and Impact of Agricultural Land-Improvement: The Post-War Trends in the Romney Marsh', *Journal of the Royal Agricultural Society of England* 145 (1984), pp. 43–56.
Siebert, E., 'Entwicklung des Deichwesens vom Mittelalter bis zur Gegenwart', in J. Ohling, ed., *Ostfriesland im Schutze des Deiches*, vol. 2 (Pewsum, 1969), pp. 77–385.
Simmons, I. G., *Fen and Sea: The Landscapes of South-East Lincolnshire AD 500–1700* (Oxford, 2022).
Skertchly, S. B. J., *The Geology of the Fenland* (London, 1877).
Slager, K., *De Ramp: Een reconstructie* (Goes, 1992).
Slicher van Bath, B. H., *Boerenvrijheid* (Groningen and Batavia, 1948).
———'Problemen rond de Friese middeleeuwse geschiedenis', in B. H. Slicher van Bath, *Herschreven historie: Schetsen en studiën op het gebied der middeleeuwse geschiedenis* (Leiden, 1949), pp. 259–80.
———'The Economic and Social Conditions in the Frisian Districts from 900 to 1550', *A. A. G. Bijdragen* 13 (1965), pp. 97–133.
Sly, R., *Soil in their Souls: A History of Fenland Farming* (Stroud, 2010).
Smits, M. B., 'Productie van marktbare gewassen', *Landbouwkundig Tijdschrift* 53 (1941), pp. 45–52.
Smout, T. C., *Nature Contested: Environmental History in Scotland and Northern England since 1600* (Edinburgh, 2000).
Soens, T., *De spade in de dijk? Waterbeheer en rurale samenleving in de Vlaamse kustvlakte (1280–1580)* (Gent, 2009).
———'Floods and Money: Funding Drainage and Flood Control in Coastal Flanders from the Thirteenth to the Sixteenth Centuries', *Continuity and Change* 26 (2011), pp. 333–65.
———'Threatened by the Sea, Condemned by Man? Flood Risk and Environmental Inequalities along the North Sea Coast, 1200–1800', in G. Massard-Guilbaud and R. Rodger, eds., *Environmental and Social Inequalities in the City: Historical Perspectives* (Cambridge, 2011), pp. 91–110.
———'The Origins of the Western Scheldt. Environmental Transformation, Storm Surges and Human Agency in the Flemish Coastal Plain (1250–1600)', in Thoen et al., eds., *Landscapes or Seascapes?*, pp. 287–312.
———'Flood Disasters and Agrarian Capitalism in the North Sea Area: Five Centuries of Interwoven History (1250–1800)', in *Gestione dell'acqua in Europa (XII–XVIII secc.)/Water Management in Europe (12th–18th Centuries)* (Florence, 2018), pp. 369–91.
———'Resilient Societies, Vulnerable People: Coping with North Sea Floods before 1800', *Past and Present* 241 (2018), pp. 143–77.
Soens, T., and E. Thoen, 'The Origins of Leasehold in the Former County

of Flanders', in B. van Bavel and P. R. Schofield, eds., *The Development of Leasehold in Northwestern Europe, ca. 1200–1600*, CORN Publication Series 10 (Turnhout, 2008), pp. 31–55.

Soens, T., P. De Graef, H. Masure and I. Jomgepier, 'Boerenrepubliek in een heerlijk landschap? Een nieuwe kijk op de wase polders als landschap en bestuur' in B. Ooghe, C. Goossens and Y. Segers, eds., *Van brouck tot dyckagie: Vijf eeuwen Wase polders* (Sint-Niklaas, 2012), pp. 19–44.

Soens, T., D. Tys and E. Thoen, 'Landscape Transformation and Social Change in the North Sea Polders, the Example of Flanders (1000–1800 AD)', *Siedlungsforschung: Archäologie – Geschichte – Geographie* 31 (2014), pp. 133–60.

Spahr van der Hoek, J. J., and O. Postma, *Geschiedenis van de Friese landbouw* (Leeuwarden, 1952).

Spufford, M., *Contrasting Communities: English Villagers in the Sixteenth and Seventeenth Centuries* (Cambridge, 1974).

Steensel, A. van, *Edelen in Zeeland: Macht, rijkdom en status in een laatmiddeleeuwse samenleving* (Hilversum, 2010).

——'Bewoning en sociale structuren', in P. Brusse and P. A. Henderikx, eds., *Geschiedenis van Zeeland. Prehistorie–1550* (Zwolle, 2012), pp. 211–25.

——'Noord-Beveland: De geschiedenis van een middeleeuws eiland', *Archief: Mededelingen van het Koninklijk Zeeuwsch Genotschap der Wetenschappen* (2023), pp. 13–120.

Steensen, T., ed., *Deichbau und Sturmfluten in den Frieslanden* (Bredstedt, 1992).

Steers, J. A., 'The East Coast Floods', *Geographical Journal* 119 (1953), pp. 280–95.

Stol, T., 'Opkomst en ondergang van de Grote Waard', *Holland* 13 (1981), pp. 129–45.

——*De veenkolonie Veenendaal: Turfwinning en waterstaat in het zuiden van de Gelderse Vallei 1546–1643* (Zutphen 1992).

——'Schaalvergroting in de polders in Amstelland in de 17^e en 18^e eeuw', *Tijdschrift voor Waterstaatsgeschiedenis* 3 (1994), pp. 13–21.

Stone, L., 'The Return of the Narrative: Reflections on a New Old History', *Past and Present* 85 (1979), pp. 3–24.

Sturmey, S. G., 'Owner-Farming in England and Wales, 1900–1950' in W. Minchinton, ed., *Essays in Agrarian History* (Exeter, 1968), pp. 281–306.

Summers, D., *The Great Level: A History of Drainage and Land Reclamation in the Fens* (Newton Abbot, 1976).

——*The East Coast Floods* (Newton Abbot, 1978).

Sundberg, A. D., *Natural Disasters at the Closing of the Dutch Golden Age: Floods, Worms, and Cattle Plague* (Cambridge, 2022).

Swart, F., *Zur friesischen Agrargeschichte* (Leipzig, 1910).

Bibliography

TeBrake, W. H., *Medieval Frontier: Culture and Ecology in Rijnland* (College Station, 1985).

——'Taming the Waterwolf: Hydraulic Engineering and Water Management in the Netherlands during the Middle Ages', *Technology and Culture* 43 (2002), pp. 475–99.

Thirsk, J., ed., *The Agrarian History of England and Wales Vol. IV, 1550–1640* (Cambridge, 1967).

——*English Peasant Farming: The Agrarian History of Lincolnshire from Tudor to Recent Times* (2nd edn., London and New York, 1981).

——*England's Agricultural Regions and Agrarian History, 1500–1750* (Basingstoke and London, 1987).

——'The Crown as Projector on its Own Estates, from Elizabeth I to Charles I', in R. W. Hoyle, ed., *The Estates of the English Crown 1558–1640* (Cambridge, 1992), pp. 297–352.

Thoen, E., 'Technique agricole, cultures nouvelles et économie rurale en Flandre au bas Moyen Age', in *Plantes et cultures nouvelles en Europe Occidentale au Moyen Age et à l'époque moderne*, Flaran 12 (1990), pp. 51–67.

——'The Count, the Countryside and the Economic Development of the Towns in Flanders from the Eleventh Century to the Thirteenth Century: Some Provisional Remarks and Hypotheses', in E. Aerts, B. Henau, P. Janssens and R. van Uytven, eds. *Studia Historia Oeconomica: Liber Amicorum Herman Van der Wee* (Leuven 1993), pp. 259–78.

——'The Birth of "the Flemish husbandry": Agricultural Technology in Medieval Flanders', in G. Astill and J. Langdon, eds., *Medieval Farming and Technology: The Impact of Agricultural Change in Northwest Europe* (Leiden, 1997), pp. 69–88.

——'Transitie en economische ontwikkeling in de Nederlanden met de nadruk op de agrarische maatschappij', *Tijdschrift voor Sociale Geschiedenis* 22 (2002) pp. 147–74.

——'Social Agrosystems as an Economic Concept to Explain Regional Differences: An Essay Taking the Former County of Flanders as an Example', in B. J. P. van Bavel and P. Hoppenbrouwers, eds., *Land Holding and Land Transfer in the North Sea Area (Late Middle Ages – 19th Century)*, CORN Publication Series 5 (Turnhout, 2004), pp. 47–66.

Thoen, E., G. J. Borger, A. M. J. de Kraker, T. Soens, D. Tys, L. Vervaet and H. J. T. Weerts, eds., *Landscapes or Seascapes? The History of the Coastal Environment in the North Sea Area Reconsidered*, CORN Publication Series 13 (Turnhout, 2013).

Thoen, E., and T. Soens, eds., *Struggling with the Environment: Land Use and Productivity. Rural History in North-Western Europe, 500–2000* (Turnhout, 2015).

——'The Low Countries, 1000–1750', in Thoen and Soens, eds., *Struggling with the Environment*, pp. 221–58.

Bibliography

———'Contextualizing 1500 Years of Agricultural Productivity and Land Use in the North Sea Area', in Thoen and Soens, eds., *Struggling with the Environment*, pp. 455–95.

Thurkow, A. J., 'De overheid en het landschap in de droogmakerijen van de 16e tot en met de 19e eeuw', *Historisch-Geografisch Tijdschrift* 9 (1991), pp. 49–56.

Tielhof, M. van, *De Hollandse graanhandel, 1470–1570: Koren op de Amsterdamse molen* (The Hague, 1995).

———'Turfwinning en proletarisering in Rijnland 1530–1670', *Tijdschrift voor Sociale en Economische Geschiedenis* 2/4 (2005), pp. 95–121.

———'Betrokken bij de waterstaat: Boeren, burgers en overheden ten zuiden van het IJ tot 1800', in E. Beukers, ed., *Hollanders en het water: Twintig eeuwen strijd en profijt* (Hilversum, 2007), pp. 61–98.

———'Forced Solidarity: Maintenance of Coastal Defences along the North Sea Coast in the Early Modern Period', *Environment and History* 21 (2015), pp. 319–50.

———'After the Flood: Mobilising Money in Order to Limit Economic Loss (the Netherlands, Sixteenth–Eighteenth Centuries', in *Gestione dell'acqua in Europa (XII–XVIII secc.)/Water Management in Europe (12th–18th Centuries)* (Florence, 2018), pp. 393–411.

———*Consensus en conflict: Waterbeheer in de Nederlanden 1200–1800* (Hilversum, 2021).

Tielhof, M. van, and P. J. E. M. van Dam, *Waterstaat in stedenland: Het hoogheemraadschap van Rijnland voor 1857* (Utrecht, 2006).

Tilly, C., *Coercion, Capital, and European States, AD 990–1992* (Cambridge, MA and Oxford, 1992).

Tocqueville, A. de, *De la démocratie en Amérique* (2 vols., Paris, 1835–40, reprint 1981).

Toussaint, B., 'Eerbiedwaardig of uit de tijd? De positie van de waterschappen tussen 1795 en 1870', *Tijdschrift voor Waterstaatsgeschiedenis* 18 (2009), pp. 40–50.

Tracy, M., *Agriculture in Western Europe: Challenge and Response 1880–1980* (London, 1982).

Tys, D., 'The Medieval Embankment of Coastal Flanders in Context', in Thoen et al., eds., *Landscapes or Seascapes?*, pp. 199–239.

Vandewalle, P., *De geschiedenis van de landbouw in de kasselrij Veurne (1550–1645)* (Brussels, 1986).

———*Quatre siècles d'agriculture dans la région de Dunkerque 1590–1990: une étude statistique* (Gent, 1994).

Vellema, D., 'De landbouw in het Lage Midden van Friesland, 1800–2020' (unpublished BSc thesis, Wageningen University, 2021).

Ven, G. P. van de, ed., *Man-Made Lowlands: History of Water Management and Land Reclamation in the Netherlands* (4th edn., Utrecht, 2004).

Bibliography

Vergouwe, A., *Emigranten naar Amerika uit West Zeeuws-Vlaanderen* (n.p., 1993).

Verhulst, A., 'L'intensification et la commercialisation de l'agriculture dans les Pays-Bas méridionaux au XIIIe siècle', in *La Belgique rurale du moyen-age à nos jours. Mélanges offerts à Jean-Jacques Hoebanx* (Brussels, 1985), pp. 89–100.

——*Précis d'histoire rurale de la Belgique* (Brussels, 1990).

——*Landschap en landbouw in Middeleeuws Vlaanderen* (Brussels, 1995).

Verhulst, A., and M. K. E. Gottschalk, eds., *Transgressies en occupatiegeschiedenis in de kustgebieden van Nederland en België* (Gent, 1980).

Vervaet, L., 'Goederenbeheer in een veranderende samenleving. Het Sint-Janshospitaal van Brugge ca. 1275 – ca. 1575' (unpublished PhD thesis, Gent University, 2014).

Vervloet, J. A. J., *Inleiding tot de historische geografie van de Nederlandse cultuurlandschappen* (Wageningen, 1984).

Vidal de La Blache, P., *Principes de géographie humaine*, ed. E. de Martonne (Paris, 1922).

Viviano, F., 'How the Netherlands Feeds the World', *National Geographic* (September 2017), pp. 58–81.

Vleesschauwer, M. de, *Van water tot land: Polders en waterschappen in midden Zeeuws-Vlaanderen 1600–1999* (Utrecht, 2013).

——*Het Vrije van Sluis: Polders en waterschappen in West-Zeeuws-Vlaanderen 1600–1999* (Utrecht, 2013).

Vos, P., *Origin of the Dutch Coastal Landscape: Long-Term Landscape Evolution of the Netherlands during the Holocene, Described and Visualized in National, Regional and Local Palaeogeographical Map Series* (Groningen, 2015).

Vos, P., and F. D. Zeiler, 'Overstromingsgeschiedenis van Zuidwest-Nederland, interactie tussen natuurlijke en antropogene processen', *Grondboor en Hamer* 62 (2008), pp. 86–95.

Vries, H. de, *Landbouw en bevolking tijdens de agrarische depressie in Friesland (1878–1895)* (Wageningen, 1971).

Vries, J. de, *The Dutch Rural Economy in the Golden Age* (New Haven and London, 1974).

——*The Economy of Europe in an Age of Crisis, 1600–1750* (Cambridge, 1976).

——'The Transition to Capitalism in a Land without Feudalism', in Hoppenbrouwers and Van Zanden, eds., *Peasants into Farmers?*, pp. 67–84.

Vries, J. de, and A. van der Woude, *The First Modern Economy: Success, Failure, and Perseverance of the Dutch Economy, 1500–1815* (Cambridge, 1997).

Walker, H. J., 'The Coastal Zone', in B. L. Turner II, W. C. Clark, R. W. Kates, J. F. Richards, J. F. Matthews and W. B. Meyer, eds., *The Earth as*

Transformed by Human Action: Global and Regional Changes over the Past 300 Years (Cambridge, 1990), pp. 271–94.

Watts, H. D., 'The Location of the Beet-Sugar Industry in England and Wales, 1912–1936', *Transactions of the Institute of British Geographers* 53 (1971) pp. 95–116.

Wells, S., *The History of the Drainage of the Great Level of the Fens called Bedford Level; with the Constitution of the Bedford Level Corporation* (2 vols., London, 1830).

Werveke, H. Van, 'La famine de l'an 1316 en Flandre et dans les régions voisines', *Revue du Nord* 41 (1959), pp. 5–14.

Whetham, E. H., *The Agrarian History of England and Wales: Volume VIII 1914–1939* (Cambridge, 1978).

White, Jr., L., *Medieval Technology and Social Change* (Oxford, 1962).

Whittle, J., *The Development of Agrarian Capitalism: Land and Labour in Norfolk 1440–1580* (Oxford, 2000).

Wickham, C., 'Social Relations, Property and Power around the North Sea, 500–1000', in B. van Bavel and R. Hoyle, eds., *Social Relations: Property and Power. Rural History in Northwestern Europe, 500–2000* (Turnhout, 2010), pp. 25–47.

Wiel, K. van der, 'De boer als Assepoester van de Zaanse geschiedenis: Het boerenbedrijf van de Zaanstreek en de invloed van de industriële ontwikkeling', in E. Beukers and C. van Sijl, eds., *Geschiedenis van de Zaanstreek* (Zwolle, 2012), pp. 219–37.

Wieringa, W. J., *Het Aduarder Zijlvest in het Ommelander waterschapswezen* (Groningen, 1946).

Wilderom, M. H., *Tussen afsluitdammen en deltadijken* (4 vols., Vlissingen, 1961–73).

Williams, C. M., and J. B. Hardaker, *The Small Fen Farm: Problems and Prospects of Intensive Arable Smallholdings on Skirt Soils in the Fens* (Cambridge, 1966).

Williamson, T., *The Norfolk Broads: A Landscape History* (Manchester, 1997).

—— *The Transformation of Rural England: Farming and the Landscape 1700–1870* (Exeter, 2002).

Wilt, C. de, *Landlieden en hoogheemraden: De bestuurlijke ontwikkeling van het waterbeheer en de participatiecultuur in Delfland in de zestiende eeuw* (Hilversum, 2015).

Winchester, A. J. L., *Common Land in Britain: A History from the Middle Ages to the Present Day* (Woodbridge, 2022).

Winiwarter, V., 'Landwirtschaft, Natur und Ländliche Gesellschaft im Umbruch: Eine umwelthistorische Perspektive zur Agrarmodernisierung', in Ditt, Gudermann and Rüße, eds., *Agrarmodernisierung*, pp. 733–67.

Bibliography

Winsemius, J. P., *De historische ontwikkeling van het waterstaatsrecht in Friesland* (Franeker, 1947).

Wintle, M. J., 'Dearly Won and Cheaply Sold: The Purchase and Sale of Agricultural Land in Zeeland in the Nineteenth Century', *Economisch-en Sociaal-Historisch Jaarboek* 49 (1986), pp. 44–99.

Wiskerke, J. S. C., *Zeeuwse akkerbouw tussen verandering en continuïteit: Een sociologische studie naar de diversiteit in landbouwbeoefening, technologieontwikkeling en plattelandsvernieuwing* (Wageningen, 1997).

Woestenburg, M., *Waarheen met het veen: Kennis voor keuzes in het westelijk veenweidegebied* (Wageningen, 2009).

Woud, A. van der, *Het landschap, de mensen: Nederland 1850–1940* (Amsterdam, 2020).

Wouda, B., ed., *Ingelanden als uitbaters: Sociaal-economische studies naar Oud- en Nieuw-Riederwaard, een polder op een Zuid-Hollands eiland* (Hilversum, 2004).

Woude, A. M. van der, *Het Noorderkwartier: Een regionaal historisch onderzoek in de demografische en economische geschiedenis van westelijk Nederland van de late Middeleeuwen tot het begin van de negentiende eeuw*, A. A. G. Bijdragen 16 (Wageningen, 1972).

Wrigley, E. A., *The Path to Sustained Growth: England's Transition from an Organic Economy to an Industrial Revolution* (Cambridge, 2016).

Zanden, J. L. van, *De economische ontwikkeling van de Nederlandse landbouw in de negentiende eeuw 1800–1914*, A. A. G. Bijdragen 25 (Wageningen, 1985).

——'The First Green Revolution: The Growth of Production and Productivity in European Agriculture, 1870–1914', *Economic History Review* 44 (1991), pp. 215–39.

——'Taking the Measure of the Early Modern Economy: Historical National Accounts for Holland in 1510/14', *European Review of Economic History* 6 (2002), pp. 131–63.

Zeischka, S., *Minerva in de polder: Waterstaat en techniek in het Hoogheemraadschap van Rijnland 1500–1865* (Hilversum, 2007).

Zeist, W. van, 'Milieu, akkerbouw en handel van middeleeuws Leeuwarden', in M. Bierma, A. T. Clason, E. Kramer and G. J. de Langen, eds., *Terpen en wierden in het Fries-Groningse kustgebied* (Groningen, 1988), pp. 129–41.

Zomer, J., *Middeleeuwse veenontginningen in het getijdenbekken van de Hunze: Een interdisciplinair landschapshistorisch onderzoek naar de paleogeografie, ontginning en waterhuishouding (ca. 800 – ca. 1500)* (Eelde, 2016).

Zonneveld, I., 'Schoen riet, rijs ende grienten: Begroeiing en benutting van de Verdronken Waard', in *Nijet dan water ende wolcken: De onderzoekscommissie naar de aanwassen in de Verdronken Waard (1521–1523)* (Tilburg, 2009), pp. 109–17.

Bibliography

Zwarts, H., 'Knowledge, Networks, and Niches: Dutch Agricultural Innovation in an International Perspective, c. 1880–1970' (unpublished PhD thesis, Wageningen University, 2021).

Zwet, H. van, *Lofwaerdighe dijckagies en miserabele polders: Een financiële analyse van landaanwinningsprojecten in Hollands Noorderkwartier, 1597–1643* (Hilversum, 2009).

INDEX

Aa, River   62
Aalburg   74–5
Aardenburg   93, 171
Achlum   229
Adolf-Hitler-Koog   254
Aduard, abbey of   63, 129
Aduarder Zijlvest   63
Afsluitdijk   249, 273
Albert and Isabella, Archdukes   144
Alblasserwaard   39, 60–1, 172
Alkemade   172
Alkemade, Floris van   94
Alkmaar   93
Altes Land   17, 49, 105, 108, 131, 182–4, 199, 209–10, 258
amphibious culture   15
Amstelland   158
Amsterdam   114, 140, 143, 158
Ancholme Level   147–8, 153
Anton Günther, count of Oldenburg   139, 144
Antwerp   81, 90, 137, 139–40, 143
agrarian capitalism   113, 121, 128, 132, 166–7, 190–1, 213, 286
agricultural regions   13, 15, 261
arable farming   49, 73, 76, 120, 129, 134, 146, 170, 177–8, 186–7, 190, 269, 279
  contraction of   38, 42, 84, 123–4, 132, 264, 267
  expansion of   70, 182, 184, 223–4, 229–30, 233, 237, 240–1, 243–4, 275, 289
arable land   32, 34, 42, 49, 74, 178, 182, 228, 237, 262, 264, 275
Aragon   8
Arkemheen   125–6, 162
Atlantic Gulf Stream   87
Aubin, Hermann   43–4

Baarderadeel   126
Badeslade, Thomas   39
Baltic Area   28, 121
Bankoff, Greg   1, 12, 201, 255

Banks, Sir John   153
Banks, Sir Joseph   233
Barendrecht   74–75
barley   28–9, 31, 70, 72–6, 90, 117, 131–2, 177, 179, 182, 184, 224, 236, 272
  see also cereal yields
Bathe, Henry de   64
Bavel, Bas van   10, 11, 13–14, 45, 68, 72, 75, 78, 119, 121–4, 166, 168, 170–2, 179, 182, 184, 240, 286
Bedford, Earl/Duke of   157, 188, 202, 211, 231, 270
Bedford Level, *see* Great Level
Bedford Level Corporation   202–3, 234
Beemster, Lake/Polder   145–6, 151
Beherdischheit   131, 185
beklemrecht   131, 185, 236
Belgium   194, 217, 246, 253, 257
Bergen (Norway)   183
Berghmans, Sander   223
Berkel (Zuid-Holland)   173
Bieleman, Jan   1, 15
Bildt, het   143, 184
Bismarck, Otto von   194
Black Death   51, 80–1, 133
Blanchard, Raoul   3
Blankenbergse Watering   114
Blijdorp, Polder   196
Boer, T.J. de   127
boezem   104, 205–6, 235, 257
Bommelerwaard   178
Bonaparte, Napoleon   216
Bont, Chris de   173
Borsselepolder   145
Boskoop   46
Boston (Lincolnshire)   58–9, 63, 233
Bottschlotter Werk   145
Boudewijn V, Count of Flanders   46
Boulton & Watt   196
Braakman   87
Brabant   23, 26, 81–3, 85, 90, 139–41, 143, 150, 198, 217–18

330

Index

Bradley, Humphrey 152
Brassley, Paul 261, 274, 279
Brazil 1
Bremen 81, 131, 183, 246
Bremen, Archbishop of 23, 47–8, 50, 108
Bremen, Archbishopric of 47, 83, 139
Brenner, Robert 14, 113, 121, 166, 190
Brouwer, Anne Gerrit 218
Bruges 70–1, 80, 88, 113–14, 224
Bruges, Franc de 62
Brümmel, Peter 199
Buitelaar, A.L.P. 47
Bunschoten 173
Burgundian Lands 82
Burgundy, Duke of 82
Butjadingen 11, 17, 83, 131, 161–2, 183, 211–12, 239–41, 244, 266, 287–8
butter 74, 85, 121, 182, 226–7, 230–1, 246, 266, 268
Buxtehude 131
Byford, Daniel 187

Cadzand 208–9, 216, 224
Calais 215
Calaisis 224
Cambridgeshire 17, 46, 49, 51, 89, 91, 134, 136, 202, 209, 242, 269
Campbell, Bruce 15, 80
Cantley 265
Canvey Island 34, 133, 146, 186, 250
capital
  markets 172
  urban 93–4, 110, 136, 145, 157, 160, 164, 169, 284
Carrington, Lord 267, 270
Catchment Boards 255
cattle
  breeding of 29, 125, 177, 181, 188, 190, 229, 231–2, 264, 285
  fattening of 116, 123, 125, 129, 146, 179, 190, 228–9, 231, 263–4, 279
  see also cows, dairy farming, oxen
Central Asia 80
cereal yields 41, 71, 78, 124–5, 170, 177–9, 221–2, 224, 228–30, 233, 235, 237, 241–3, 262, 270–1
Charles I, King of England 153, 164, 231, 285

cheese 121, 182, 189, 226–7, 230–1
China 4–5
Cirksena, Ulrich 83, 105
climate change 21–3, 65, 79–80, 87, 91, 137–8, 194, 293
Commission of Sewers 7, 64–5, 78, 108–10, 153, 255, 283
Common Agricultural Policy 246, 291
common land/commons 34, 36, 67–8, 76–8, 89, 134, 148, 150, 152–4, 156, 164, 167, 184, 187, 283, 285
commoners 148–50, 153–4, 156, 164, 187, 285–6
contentement 151
cope 45–9, 51, 71, 211
Cortswagers, Lijsbeth 176
cows 71–2, 74–5, 84, 105, 116, 120, 128–9, 170, 175–6, 181, 184, 226, 229, 231–2, 249, 273, 278–9
credit 10, 82, 160, 176, 190
Cromwell, Oliver 231
Cronshagen, Jessica 240
crop rotation 70, 241, 270
Cruquius steam pumping station 197
Curtis, Daniel 81

Dagebüller Bucht 145
dairy farming/production 38, 70, 85, 116, 129, 146, 170, 175–6, 179, 182, 190, 205, 223–4, 226–8, 231–2, 242, 246, 264, 266
Dam, Petra van 15, 94, 172
Danelaw 49
Darby, Henry Clifford 17, 57, 188, 203–4, 231–2
Dean, Arthur 196
Deeping Fen 233–4
Deeping St. Nicholas 234
Deichacht 7, 105
Deichverband 7, 257
Delfgauw 73
Delfland 93, 95, 103, 172–3, 175–6, 180, 231
Delft 46, 74, 94–5, 103, 140
Delta Works 259
Denmark 13, 17, 23, 139, 194, 210, 245–7, 266
Dieleman, Petrus 256

331

Index

Dienst Zuiderzeewerken 248
dijkgraaf 61, 155, 216
dike construction 32–3, 41–2, 91–2, 143, 248–51
Dirk III, Count of Holland 46
Dissolution of the Monasteries 168
Dithmarschen 176, 24, 56, 83, 130, 154, 182, 240, 278, 286
Doctor Cassandra 249
Dollard 86, 107
Dordrecht 73, 88, 143, 169
drainage 3–4, 147–9, 157–60, 195–8, 251–3, 268
drainage company 155–7
drainage mill 39, 103–5, 109–10, 120, 125, 141–2, 157–60, 179, 182, 188, 190, 197, 202–3, 205, 214, 235, 289
  introduction of 93–7
  investment in *see* investment
Dutch River Area 17, 26–7, 44, 49–50, 113, 135, 244, 283
  drainage of 32, 124–5, 243
  farming in 121–5, 177–8, 190, 228–30, 263, 270–1, 276
  water management in 198–200, 217
Duyverman, Johannes Jacob 40

East Anglia 265, 269
East and West Fen 148, 167
East Sussex 1
Eckwarden 239
Eiderstedt 182, 264, 278
Eiesluis, Watering 113
Elbe, River 3, 26, 47–9, 66, 100, 105, 108, 132, 198–200, 211–12, 246, 258, 264
Elbe-Weser-Triangle 267
electric pumping station 252
elites 11, 116, 139, 161–5, 194, 199, 208, 213–15, 224, 240, 256, 287
Elizabeth I, Queen of England 152–3
Elvin, Mark 4–5
Ely 269
Emden 131
emmer wheat 28, 73–5
Ems 3
English Fens 3, 147, 203, 205, 220–1, 230, 236, 241, 243–4, 261, 265, 276, 288–9, 291
Epstein, S.R. 192
Essex 34, 88, 265, 269
exploitation of wetlands 31–2, 35, 282
exports 81, 132, 182, 226–30, 241, 263, 266, 275

Faipoult, Guillaume Charles 215–16
fallow/fallowing 38, 70, 184, 221, 225
farm buildings 72, 123, 130–1, 142, 159, 181, 224, 243, 253
farm size 69–71, 115–17, 122–3, 128–30, 134–5, 169–70, 175–8, 181–3, 185–6, 189, 222–3, 234, 240
farmer, *see* owner-occupier, tenant farmer
fens/fenlands
  drainage of 147–9, 202–4, 233–4, 251–2, 285
  oxidation and shrinkage of 36–41, 73, 83–6, 283–4
fertilisation 269
  artificial 225, 260, 274
  claying 233–4, 237, 241
  manuring 180, 221, 225
  marling 123–4
  sediment 126, 149, 179, 241, 272–4
  terpaarde 230
  see also leguminous plants
Ferwerderadeel 126–7
field drainage 3–4, 21, 234, 237–8, 251–2, 259, 271, 279
Finsterwolde 238
Flanders 23–4
  see also Flemish Coastal Plain
flax 20, 28–9, 31, 73–4, 76, 96, 135, 170, 183, 188, 191, 223
Flemish Coastal Plain 17, 25, 28, 86–7, 91–3, 135, 177–8
  farming on 69–71, 78, 116–17, 169–70, 181, 190, 222–5, 242, 261–2
  flooding of 86, 91, 167, 245
  property relations on 14, 44, 46, 92, 113–5, 168
  water management on 100–1, 215, 217, 283

332

Index

floods 41–2, 86–8, 138, 162, 183, 206–7, 210–11, 213, 248–51
fodder/fodder crops 124, 225, 237, 263, 269, 291
France 1, 3, 13, 23, 44, 121, 140, 167, 215, 217, 246, 262, 265
freedom 19, 24, 43–4, 48–53, 67, 191
Friesland
 central 129, 160, 179, 181, 205, 235, 253, 275, 285, 288
 farming in 28–9, 126–8, 178–80, 229–30, 235–6
 water management in 56–7, 65, 106–7, 153, 159–60, 197, 205–6, 253, 284–5
 see also Wadden Coast
fruit cultivation 177, 183, 228–9, 263, 265, 271, 275

Galicia 267
Galloway, James 133
Gasser, Adolf 43
Gelderland 201, 218, 271
Gelderse Vallei 179
gemeenmaking 100–1
Germany 194, 246–7, 250, 253, 257–8, 276
Ghent 26, 71, 88, 115, 224
Gibbs, Joseph 196
Goeree 207
Goes 208
Gouda 72–3, 75
grass *see* hay, meadow, pasture
Grebbe Sluice 6, 9–10
Great Depression 246, 268
Great Level 40, 148–9, 152–3, 156, 187–9, 202, 234, 252
Great Ouse, River 204, 251
Green Heart of Holland 4
Greetsiel 212
Griffiths, Elizabeth 152
Grimsby 270
Groningen
 city of 81
 farming in province of 130, 180, 236–8, 242–4, 264–5, 288
 flooding of province of 138, 163, 211
 property relations in province of 129, 131, 184–5

water management in province of 97, 106–7, 111, 161–4, 199, 201, 213–14, 219, 253–4, 256, 287–8
 see also Wadden Coast
Grote Waard 87–9, 91
Guelders 17, 23, 44, 60–2, 78, 82–3, 122, 125–6, 162, 166, 177–8, 243
Guelders River Area, *see* Dutch River Area

Haarlem 119, 140, 196
Haarlemmermeer 86, 194, 196–7, 220, 248, 290
Habsburg, House of 82–3, 106–7, 138–9, 159, 167–8, 170, 172
Haddenham Level 202
Hadeln 17, 105, 108, 131, 200–1, 240, 250, 258, 264, 267, 278, 286
Ha(e)sdonck, Jan van 152
Hallam, Herbert E. 16, 51, 191
Hallig 90, 184, 251, 258, 271–2, 274, 292
Hamburg 81, 183, 246, 258
Hannover, Kingdom of 194, 212
Hardin, Garrett 68
Harlebucht 86
Harlingen 229
Harlingerland 212
Hatfield Chase 17, 147, 149, 153, 156, 187–8, 204, 234, 253
hay 13, 32, 72, 84, 129, 179, 205, 220, 227, 229, 235, 272–4, 276, 285
Heemraadschap 7, 60, 64, 103, 136
Heerenveen 205
Hellevoetsluis 207
hemp 28, 73–4, 96, 135, 188
Hennaarderadeel 126–7
Hermann-Göring-Koog 254
Heusden, Land van 74
Hier 124
Hills, Richard L. 231
Hipkin, Stephen 185–6
Hogebeintum 29
Hohenkirchen 130
Holland (Lincolnshire) 46
Holland, 17, 23, 82
 Count of 45, 48, 61, 63, 69, 73, 84, 93
 farming in 71–5, 120–1, 175–7, 226–8, 266, 286

333

Index

flooding of 249–50
property relations in 44–7, 118–20, 136, 169, 171–4
water management in 60–1, 103–4, 111, 142–3, 151, 195–7, 207, 283, 285
see also fens/fenlands
Holland–Utrecht Plain 27, 33–4, 40, 43, 45–6, 69, 76, 94, 97, 136, 147, 233
Hollandse IJssel 84
Holme Post 37–8
Holstein, Duchy of 23, 83, 194, 245
Holsteinische Elbmarschen 132, 182, 264
Hooge 271–2
Hoogheemraadschap 110, 204, 218
Hoorn 120
Hoppenbrouwers, Peter 42, 122
hops 178
horses 71–2, 74–5, 78, 132, 141, 182–3, 188, 210, 222, 230, 249, 276
breeding of 123, 129, 177, 229, 263
horticulture 38, 120, 224, 226, 246, 263, 265, 271, 277, 291
Hoyle, Richard W. 153
Hudig, Joost 40
Humber, River 27

Iberian Peninsula 50
Idaarderadeel 179
IJssel 17, 126, 178
IJsselmeer 248, 273
IJsselmonde 95
Industrial Revolution 2, 192–3, 204, 248
industry 20, 81, 85, 94–5, 121, 132, 140, 185, 226, 263, 265, 273
institutions 6–12, 43–51, 55–68, 99–109, 150–6, 158–63, 198–210, 212–9, 255–8, 284–9
intensification 31, 238, 242, 288
Internal Drainage Board 192, 202–3, 219, 255
investment
in drainage mills 42, 94–7, 99, 158, 164, 172, 176, 180, 284
in farming 67, 180, 188, 237, 241, 244, 268, 271

in land reclamation 54, 144, 148, 155, 157, 254, 286
in office holding 185, 287
in steam pumps 196–7, 210
in water management 4, 14–15, 82, 88, 91, 93, 100, 115, 133, 190, 198, 207, 209, 212, 216, 236, 256, 259
Isle of Axholme 17, 134–5, 187, 234, 267
Isle of Ely 134, 269, 275, 277
Isabellapolder 171
Italy 8, 10

Jabben, Umme 130
Jadebusen 86
James I, king of England 153
Jansz, Dirck 184
Jever 17, 83, 130, 183, 212, 240–1
Jutland 17, 23, 25

Kabeldeichung 55, 198, 258
see also maintenance in kind
Kampen 126, 272–3
Kampereiland 126, 178–9, 272–3
Kats 277
Katwijk 118
Kehdingen 17, 105, 108, 131, 154, 199, 258
Kent 17, 27, 64, 88
Ketelswarf 184
Kieldrecht 90
King's Lynn 265
King's Lynn Conservancy Board 251
Knibbe, Merijn 180, 230
Knottnerus, Otto 130, 240, 290
Kole, Heleen 102
Kommuniondeichung 100, 258
see also monetisation
Kraker, Adriaan de 138
Krimpenerwaard 118–19

labour productivity 124, 170, 178, 222, 225, 240, 260, 276, 288
labourers 33, 90, 112–13, 132, 141, 148, 169–71, 176, 180, 190, 209, 222, 239–40, 263, 276, 286
seasonal 170, 177, 190, 267, 288, 291
lake drainage 141–3, 195–7, 235

Index

land consolidation 253, 271, 273–4, 276, 278, 289
Land Drainage Act 1930 219, 255
landownership, *see* Property relations
land productivity *see* cereal yields
land registration 101, 110
Landgemeinde 24, 54–5, 77, 83, 105–8
Langen, Gilles de 128
leasehold 66, 68, 115–16, 118, 122, 128, 131
Leeuwarden 179
Leeuwarderadeel 126–7
Leeuwenhorst, abbey of 174
leguminous plants 70
  beans 70, 96, 224
  clover 237, 241
  peas 70, 188, 265
  vetches 70
Leiden 94, 96–7, 118–19, 140, 158, 172, 174, 196
Lek, River 119
Lely, Cornelis 248, 251
Lemmer 153
Leybucht 86
Lille 71
Lincolnshire 16–7, 27, 33–4, 46, 49, 51, 65, 69, 76, 91, 134–6, 146–8, 186–9, 233–4, 250, 269
Linden, Hendrik van der 43–9, 73
Lindsey 146, 187
Little Ice Age 79, 137, 194
Littleport Fen 204
London 20, 24, 81, 132, 134, 146, 234, 250, 265–6
Lorenzen-Schmidt, Klaus-Joachim 191
Lower Saxony 141, 246, 250, 257–8
Lowestoft 265
Lyon, Bryce 43

Maasland 95, 173, 176
Macdonald, Sir Murdoch 251
madder 70–1, 78, 118, 170, 191, 223, 225–6, 244, 261–2, 288
maintenance in kind 100–1, 109–11, 162, 192, 198–201, 209, 258, 287–8
Maldon 265
manorial system 44, 46, 49–52, 78, 122

Mansholt, Sicco 254, 260
March 188
Mare 90
Mariengaarde, abbey of 127
Mariënweerd, abbey of 123
Marken 90, 273
market gardening *see* horticulture
markets
  capital 172
  land and lease 67, 121–3, 135, 69, 172, 180, 186, 240, 268, 286, 288
  product 34, 69, 71, 82, 112, 120–1, 128, 132, 150, 170, 188, 190–1, 226–7, 229–30, 234, 246, 265–6
Mastenbroek 125–6
Maurice, Prince of Orange-Nassau 144
meadow 32, 97, 120, 180, 187, 224, 273
meat 30–1, 69, 74, 125, 129, 132, 136, 185, 190–1, 193, 228, 268, 279
Mechelen 93, 143
Meuse, River 3, 17, 20, 26, 85, 248
Middelburg 117, 143
Midwolda 181
mill polder 104–5, 158–60, 176, 203, 206, 210
mixed farming 29, 72–4, 84, 123–4, 131–2, 134, 177–8, 182–3, 225–6, 228, 234, 237–8, 240–1, 261–5, 269–71
modification of wetlands 31–2, 35–6, 89, 272, 273–4, 282–3
Mol, Hans 128
monetisation of maintenance 99–102, 110–11, 136, 162–3, 192, 198–9, 211, 258, 287
  see also gemeenmaking, Kommuniondeichung
Monster 173
Morgan, John 109

Morton's Leam 232
Namibia 254
Nene, River 204
Netherlands 194, 217–18, 246–50, 252–3, 255–7, 270, 275
Nieuwkoop 46
Nijmegen 179, 273
Nissen, Sönke 254

335

Index

nobility 14, 92, 117–18, 122, 127, 129, 131, 138, 167–8, 174, 180
Noord-Beveland 117, 277–8
Nord, Département du 215
Nordfriesland 17, 31, 50, 90, 92, 138, 145, 162–3, 182, 184, 201, 254, 267, 271, 292
Nordmarsch-Langeneß 184
Nordstrand 8, 130, 138, 144, 154, 200, 209
Norstrandischmoor 271, 292
Norfolk Broads 27, 51
North, Douglass 11

oats 70, 73–6, 90, 96, 124, 131–2, 179, 182, 184, 187–8, 229–30, 233, 236, 272
octrooi 151, 285
Oegstgeest 118, 174, 176
Ogilvie, Sheilagh 2, 10–11, 279
Oktroy 154
Oldambt 181–2, 211, 237
Oldenbarnevelt, Johan van 151
Oldenburg 23, 83, 131, 139, 144, 154, 156, 183, 194–5, 218, 241, 246
Oldenburger Wesermarsch 266
oligarchisation 160–1, 164, 286
  see also elites
Ommelanden 60, 106, 287
Ooijpolder 179, 273–4
Oostburg 113, 115
Oostergo 24, 107
Oosterwolde 125
Oste, River 241
Osterstade 258
Ostfriesland 17, 67, 183, 105–6, 111, 130–2, 182, 184–5, 195, 212, 218, 240–1, 264, 278, 286
Ostrom, Elinor 8
Otterndorf 108
Oude Rijn, River 28, 96–7
Oude Yevenewatering 113–15
Over-Betuwe 117–18, 228
Overijssel 17, 61, 78, 83, 125, 166, 177, 218
owner-occupier 112–13, 116–17, 122, 127, 125–30, 178, 183–4, 190, 213, 224, 236, 240, 270, 282

oxen 75, 123, 125–6, 129, 132, 134, 146, 174, 179, 182, 279

Papendrecht 73
Pas-de-Calais 215, 262
pasture 32, 34, 90, 120, 125–6, 134, 147–8, 170, 174, 180, 184, 187, 205, 224–5, 229–30, 241, 272–3, 276, 282
patent 151–2, 154, 156, 285
path dependence 11
peat digging/mining 82, 86–7, 172–3, 195, 235
Pellworm 201
Pevensey Levels 1
Pfanddeichung *see* Kabeldeichung
Philippines 201
pigs 44, 67, 74–5, 249, 291
pileworm 213
Poland 41, 267
polder 7, 104
  see also mill polder
Pollution of Surface Waters Act 1970 256
population 20–1, 80–2, 247, 267–8
potatoes 229, 233–4, 236, 243, 265, 267, 269, 275, 277
Priester, Peter 222, 237, 243
privileges 108, 138, 192–3, 208–9, 212, 216–17, 220
property relations 66–8, 113–19, 122–3, 126–33, 167–74, 184–5, 189, 224, 228, 236–7, 239
  see also leasehold, owner-occupier, tenant farmer
Prussia 194, 246, 272
Putnam, Robert 9

rapeseed 170–1, 187–8, 191, 241
rates 6–8, 64, 100–1, 104, 106, 109–11, 156, 158, 160–3, 165, 198, 202–3, 207, 212, 226, 257, 284, 287
  see also monetisation
Ravensdale, J.R. 189
reclamation 31–6, 140–50, 195–9, 248, 253–4
  unintended consequences of 36–42, 83–90
Reformation 129, 174
Regional Water Authorities 255

Index

Regional Water Authorities Act 1992   257
regression phase   22
Reijerwaard   95
Reimerswaal   88
Rennie, John   233
Rhenen   6
Rhine, River   3, 6, 17, 26, 84–5
Rhineland   44, 129
Ricardo, David   2
Rijnland   33, 60, 84, 95–6, 103, 118–19, 158, 172–4, 218
Rijnsburg   118, 174, 176
Rijnsburg, abbey of   174, 176
rinderpest   226–7, 229
Rippon, Stephen   31
Ritzebüttel   258
River Authorities/Boards   255
Robles, Caspar de   107
Romboutswerve, Watering   114
Romney Marsh   17, 27, 46, 49, 64–5, 132–6, 185–6, 190, 275, 283
Ronde Hoep   158
Roosen, Joris   81
Rotterdam   187, 196, 207
Rudolf II, Count of Stade   49
Rungholt   86
Russia   195
Rüstringen   24
rye   28, 72–3, 75, 132, 184

Sachsen-Lauenburg   108
salt production   30–1, 85, 87, 91
Sangatte   17
Saxony   44
Sayers, Dorothy L.   219
Schauung   56
Scheldt, River   3, 17, 62, 85, 93, 140, 142, 248
Schenk van Toutenburg, George   107
Schermer, Lake   143
Schiphol   290
Schleswig, Duchy of   83, 131–2, 154, 194, 245
Schleswig-Holstein   26, 81, 83, 132, 138–9, 182, 218, 246, 251, 254, 257, 272
Schleswig-Holstein-Gottorf, Duchy of   139, 144, 154

Schokland   90
schouw   56
Schouwen   91, 117, 198, 200–1, 210
Schweiburg   144
seepage   42, 125, 178, 228–9
serfdom/serfs   44, 46, 48, 50–2, 282
sheep   29–30, 34, 36, 63, 67, 74, 76, 132, 134, 146, 184–6, 188–9, 227, 231–2, 275, 282
Sielacht   105
Sint-Truiden, abbey of   74
Skertchly, Sidney B. J.   40
Slicher van Bath, Bernard Hendrik   28, 43
sluices
   construction of   99, 212–13
   maintenance of   58, 63, 65, 100, 105–6, 160, 205–6, 235, 257
Sneek   205, 235
social agrosystems   13, 15
social capital   9–10, 12
Soens, Tim   4, 14, 17, 92–3, 114, 116, 239
soil fertility   25–6, 73, 178
soil types   25–6, 94–5, 118, 124, 159, 170, 173, 178, 183–4, 224, 228, 264
Sönke-Nissen-Koog   254
South Lincolnshire Smallholders' Association   267
Spalding   267, 270
specialisation   20, 71, 175–7, 179–80, 190, 227, 230, 234, 243, 266, 282, 286, 289
St. Peter's Abbey Ghent   115
Stade   49, 108, 131
Stadland   83, 131, 183
Staring, Winand   228
Starnmeer   146
state formation   14, 23, 82–3, 139–40, 193–4, 245–6
States-General   207–9
steam pumping station   196–8, 204, 210
Stedingen   24, 48
storm surges, *see* floods
Stuart, House of   17
Studler van Zurck, Anthony   144
sugar beet   225–6, 229, 243, 261–7, 275, 278

337

Index

Summers, Dorothy 192, 231
Sussex 88
Swart, Friedrich 130, 260
Sweden 139
Switzerland 3, 27

Tambora, Volcano 194
taxation/taxes 50, 82, 119, 126–7, 130, 207, 223, 226–7
  exemption of 151, 163, 207, 285
  see also rates
technological lock-in 5
tenant farmer 56, 115, 121, 123–4, 128, 132, 169, 183, 185, 190, 223, 240, 243–4
Ter Aar 172
terp 28–30, 33, 35, 230
territorial lords 24, 50, 52, 61, 108, 111, 131, 154, 282–3
Thames, River 17, 27, 36, 57, 88, 91, 133–5, 186, 250
Thames Barrier 250, 259
The Hague 103, 119, 139, 143, 207–8
Thirsk, Joan 13, 16, 134, 231, 234
Tholen 208
Thorbecke, Johan Rudolph 218
Thorney, estate 188–9, 231–2, 234, 270
Tielhof, Milja van 102, 172
Tiengemeten 290
Tilly, Charles 23
tithes 66, 73, 84, 93, 95–6
Tocqueville, Alexis de 194
Tönning 194
trade 20, 30–1, 58, 74, 81, 86, 125, 140, 144, 188
  free trade 246
  trade routes 150
transformation of wetlands 4, 31–2, 35, 76, 126, 282
  see also reclamation
transgression phase 22
transhumance 31
 Trent, River 253
trust 9–10, 163, 192, 204–5, 209, 287

underdrainage *see* field drainage
United East India Company (VOC) 142

United Kingdom 246–7, 250, 252–3, 255, 265, 268, 275
United States 1, 267
Upwell 234
urbanisation 20, 23, 81, 111, 193
urban investors/investment *see* capital, urban
Utrecht
  bishop of 23–4, 45–6, 50, 126
  city of 80
  territory of 17, 33, 36, 44, 47, 49, 60–2, 65–6, 71, 75, 78, 82–5, 89, 106, 110–11, 122–3, 173, 266

Valencia 8
Valkenburg 118
Veen, Johan van 248, 251, 293
Velsen, Cornelis 40
Venny, Abraham 188
verhevening 55
  see also maintenance in kind
verhoefslaging 55
  see also maintenance in kind
Vermuyden, Sir Cornelius 147, 202, 251
Veurne, Castellany of 115–16, 170, 222–4, 242
Vijf Deelen Dijk 107
Vlaardingen 46, 73
Voorne 208
Vries, Jan de 1, 120, 166, 175–6, 179

Waal 17, 217, 219
Wadden Coast 17, 23, 26, 28–9, 44, 50, 60, 83, 105, 113, 126, 143, 154–5, 166, 179, 290
  farming on 129–31, 184, 190, 229, 239, 2645
  flooding of 138, 210, 239
  water management on 62, 143, 156, 161, 164, 198–9, 254
Wadden Sea 26, 65, 78, 81, 86, 99, 130, 182, 184, 248
Wageningen 6, 9
wages/wage costs 121–3, 178, 199, 222, 230, 267, 276, 279, 288
Wainfleet 57
Walcheren 60, 81, 101, 117
Warft 28, 90, 184, 272

338

Index

Wash 27, 253–4
Wassenaar, Lord of 174
Waterbeach 232
watering 7, 60–2, 224, 257, 283
Wateringen 175
waterschap 7
Wells, William 234
werf/werve 28
Weser, River 3, 26, 47–9, 144, 156, 258
Westergo 24, 107
Westerkwartier 129
West-Friesland 39, 84, 96
Westland 291
Westphalia 132, 177, 288
Westpolder 254
wheat 70, 73, 117, 124, 134, 188, 223, 233–4, 236, 241, 243–4, 262, 269, 275, 277, 288
  see also cereal yields
Whittle, Jane 167
Whittlesey Mere 38
Wiemsdorf 239
Wierde 28
Wieringen 233, 248
Wildmore Fen 148
Willem I, King of the Netherlands 218
Willem II, Count of Holland 60
William the Conqueror 63
Willingham 134–5, 189
Wilt, Carla de 173
Wimsey, Lord Peter 219
Wing, John 232
Winiwarter, Verena 279, 289
Witham, River 233
wood 3, 21, 31, 56, 67, 72, 89, 92, 99, 143, 145, 212–13, 224, 241, 283

wool 20, 30–1, 69, 76, 132, 136, 185, 227
Wulpen 93
Würden 258
Wursten 17, 211
Wurt 28

Yarmouth 265
Yorkshire 1, 17, 88, 147
Young, Arthur 40, 203–4, 234
Ypres 71, 80

Zanden, Jan Luiten van 72, 222
Zeeland 17
  farming in 117, 177–8, 222–5, 228, 236, 261–3, 271, 276, 286
  flooding of 88, 90, 167, 249–50, 256
  property relations in 44, 50, 92, 102, 110, 116–17, 168–9, 190
  water management in 41, 60–1, 65, 78, 92–3, 99–100, 110, 141, 144, 163–4, 198, 207–9, 217–18, 256, 283, 285
Zeeland Flanders 26, 70, 91, 113–14, 116, 145, 150, 168, 207–9, 215–16, 222–5, 242–3, 256, 267
Zeevang 120
Zeischka, Siger 96
Zierikzee 117, 208
zijlvest 62–3, 65, 105–7, 111
Zuid-Beijerland 169
Zuidplas 196
Zuider Zee 17, 35, 90, 107, 113, 125–6, 206, 214, 235, 248, 253, 273
Zuider Zee Works 248, 259
Zuienkerke 70
Zwijndrechtse Waard 155

Boydell Studies in Rural History

*The Real Agricultural Revolution: The Transformation of
English Farming, 1939–1985*
Paul Brassley, David Harvey, Matt Lobley and Michael Winter

Agricultural Knowledge Networks in Rural Europe, 1700–2000
Edited by Yves Segers and Leen Van Molle

Landless Households in Rural Europe, 1600–1900
Edited by Christine Fertig, Richard Paping and Henry French

Agriculture, Economy and Society in Early Modern Scotland
Edited by Harriet Cornell, Julian Goodare and Alan R. MacDonald

The Tithe War in England and Wales, 1881–1936: A Curious Rural Revolt
John Bulaitis

Printed in the United States
by Baker & Taylor Publisher Services